P9-CAT-038

MACHINERY FOR HORTICULTURE

BRIAN BELL and STEWART COUSINS

Farming Press

First published 1991

A catalogue record for this book is available from the British Library

ISBN 0 85236 231 5

Published by Farming Press Books
4 Friars Courtyard, 30–32 Princes Street
Ipswich IP1 1RJ, United Kingdom

Distributed in North America
by Diamond Farm Enterprises
Box 537, Alexandria Bay, NY 13607, USA

Cover design by Andrew Thistlethwaite
Phototypeset by Galleon Photosetting, Ipswich
Printed and bound in Great Britain by Butler and Tanner, Frome, Somerset

CONTENTS

A colour section appears between pages 152 and 153

INTRODUCTION

This book is designed to give the reader a sound knowledge of the wide range of tractors and machinery used in the horticultural industry.

There are sections on tractors and self-propelled machinery, cultivation, drilling and planting equipment, also crop treatment and harvesting machinery. Glasshouse, grass cutting and turf care equipment together with irrigation and machinery for estate and grounds maintenance, forms an important part of this book. The final section deals with the horticultural workshop and sources of power including oil, electricity, gas and solid fuels. Each chapter is concluded with some suggestions for student practical work and a reminder of relevant safety matters.

Readers interested in learning more about workshop processes and techniques will find more detailed information in *Farm Workshop*, also written by Brian Bell.

Machinery for Horticulture will be an invaluable guide for students taking both full-time and part-time college courses in preparation for the various levels of the new National Vocational Qualifications (NVQ) in a wide range of horticultural subjects. The book will also prove useful to those embarked on an Open Learning course or for the private study of those who wish to learn more about modern horticultural machinery and equipment.

It is not possible to give specific information about any particular horticultural tractor or machine in a book of this type. Average figures are given to indicate typical dimensions, settings and adjustments. However, the operator's instruction manual should be considered essential reading for anyone using specialist horticultural equipment.

Reference is made to Health and Safety Regulations throughout this book which summarises their main requirements. However the reader is advised to study the relevant information leaflets and in particular those concerning the safe use of pesticides.

Dimensions and settings are given in metric units and approximate imperial equivalents are also cited in some cases. A list of metric conversions can be found on page 289.

ACKNOWLEDGEMENTS

We are grateful for the help so freely given by the many individuals and companies involved in the manufacture of horticultural machinery and equipment during the preparation of this book. Many of the illustrations have been provided by these companies, and this is acknowledged in the caption of the plate or figure concerned. Special thanks are due to John Blyth, John Briscoe, Delia Chinnery, Ian Carr, David Jenkins, Bob Rendall and Alec Williamson.

Chapter 1

TRACTORS AND POWER UNITS

Various types of tractor have been used for horticultural work since the early part of this century. They include simple pedestrian controlled machines, medium and low power agricultural tractors and small crawler tractors. In the early days of horticultural mechanisation, tractors and pedestrian controlled cultivators were mainly driven by small air cooled petrol engines. Modern horticultural tractors can be sub-divided into pedestrian controlled and ride-on machines. Many large tractors have air cooled diesel engines with one or more cylinders; others have multi-cylinder water cooled diesel or petrol engines.

PEDESTRIAN CONTROLLED TRACTORS

Most pedestrian controlled horticultural tractors are primarily designed as self-propelled rotary cultivators, but are also capable of using a range of toolbar implements such as cultivator tines, ridging bodies or a plough. Small trailers, often with a seat attachment for the operator, are also available for larger models of self-propelled cultivators.

Pedestrian controlled tractors have either one or two wheels. Single wheel models require the operator to keep the machine balanced; this can prove heavy work in difficult

PLATE 1.1 *Compact tractors at work with a rotary cultivator and a transplanter.* (Ford New Holland)

PLATE 1.2 *Pedestrian controlled rotary cultivator.* (Dowdeswell)

conditions. The handlebars can be adjusted to allow the operator to walk behind the machine or can be offset to either side when it is undesirable to walk on the land being cultivated.

The various types of pedestrian controlled grass cutting machines, available in both hand pushed and self-propelled versions, are another major group of powered horticultural equipment. Here again, the larger models often have a seat for the operator, carried on a roller, towed behind the machine. Grass cutting machinery is dealt with in detail in Chapter 14.

Pedestrian controlled horticultural equipment, including self-propelled rotary cultivators, usually has single cylinder air cooled petrol or diesel engines. Some models have electric starting but it is more common for the engine to be hand started with a recoil starter.

Ride-on Tractors

Ride-on tractors used in horticulture range from simple three or four wheeled machines designed for the large garden, up to medium powered tractors of around 30 kW (40 hp) which are used for such work as field scale fruit and vegetable production, transporting materials and maintenance of parks and sportsgrounds.

Ride-On Garden Tractors

The smaller ride-on garden tractors, designed as single purpose ride-on domestic rotary grass mowers with a cutting width of 650 to 750 mm (25 to 30 in), have an air cooled

PLATE 1.3 *Pedestrian controlled rotary cultivator with the handlebars offset so that the operator does not have to walk on cultivated ground.* (Kubota)

PLATE 1.4 *Cylinder lawn mower with a seat for the operator.* (Ransomes, Sims and Jefferies)

PLATE 1.5 *Ride-on garden tractor with mid-mounted rotary mower.* (John Deere)

PLATE 1.6 *A garden tractor with a 12 kW (18 hp) diesel engine, mid-mounted rotary mower and grass collector. The tractor has four wheel steering for added manoeuvrability when working in confined spaces.* (Kubota)

petrol engine developing from 4.5 to 7.5 kW (6 to 10 hp) depending on the model. Some models have electric starting. They are equipped with a simple gearbox which provides from three to five forward gears and one reverse.

Larger ride-on garden tractors of the type shown in Plate 1.4, with a 6 to 9 kW (8 to 12 hp) engine and a cutting width from 750 to 1,060 mm (30 to 42 in), give the output

PLATE 1.7 *A turf care tractor with a rear hydraulic lift linkage and articulated steering. It can also be used with front-mounted equipment, such as a mower or rotary brush, driven by a vee-belt from a pulley under the front of the tractor.* (Ransomes, Sims and Jefferies)

needed for cutting large areas of grass. These machines have electric starting and charging systems. Manual gearbox models usually have five or six forward speeds and a reverse. More expensive versions have change on the move manual gearboxes or hydrostatic transmission.

Many of these machines have a grass collection system with a large capacity grass bag carried on the back of the tractor or towed behind it. Other equipment such as small trailers, leaf collectors and lawn rollers can be attached to the drawbar to speed up maintenance work in the larger garden.

There is also a range of ride-on mowers for professional grounds maintenance work. A water cooled petrol or diesel engine of up to 16 kW (22 hp) or more provides the power for these machines which have hydrostatic transmission, four wheel drive and a three point linkage hydraulic system. Some ride-on mowers also have four wheel steering. These highly sophisticated and expensive models have provision for attaching power driven implements to both front and rear, and in addition to operating hydraulic motor driven rotary or cylinder mowers, they can be used

PLATE 1.8 *This 9.4 kW (12.5 hp) radio controlled mower with a three cylinder diesel engine and four wheel drive will operate on both uphill and downhill graded slopes up to 40 degrees where it would be unsafe for a driven tractor to be used. It has a radio control radius of 100 metres and has several safety features. The engine cuts out if the radio control unit fails and automatic brakes which engage whenever the mower or the engine stops. The engine also stops automatically if the oil pressure becomes too low or the cooling system temperature is too high.* (Kubota)

with a range of turf care and estate maintenance equipment.

Ride-on front mowers with either a rotary or flail cutting mechanism carried on a hydraulic linkage provide high output cutting for the maintenance of large areas of grass on estates and public parks. These machines are very manoeuvrable with rear wheel steering, diff-lock, independent brakes and hydrostatic transmission system. A typical front mower has a three cylinder diesel engine which develops 16.5 kW (22 hp).

Compact Tractors

Compact tractors (Plate 1.1) are based on the design of agricultural models. They come within a power range of approximately 11 to 17 kW (15 to 25 hp) and have come into widespread horticultural use in recent years.

PLATE 1.9 *A diesel engined four wheel drive front mower with hydrostatic transmission. The side discharge rotary mower deck is raised and lowered hydraulically.* (Kubota)

PLATE 1.10 *Compact tractor mini-digger with four wheel drive. The driver's seat can be adjusted to face the front or rear.* (Kubota)

Compact tractors have multi-cylinder air or water cooled diesel engines with a hydraulic system and a choice of either manual or hydrostatic transmission. Some have two wheel drive but four wheel drive is more common.

Health and Safety Regulations require that all tractors weighing more than 560 kg must have a safety frame or cab to protect the driver should the tractor overturn. A further regulation requires a 'quiet cab' if the noise level is above 85 decibels. Most compact tractors do not reach this noise level and do not require a quiet cab; however for most operations they do require a safety frame. When working in confined spaces such as buildings or an orchard, it is permissible to remove or fold down the safety frame.

The compact tractor also provides the power unit for mini-diggers, whose compact design and ability to work in confined spaces have made them very popular for landscape construction work with the rear trencher and front loader bucket.

All Terrain Vehicles (ATVs)

ATVs, as they are popularly known, are used by some growers to provide the power for horticultural operations, especially where the light weight and high speed of an ATV are of particular benefit. ATVs have a single cylinder air cooled engine with a power output varying from 7 to 18 kW (9.5 to 24 hp) or more depending on the model.

As well as its use for personal transport and conveying materials at high speed, an ATV is useful for a variety of other tasks. These include spreading fertiliser or broadcasting seed with an electro-broadcaster

PLATE 1.11 *An ATV pulling a trailed fertiliser broadcaster. A single cylinder air cooled engine is used to drive the broadcaster.* (Kawasaki)

towed behind or mounted on the ATV and operated from its 12 volt power outlet socket. ATVs can also tow small sprayers, power brushes, grass mowers, etc. where full benefit can be gained from the vehicle's low ground pressure of about 0.15 bar (2.5 psi). Where power is required to operate trailed implements, the choices include land wheel drive, electric motor drive from the ATV battery or a separate single cylinder, air cooled engine fitted to implements such as a grass mower.

Rowcrop Tractors

Tractors up to about 37 kW (50 hp) used for agricultural rowcrop work also provide the power for large scale horticultural crop production, maintenance of sportsgrounds and parks and general transport work.

Rowcrop tractors have a small turning circle, easily adjusted wheel track settings and they offer good visibility of the crop or ground being worked. Fruit and vegetable growers sometimes use rowcrop tractors with a high ground clearance kit which makes it possible for the tractor to straddle tall rowcrops. Some high clearance tractors have sufficient clearance to be driven over rows of fruit bushes.

Rowcrop tractors in the 22 to 34 kW (30 to 45 hp) power range are also used to operate multiple gangs of hydraulically or land wheel driven cylinder mowers on sportsgrounds and parks. These tractors have the hydraulic capacity to lift, carry and drive high speed grass cutting equipment.

PLATE 1.12 *Rowcrop tractor with a 34 kW (45 hp) diesel engine. It is a simple task to adjust the track width of this two wheel drive tractor.* (Massey-Ferguson)

Another type of rowcrop tractor sometimes used by the vegetable grower is the self-propelled toolbar or tool carrier. It consists of a tool frame carried on three or four wheels which can be fitted with a variety of implements including cultivator tines, hoe blades and ridging bodies. The implements are simple to attach and the self-propelled toolbar gives the driver an excellent view of the rows.

Vineyard Tractors

Vineyard tractors are another type of rowcrop tractor which has been modified for working between narrow rows of fruit bushes, fruit trees, vines, etc. where the crop is too tall for the tractor to straddle the rows. Some tractor manufacturers offer a vineyard version of certain models in their range. Vineyard tractors may have two or four wheel drive, and they often have extra deflector shields fitted around the wheels to reduce the risk of damage to the trees or bushes. A typical vineyard tractor has a minimum overall width of 1.33 m (52.4 in).

PLATE 1.13 *This four wheel drive vineyard tractor is designed to work between narrow rows of fruit and vines. It has extended mudguards to reduce the risk of damaging the trees and bushes.* (Massey-Ferguson)

Tracklayers

Tracklayers have few horticultural applications but are used for drainage work and moving large quantities of soil for landscape construction projects. A number of small tracked tractors are in production, some of which are built as small digger loaders. More powerful tracklayers are used in the construction industry and in agriculture. Tracklayers cannot be driven on public highways except when fitted with track plates which prevent damage to the road surface. Some of the latest tracklayers have rubber tracks which can be used on public roads and are capable of quite high road speeds.

The great advantage of the tracklayer is its very low ground pressure compared with that of a wheeled tractor. This means that soil compaction is kept to a minimum.

Forklift Trucks

There are two main types of forklift truck: industrial and rough terrain. Both have many uses including loading and unloading lorries, stacking materials and moving them from one site to another. A range of attachments such as pallet forks, buckets and root forks is available.

Drivers must be trained in the operation of forklift trucks. New users, and those who use a new machine of a different type, must attend an operator instruction course.

Industrial forklifts are designed to work on hard surfaces, such as storage areas, and in pack houses, glasshouses, etc. They have small wheels and may be powered by an internal combustion engine, gas or electricity. Petrol and diesel engined models are not suitable for working in buildings because of their toxic exhaust gases. Electric forklifts require regular recharging of their batteries; this is usually done overnight so that the machine is ready for work the next morning.

Rough terrain forklift trucks are, as their name suggests, designed to work on uneven surfaces. Most models have four wheel drive and hydrostatic transmission. The mast type is similar in design to an industrial forklift, but has limitations in reach and height of lift. Telescopic machines have a single telescopic lifting arm which can be extended from the driving position to give greater lifting height and reach than mast machines.

Driving Controls

There are many instruments and driving controls on ride-on mowers and horticultural tractors. It is important to have instruction on the driving controls and if possible to read the section on controls in the operator instruction book before driving an unfamiliar tractor.

The basic models of tractor and self-propelled turf care machinery have a simple set of controls and instruments; the more expensive power units may be equipped with a computerised instrument panel and even have the luxury of an air conditioned cab.

The **throttle** is used to set the engine speed. All tractors have a hand throttle and many also have a foot throttle. The hand throttle lever is used to set the engine speed, which is maintained by the engine governor for continuous work such as cultivating, ploughing and grass cutting. The foot throttle can override the hand throttle and is used where frequent changes of engine speed are needed, for example when using a loader or driving on the public highway.

The **clutch pedal** is on the left-hand side of the tractor. It must not be used as a foot rest when driving as this will cause undue wear on the clutch thrust bearing. The clutch engages and disengages the drive from the engine to the gearbox and power take-off shaft.

Some tractors have a single stage clutch which controls the drive to the transmission system and the power shaft. Others have a dual clutch which gives separate control of the transmission and power shaft. The pedal is pushed half-way down to stop the wheels and fully down to stop the power take-off.

PLATE 1.14 *The main driving controls of a compact tractor. The clutch pedal is on the left, the main gear lever is below the steering wheel and the range gear lever (not shown) is to the left. The brake pedals are on the right with the stop control knob above them. The ignition switch is left of the stop control with the hand throttle above it. The power take-off lever is bottom right, below the steering wheel.* (Massey-Ferguson)

Tractors not fitted with a dual clutch may have an independent clutch to control the drive to the power take-off shaft. A single stage clutch, operated by the pedal, controls drive to the wheels and a completely separate clutch, controlled by a hand lever, engages and disengages the drive to the power shaft.

The **brakes** have three functions. These are:

- To stop the tractor.
- To park.
- To assist with steering, especially on headlands and in confined spaces.

There are two brake pedals, one for each rear wheel. They can be used independently for fieldwork to assist steering and are locked together for road work. It is very dangerous

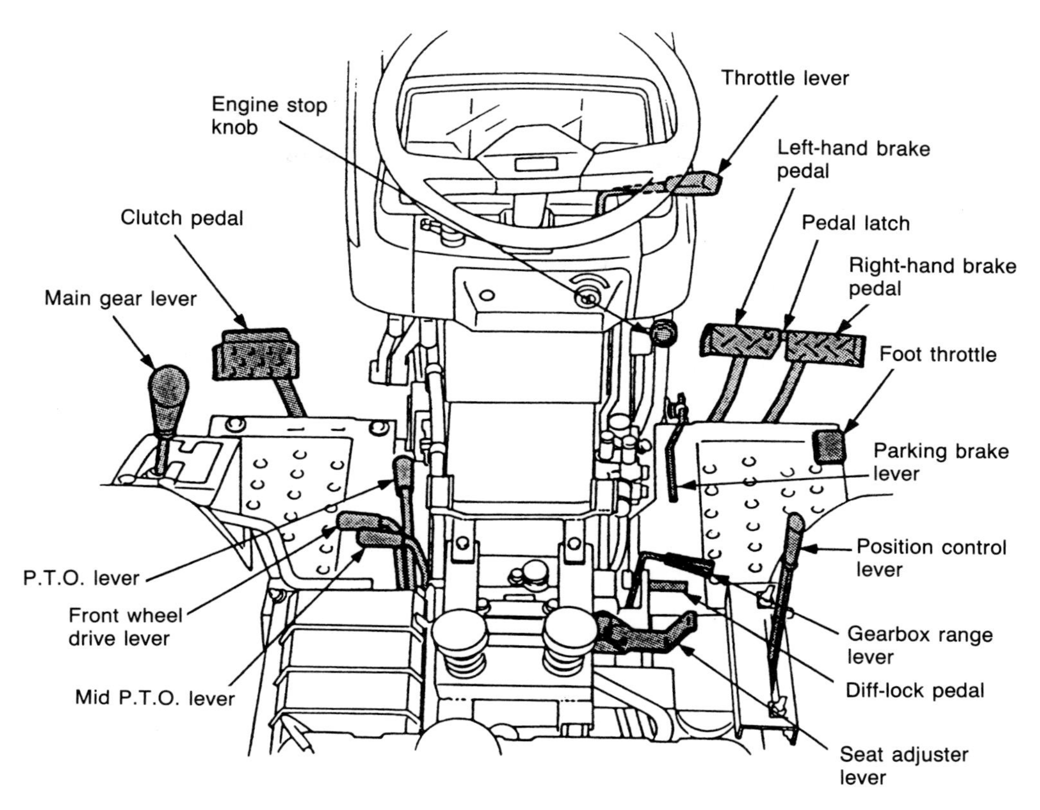

FIGURE 1.1 The driving controls. *(Kubota)*

to drive a tractor on the road without first locking the brake pedals together. For parking purposes, some tractors have a hand brake; others have a pedal latch mechanism which is used to lock the foot brakes after they have been applied.

The **differential lock**, or diff-lock pedal, is used to overcome the problem of wheelspin which occurs when working in difficult soil conditions. The differential allows the driving wheels to rotate at different speeds when cornering. When the diff-lock is applied, the differential cannot function and wheelspin is reduced or eliminated. It is not possible to drive round a corner with the diff-lock engaged. It is designed to disengage automatically but can be released by applying either the clutch or one of the brake pedals if it fails to disengage when required.

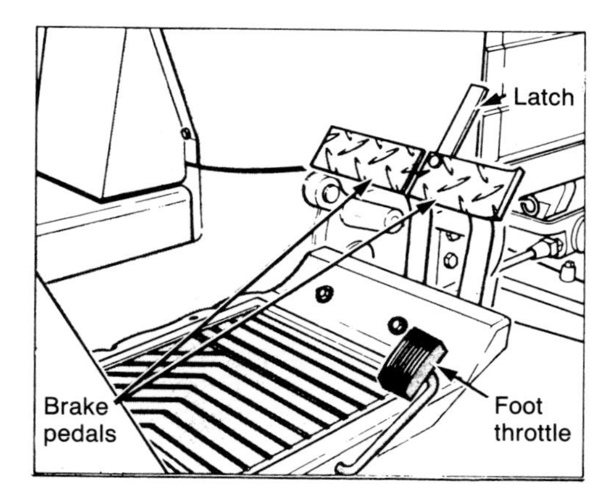

FIGURE 1.2 Independent brake pedals must be latched together before taking a tractor on the road. *(Ford New Holland)*

The **gear levers** are used to select the required forward, reverse or neutral gear. Most tractors have two levers: the range lever selects high or low ratio (some models have high, medium and low ratios), and the main lever is used to engage the required speed within the chosen range. A typical compact tractor has three forward and one reverse gears, which are doubled up to six forward and two reverse by selecting either high or low ratio with the range lever.

Tractors and turf care equipment with hydrostatic transmission have a gear lever which gives either two or three speed ranges. Forward or reverse speed is varied by either a hand lever or a foot control pedal. The pedal or lever is moved in one direction to vary forward speed and in the opposite direction to alter reverse speed.

The **power take-off** drive is engaged with a hand lever. Some tractors require the transmission or dual clutch to be disengaged before using the power take-off lever. Others have an independent clutch system allowing the power shaft to be engaged or disengaged without using the main clutch pedal.

Many tractors have two or more power take-off shaft speeds selected either with the main lever or by a separate speed range lever.

The **hydraulic system** has numerous controls and each manufacturer has special design features. The basic controls are:

- Position control used for implements such as a sprayer, which needs to be held on the hydraulic linkage at a fixed height above the ground.
- Draft control used for implements which work in the soil such as a plough or cultivator. The system maintains a regular working depth without the need for depth wheels on the implement.
- Auxiliary service control used to operate external rams on front loaders, tipping trailers, etc.

Rowcrop and general purpose tractors have all three hydraulic services as standard equipment. As most compact tractors do not have draft control, the working depth of soil engaging implements is controlled by depth wheels.

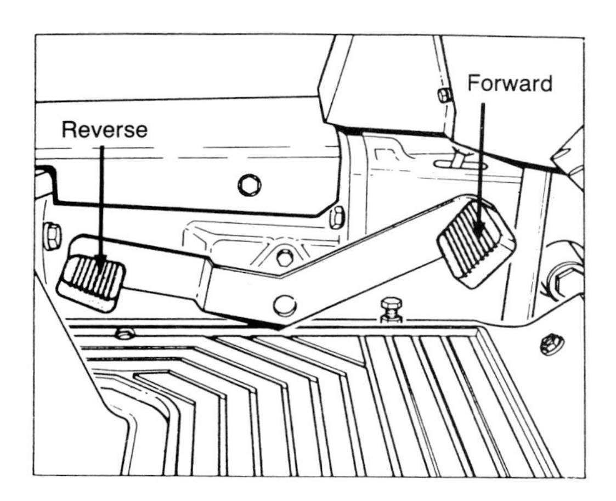

FIGURE 1.3 Heel and toe pedal for controlling a compact tractor hydrostatic transmission. *(Ford New Holland)*

The Instrument Panel

Tractors and self-propelled grass care equipment have various switches, gauges and warning lights on the instrument panel, as all models need a minimum number of instruments to monitor and control the basic functions of the machine. A comprehensive set of switches, gauges and warning lights will be found on the more expensive tractors and self-propelled turf care equipment.

Warning lights for low oil pressure, ignition, parking brake, headlamp main beam, direction indicators and hazard lights are standard on most machines. The more sophisticated instrument panel will also have warning lamps to indicate low hydraulic oil pressure, engine airflow restriction, low fuel level and that the parking brake is on.

All warning lamps are there to draw the attention of the driver to a fault in the various systems or to indicate, for example, that the headlamps are on main beam. When one of the system warning lamps is on, especially those indicating low oil pressure and air flow restriction, or when the ignition warning lamp shows that the battery is not charging, it is very important to stop the engine and not re-start it until the fault has been put right.

The **gauges** found on most instrument

PLATE 1.15 *The instrument panel of a rowcrop tractor. At the top from the left are the engine rev counter, the warning light panel, fuel gauge and temperature gauge. The horn and indicator switch is left of the steering wheel with the throttle and ignition switch on the right. At the bottom from the left are the stop button, hazard warning switch and light switches.* (Massey-Ferguson)

panels include cooling system temperature, engine oil pressure and fuel level gauges. Some models will also have an ammeter which indicates battery charging rate.

The **ignition switch** usually incorporates the starter switch, and on diesel engines it generally controls the pre-heater unit for cold starting. Petrol engines are stopped by turning the ignition key to the 'off' position. Some diesel engines are also stopped with the ignition key, which, by means of a small solenoid, cuts off the fuel supply to the injectors.

The **proof meter** indicates engine crankshaft speed in rpm. It also records the engine working hours, usually calculated at an engine speed of around 1,600 rpm. At full throttle, more hours will be recorded than shown by the driver's watch. However, some proof meters record the hours the engine runs regardless of engine speed.

The proof meter indicates the engine speed required to achieve standard power take-off speeds which are necessary to operate power driven machines. Some of them also measure forward speed in mph and km/h in most forward gears. It is more common for a speed chart near the steering wheel to be used in conjunction with the engine speed to find the forward speed in each gear. This information is important when using sprayers, fertiliser spreaders, etc.

Other important items on the instrument panel include the horn button, light switches and traffic indicators. Tractors with cabs have heater and radio controls and the expensive models may also have air conditioning switches.

Electronic instrument panels are found on some horticultural tractors and self-propelled grass machinery. A typical example has eighteen electronic instruments, which monitor ignition, engine coolant temperature, engine oil pressure, parking brake, low fuel level, etc. Bar graphs indicating engine speed in rpm, engine oil pressure, coolant temperature and fuel level replace dial gauges.

Starting a Tractor Engine

The driver must always be in the driving seat before attempting to start the engine. Most tractors have a safety start device requiring either the gear levers to be set in neutral or the clutch pedal to be pressed fully down before the starter can be operated. Many models of ride-on grass mowers have a safety switch under the seat which isolates the starter circuit. It requires the weight of the driver on the seat to de-activate this switch. On aged models without a safety start switch, take extra care to ensure that the gears are set in neutral and the clutch pedal is pressed down before operating the starter.

To start a diesel engined tractor:

1. Check that the gear levers are set in neutral. It is a good idea to push the clutch pedal down too as this reduces the load on the starter motor. The clutch must be disengaged if the tractor has a clutch operated safety start switch. The range lever must be set in neutral on tractors with hydrostatic transmission.
2. Set the throttle lever between half and three-quarters open. When starting a cold engine with a turbocharger, do not allow the engine to run at more than half throttle for the first two minutes to give it time to warm up.
3. Set the stop control handle in the run position.
4. Turn on the ignition switch.
5. Operate the starter switch, usually by further movement of the ignition switch key. If the engine fails to start after about 10 seconds, release the starter switch and allow the starter to come to rest; then try again after a few seconds. Excessive use of the starter without rest can damage the battery and the motor.
6. Most tractors have some form of cold starting aid. Indirect injection diesel engines fitted to compact and other models of tractor have glow plugs which heat the air in each cylinder or a heater unit which warms the air in the inlet manifold.
   Before starting an indirect engine from cold, use the ignition key to operate the heater units for about 10 seconds before engaging the starter. Engines with glow plugs have an indicator lamp on the instrument panel to show the driver when the engine is ready for starting.
   Direct injection engines may have an excess fuel button on the injection pump which is used to supply extra fuel to the cylinders when starting cold engines on winter days.

To start a petrol engined tractor or mower it is necessary to close the choke when the engine is cold. The ignition and starter switches—usually combined—should then be operated until the engine starts. Do not run the starter for more than a few seconds; if the engine does not start, allow the starter to come to rest before trying again. When the engine has run for a short while, the choke should be returned to the normal running position.

Small engines with a recoil starter will require the choke to be closed and the ignition isolator switch turned on before starting a cold engine. Make sure that the fuel tap is switched on.

Stopping the Engine

To stop a diesel engine, pull out the stop control handle. This will cut off the fuel supply to the engine. The ignition switch is used to stop the diesel engine on some tractors and forklift trucks. The switch operates a solenoid which cuts off the fuel supply to the engine.

Turbocharged engines should be run at tick-over for a couple of minutes before being stopped to allow the turbocharger to cool down.

When the engine has stopped, remember to apply the handbrake and set the gear levers in neutral or park position before leaving the tractor.

Petrol engines are stopped by switching off the ignition so that no sparks occur at the sparking plug or plugs. All engines, even small air cooled engines on pedestrian controlled lawn mowers, must have the stop switch clearly marked.

Using Tractors and Turf Care Machines

On the Road

The Road Traffic Regulations are very complicated, and the information given here is merely a guide. You must read the regulations in order to understand the full requirements of the law.

The Highway Code must be obeyed whenever a self-propelled machine is driven on the public highway. This also applies to self-propelled machines controlled by a pedestrian, which need to be licensed and insured for road use.

The maximum permitted road speed for tractors and self-propelled turf equipment is 20 mph (12 mph for very wide machines).

When driving a tractor or other horticultural machine on a public road, remember these points:

- Make sure the brakes are in full working order and if the tractor has independent brake pedals, these must be locked before driving on the road.
- The steering and tyres must be in good condition. Any lights must be in working order at all times.
- An amber flashing beacon must be used when driving along a dual carriageway with a tractor or horticultural machine with a maximum speed of less than 25 mph.
- Suitable mirrors must be fitted to give the driver a clear view of traffic behind, even when pulling a loaded trailer.
- Very long or wide loads or implements require wide load marker boards.
- Tractors and other machines with cabs must have a clean windscreen, and washers and wipers must be in working order.

Some other important points to remember when driving a tractor or turf care machine on the road are:

- The machine must have a road tax disc or tax exemption certificate. There must be a valid certificate of insurance and the driver must have a licence too! The registration number plate must be fully visible from the rear of the vehicle.
- Do not allow mud or other material to fall on to the road. If it does, the law requires that it must be removed to avoid causing a hazard for other road users.
- When towing a laden trailer, make sure the load is secure.

As well as the requirements of the Road Traffic Regulations, there are other requirements under the Health and Safety at Work Act, the Avoidance of Accidents to Children Act and the Agricultural Field Machinery Regulations which also apply to horticulture. The official leaflets give exact information, and the following notes serve only as a guide:

- Children under thirteen must not ride on or drive a tractor or self-propelled horticultural machine. (In Norfolk the minimum age for driving a tractor is fourteen.) They may ride on an empty or partly loaded trailer provided that all four sides project above the floor or load.
- Passengers must not ride on the drawbar or linkage of a tractor or other machine when towing is taking place.
- A tractor or other self-propelled machine may be started or set in motion only when the operator is in the normal driving position.
- The driver must not leave the driving seat

while the machine is in motion except in an emergency.
- Drawbar pins must be secured with a safety clip to prevent them jumping out when towing.
- All cutter bars must be protected with a guard when not in use, either in store or when being transported on a tractor.

Children are killed in accidents involving machinery every year, so always take extra care when children are playing nearby. When you need to back a trailer, check that there is no one behind before reversing.

Quiet cabs isolate the driver from the noise of children playing, tractors, other workers and their equipment, so when driving in a quiet cab take extra care driving around buildings and be prepared for an unexpected hazard just round the next corner.

Safe Driving in the Field

- Take special care when driving on sloping ground. Select a low gear before driving downhill and never change gear on a slope except when the tractor is stationary.
- Check that nobody is in the way when moving off, lowering an implement or engaging the power take-off drive.
- Do not drive close to the edge of ditches or dykes.
- Drive at a safe speed at all times. Good drivers do not have to apply their brakes fiercely as they are in full control of the machine.

Safe Parking

Always apply the handbrake when parking a tractor or other self-propelled machine. The gear lever should be left in neutral and the ignition key removed. Always lower implements and loaders to the ground before leaving a parked tractor.

Safe Hitching

- Back squarely up to an implement when preparing to attach it to the three point linkage.
- Avoid standing between the tractor and the implement when hitching; always work from the side. This is very important if someone else is reversing the tractor. Remember that the correct sequence for attaching the three point linkage is left—right—top.
- Never use a finger to check the top link pin hole alignment—you could lose your finger.
- Never tow a trailer with a ring hitch from the swinging drawbar, as this practice is likely to shear the drawbar pin and the results will be disastrous.
- Do not tow from the top link position because the tractor may somersault backwards. The only safe towing point is the drawbar.

All workers have a legal responsibility to make full use of guards and other protective devices provided on tractors and machinery. Broken and damaged guards must be reported to the employer so that a replacement can be fitted. Personal protection equipment such as ear defenders, steel toecap boots, face shield, etc., must be worn when using certain items of equipment such as strimmers, hedge trimmers, shredders and chain saws. Study the various safety leaflets to make sure that you do not break the law when using potentially dangerous machines.

Health and Safety Regulations require all workers to ensure that their actions, while at work, do not affect the health and safety of others.

Suggested Student Activities

1. Read the instruction book for any tractor or pedestrian controlled cultivator you are required to use.
2. Get to know the driving controls on different tractors and garden machines.
3. Study the Highway Code with particular reference to the sections dealing with tractors and pedestrian controlled equipment.
4. Study the many Health and Safety leaflets published for your information.

Safety Checks

Do not forget that the law requires the driver to start the engine of a tractor or self-propelled machine from the driving seat. The driver may not leave the seat while the machine is moving, except in an emergency.

Keep your feet well clear of the blades when starting a rotary mower, and protect your toes with steel toecap boots.

Chapter 2

ENGINES

Horticultural tractors, lawn mowers and self-propelled rotary cultivators are usually powered by four stroke petrol or diesel engines. Two stroke petrol engines are used for chain saws, small, hand held garden equipment and some lightweight rotary mowers. Two and four stroke engines are called internal combustion engines because the heat required to make them work is produced inside the engine cylinder. The steam engine, an important source of power earlier this century, is an external combustion engine. The heat is created by burning coal to produce the steam which is passed to the cylinder.

Petrol engines are still in common use in horticulture, although most four wheel tractors and an increasing number of pedestrian controlled horticultural machines are powered by diesel engines.

The Four Stroke Engine

A four stroke engine has one working or power stroke in every four. A two stroke engine has every second stroke as a power stroke. The basic principle of a four stroke engine is that by burning a mixture of fuel and air above a piston in a cylinder, heat is produced. Expansion of the gases caused by this heat forces the piston downwards, turning the crankshaft. Four stroke engines used in horticulture may have one, two, three or four cylinders depending on size of machine. Large four wheeled tractors, which are of limited horticultural use, often have six cylinder engines.

Each cylinder has two valves. The inlet valve allows air into the cylinder of a diesel engine and a mixture of fuel and air into the cylinder of a petrol engine. The exhaust valve releases waste gases to the exhaust pipe after the fuel and air have been burnt. The valves are opened by the push rods and rocker arms which are operated by the camshaft. The camshaft is driven by and timed with the crankshaft, ensuring that the valves are opened at the correct point during the four stroke cycle. The valves are closed by strong springs. The timing of the valves in relation to the movement of the piston is described later in this chapter.

Most multi-cylinder engines have overhead valves. This means that the valves are in the cylinder head above the pistons. Some overhead valve engines have a cross flow

PLATE 2.1 *A three cylinder four stroke diesel engine with a power output of up to 26 kW (35 hp). The fuel system, oil filter, crankshaft pulley and fan belt which drives the alternator and cooling system fan can be seen.* (Perkins)

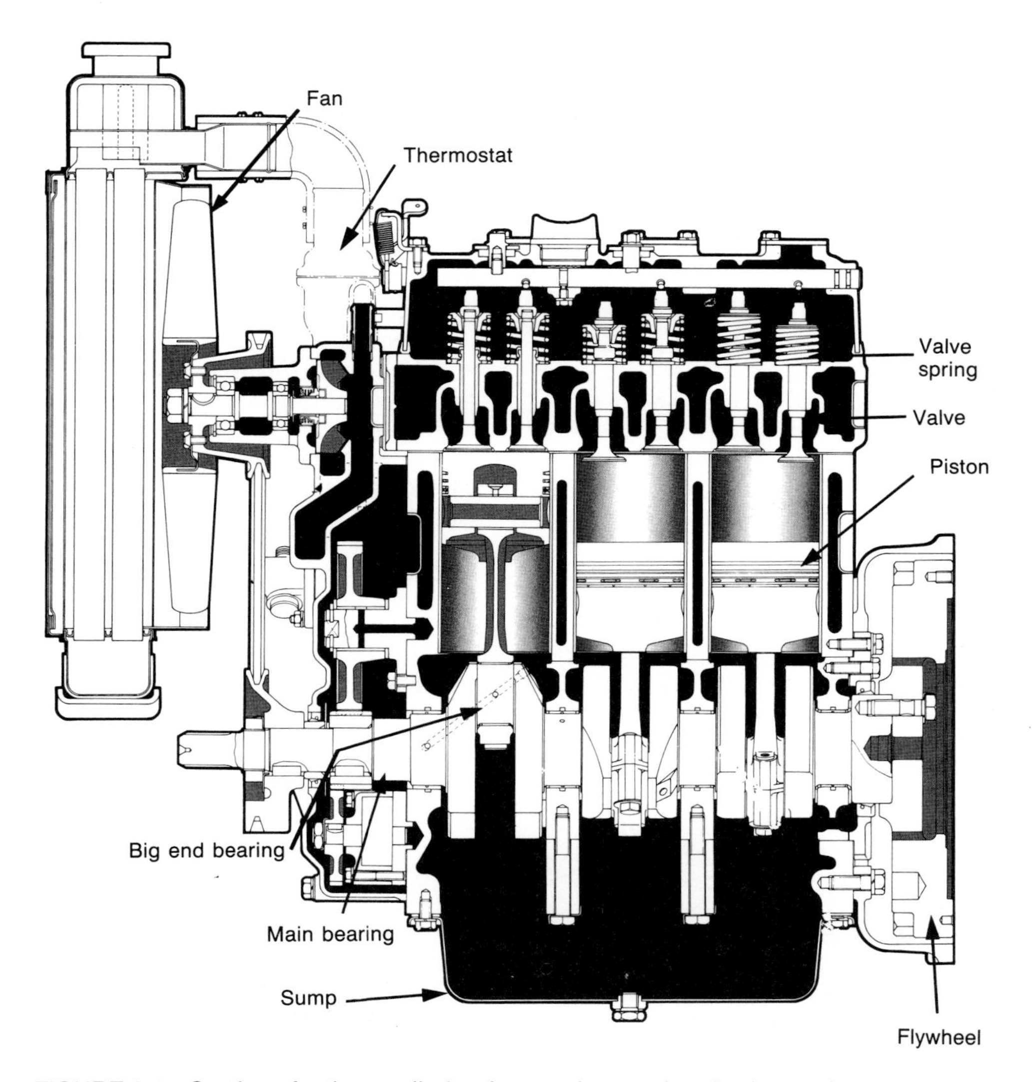

FIGURE 2.1 Section of a three cylinder, four stroke, overhead valve engine. *(Kubota)*

cylinder head. This means that the inlet and exhaust manifolds are on opposite sides of the engine. This design has the advantages of a quick exit for exhaust gases and a rapid entry for the incoming supply of air or mixture of air and fuel, depending on the type of engine.

Some single cylinder engines have side valves. These engines are used to provide the power for small pedestrian controlled machines including lawn mowers and rotary cultivators. The valves are in the cylinder block of a side valve engine, close to the piston, and move upwards when they are opened by the camshaft. Overhead valves are more efficient than side valves, but the simple design of the side valve mechanism makes it ideal for small petrol engines.

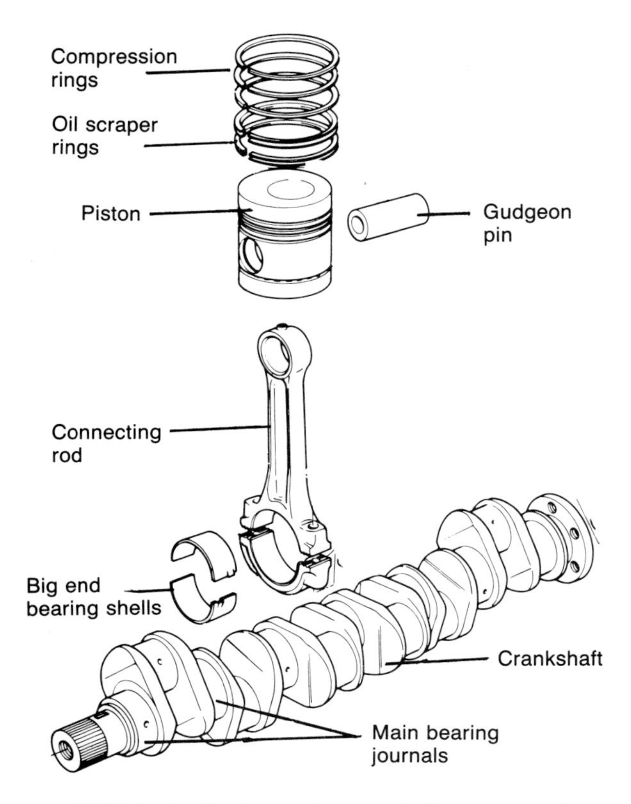

FIGURE 2.2 Piston and crankshaft components.

The piston is attached to the connecting rod (con rod) with a gudgeon pin, which runs in the little end bearing. The con rod is held on the crankshaft by the big end bearing and cap. Each piston has a set of piston rings. The top ones are compression rings, which act as a seal between the piston and the cylinder wall, to prevent loss of pressure above the piston on the compression stroke. The bottom ring is an oil scraper ring. Some engines have two of these, and their purpose is to scrape oil from the cylinder walls to prevent it getting above the piston into the combustion chamber. When an engine cylinder and piston rings are worn, oil will get past the piston and foul the sparking plugs.

The crankshaft is secured in the cylinder block by the main bearings and caps. A single cylinder engine has two main bearings, while four cylinder diesel engines and many four cylinder petrol engines have five. Bolted to one end of the crankshaft is a heavy flywheel, which stores energy and keeps the engine running smoothly between the power strokes.

The timing gear at the front of the crankshaft drives the camshaft which opens the valves at the correct point on the four stroke cycle. The crankshaft timing gear also drives the fuel injection pump on a diesel engine, and is timed to inject the fuel into the cylinder at the correct point on the compression stroke.

On petrol engines, the camshaft also times the delivery of the spark to the sparking plug at the correct point in the cycle.

Some timing systems are arranged through a series of gears, while others are chain driven from the crankshaft.

Diesel engines are known as compression ignition engines because the fuel is ignited by the immense heat created in the cylinder when the air is compressed. Petrol engines are known as spark ignition engines because the mixture of air and fuel in the cylinder is ignited by an electric spark. All internal combustion engines have a cooling

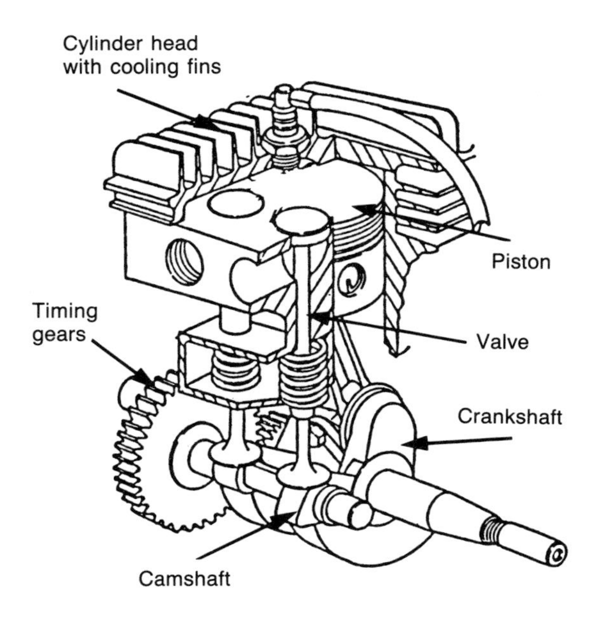

FIGURE 2.3 Layout of a single cylinder, overhead valve, air cooled engine.
(Briggs and Stratton Corp.)

system to remove unwanted engine heat. Most multi-cylinder engines are water cooled, while the majority of single cylinder engines are air cooled. Lubrication is vital so all engines have a lubrication system to ensure a constant flow of oil to all moving parts. The air cleaner is another essential part of an engine. It provides a constant, dust free supply of air to the cylinders.

Spark ignition engines have a carburettor to supply the fuel and an ignition system, either a magneto or coil ignition, to provide an electric spark which ignites the mixture of fuel and air. Compression ignition engines have a fuel system which provides precise amounts of fuel to the cylinders.

The following pages explain the working principles of a four stroke engine.

The Four Stroke Diesel Engine

The diesel or compression ignition engine relies on immense heat inside the cylinder to burn the fuel as it is injected into the combustion space. The high temperature is achieved by compressing the air in the cylinder to a very high pressure, approximately 34 bar (500 psi) giving a temperature of approximately 550°C. Some high performance engines have much higher compression pressures with the temperature in the cylinders approaching 900°C.

The four stroke diesel works in this way:

Induction stroke The piston moves downwards and sucks air from the air cleaner into the cylinder through the open inlet valve.

Compression stroke Both valves are closed and the rising piston compresses the air in the cylinder to a very high pressure, creating a temperature of 550°C or more. Just before the piston reaches the top of the compression stroke, a fine spray of diesel fuel is injected into the cylinder above the piston. The fuel burns instantly and the resultant heat causes rapid expansion of gases above the piston.

Power stroke The expanding gases in the cylinder drive the piston down, turning the crankshaft. This is the working stroke of the four stroke cycle.

Exhaust stroke The exhaust valve opens at the end of the power stroke. The waste gases are driven from the cylinder by the rising piston. Just before the top of this stroke, the exhaust valve closes and the inlet valve opens in readiness for the next induction stroke.

A single cylinder engine on a pedestrian operated cultivator or mower will run at a speed of 3,000 rpm or more. At this speed, the piston will make 6,000 strokes per minute. Each valve will open and close 1,500 times and there will be 1,500 separate injections of diesel fuel into the cylinder every minute.

The Four Stroke Petrol Engine

The petrol engine operates at lower cylinder pressures and temperatures than a diesel engine. The fuel is ignited by an electric spark after the mixture of fuel and air has been compressed in the cylinder. A four stroke spark ignition engine works in this way:

Induction stroke A mixture of fuel and cleaned air is drawn into the cylinder through the open inlet valve as the piston moves downwards.

Compression stroke Both valves are closed and the rising piston compresses the fuel and air in the cylinder. Just before the piston reaches the top of the stroke, an electrical spark ignites the compressed mixture of fuel and air.

Power stroke The expanding gases drive the piston downwards, turning the crankshaft. The exhaust valve opens at the end of the power stroke.

Exhaust stroke The waste gases are driven out of the cylinder through the open exhaust valve by the rising piston. The exhaust valve closes and the inlet valve opens as the piston reaches the top of the exhaust stroke ready for the next induction stroke.

When at the top of its stroke, the piston is said to be at top dead centre (TDC). The lowest position of the piston in the cylinder is known as bottom dead centre (BDC). The letters TDC and BDC will be found in engine

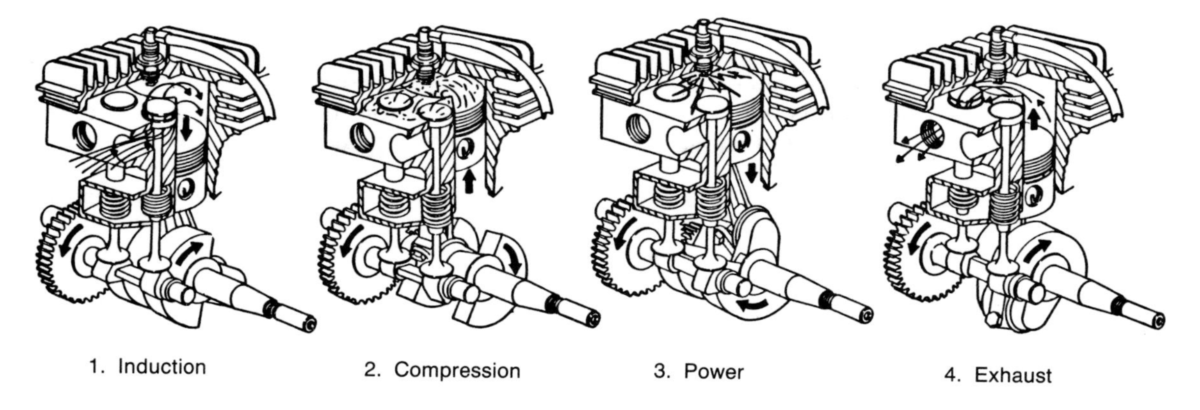

FIGURE 2.4 The working principles of a four stroke petrol engine. *(Briggs & Stratton Corp.)*

instruction manuals in sections dealing with engine timing.

Firing order
The firing order is the sequence in which the cylinder receives its spark or injection of fuel. An engine is designed with a set firing order. Any model of engine can have only one firing order. The firing order for an engine with more than two cylinders is given in the operator's manual and it may also be stamped on the engine. The firing orders for three, four and six cylinder engines are:

Three cylinder	1, 2, 3
Four cylinder	1, 3, 4, 2 or 1, 2, 4, 3
Six cylinder	1, 5, 3, 6, 2, 4 or
	1, 4, 2, 6, 3, 5

The number one cylinder is usually at the front of the engine.

Compression ratio
This is the ratio, or relationship, between the volume in the cylinder when the piston is at the bottom of its stroke (BDC) and the volume in the cylinder when it is at the top (TDC). The higher the compression ratio, the greater will be the pressure in the cylinder at TDC.

A spark ignition engine has a compression ratio of about 7 : 1. This means that the volume at TDC is one-seventh of the volume in the cylinder at BDC. A typical compression ratio for a compression ignition (diesel) engine is 16 : 1. Compression ratios may be as high as 20 : 1 on some of the smaller diesel engines.

THE DIESEL ENGINE

A diesel engine has a more robust design than a petrol engine. It must withstand much higher working temperatures and pressures resulting from the high compression ratio. The sturdier design makes the diesel engine more expensive to manufacture but it is preferred for the larger tractors used in horticulture because of its greater efficiency and operating economy compared with a petrol engine of similar size. Diesel engines are becoming more widely used to power ride-on grass mowers and pedestrian operated machinery, especially rotary cultivators. Diesel engine exhaust gases are also considered less harmful to the environment.

Direct and indirect injection
There are two types of diesel engine used in horticulture. The direct injection engine has the fuel injected into the hottest part of the cylinder, directly above the piston. This gives the direct injection diesel engine the advantage of good cold starting performance, even on a winter morning. Direct injection engines in horticulture are mainly restricted to compact tractors and models produced mainly for agricultural use.

The indirect injection engine differs in that the fuel is injected into a pre-combustion chamber at one side of the main combustion space above the piston. This is more efficient than a direct injection engine in mixing the fuel with the air in the cylinder.

Indirect diesel engines can be a problem to start in cold weather, but this difficulty is overcome by using the engine's cold starting device on winter days.

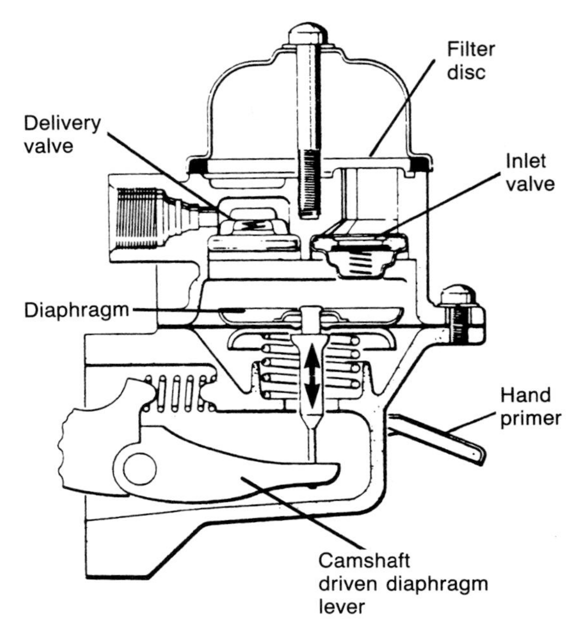

FIGURE 2.5 Diesel fuel lift pump. *(Perkins Engines)*

The Diesel Fuel System

A four stroke diesel engine requires a precise amount of spotlessly clean fuel injected into the cylinder, in an atomised form, as the piston nears TDC on the compression stroke. This is the function of the diesel fuel system.

The fuel flows from the tank through a tap to the lift pump and on to the fuel filter. Some engines do not have a lift pump but rely on gravity feed. The fuel tap is always left on to prevent air bubbles collecting in the system, as air in diesel fuel will cause misfiring. A build-up of air is likely to stop the engine which cannot then, as a general rule, be restarted until the air has been removed or bled from the system. The tap should be turned off only when servicing the fuel system.

The *lift pump,* when fitted, is driven by either the engine camshaft or the fuel injection pump. It supplies fuel under slight pressure to the filter. Some lift pumps have a fine filter to collect dirt and water from the fuel. Many have a hand priming lever used to bleed the system when unwanted air is present in the fuel.

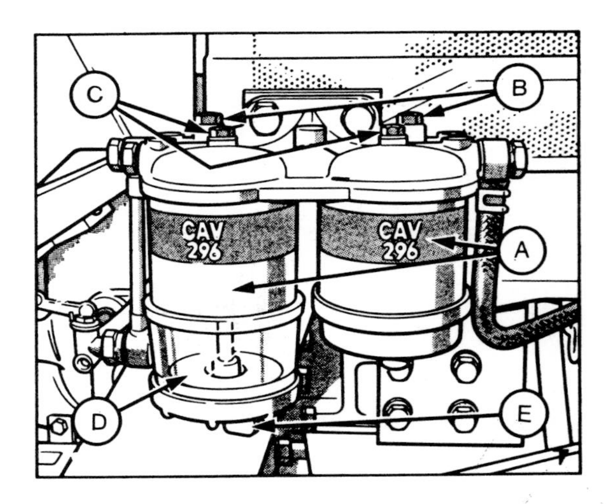

FIGURE 2.6 Diesel fuel filters. The filter elements (A) are secured by a single bolt (C). The first filter has a glass bowl (D) beneath it which collects any dirt and water in the fuel. The tap (E) is used to drain water from the bowl. Each filter unit has a bleed screw (B). *(Ford New Holland)*

The *fuel filter*—some engines have two—removes all traces of dirt from the fuel. It is usual for the filter to have a water trap, with a drain plug, which collects water in the fuel, caused by condensation in the tank. Water is heavier than diesel fuel and will drop to the bottom of the filter unit into the water trap from where it can be drained off. The fuel filters must be renewed periodically, usually after 600 hours of service. Each filter unit has a bleed screw for removing all traces of air after fitting a new filter element.

The *injection pump* is driven by the engine

timing gears. It delivers minute quantities of fuel to the cylinder at a pressure of approximately 175 atmospheres. (One atmosphere equals 14.7 psi, so a pressure of 175 atmospheres is equivalent to 2,500 psi.) There are two types of injection pump: in-line and rotary (or DPA).

The *in-line pump* has a separate pumping element for each injector. These elements are operated by a camshaft inside the pump housing which also controls the timing of fuel injection to the cylinders. An in-line injection pump on a three cylinder engine will have three pumping units. It is set by the manufacturer to ensure that each cylinder receives equal amounts of fuel. A worn in-line pump will not be able to supply equal quantities of fuel to each cylinder, thus causing uneven running and wasting of fuel.

The *rotary pump* has one pumping unit with a distributor head which delivers fuel to the injectors in the firing order. The advantage of the rotary pump is that it always delivers equal quantities of fuel to each cylinder, even when the pumping unit is worn.

The throttle setting regulates the amount of fuel injected into the cylinders by the injection pump and in this way controls engine speed. In addition to the usual hand throttle lever used to select a constant engine speed, some engines have a foot throttle. This overrides the hand throttle and is used to alter the engine speed as required.

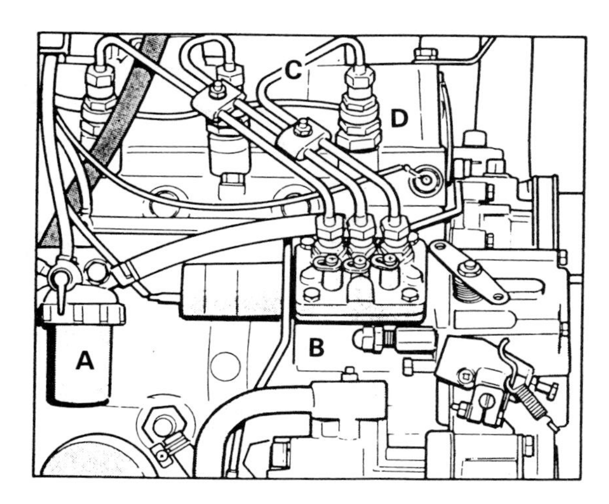

FIGURE 2.7 The diesel fuel system for a three cylinder engine. The filter (A) supplies clean fuel to the injection pump (B) which delivers minute quantities of atomised fuel through heavy duty steel fuel pipes (C) to the injectors (D).
(Ford New Holland)

A *governor* is built into the injection pump to maintain the engine speed selected with the hand throttle. The governor enables the injection pump to increase or decrease the amount of fuel supplied to the engine. When the load on the engine is increased, the crankshaft speed will fall away. To overcome this speed and power loss, the governor will increase the quantity of fuel supplied to the cylinders by automatically changing the injection pump setting. When the engine load is reduced, it will run faster than the pre-set speed. In this situation, the governor reduces the fuel supply to bring the engine back to the selected speed.

The *injector*, or atomiser, sprays a fine mist of fuel into the cylinder when the piston is a few degrees before TDC on the compression stroke. The hole, or holes, in the tip of the injector are very small, often no more than 0.2 mm in diameter. More fuel is supplied to the injectors than is actually injected into the cylinders. The extra fuel is used to cool and lubricate the injectors, and it then returns to the tank or filter unit by means of small fuel pipes called leak-off pipes.

Failure to supply clean fuel to the injectors will increase their rate of wear. When worn, the injectors will put too much fuel into the cylinders,. This wastes fuel and shortens the efficient working life of the engine. A sure sign of worn injectors is a continuous cloud of black smoke from the exhaust pipe. The injectors should be checked periodically and replaced when worn. This is a task for a trained mechanic.

Care of the Diesel Fuel System

Water and dust are the main enemies of a diesel fuel system. By filling the fuel tank at night, after the day's work is done, condensation inside the tank will be reduced and there will be less water to contaminate the fuel. Always use clean fuel. Fill the tank through a funnel with a fine mesh filter in the

PLATE 2.2 *Diesel engine fuel system showing the fuel sediment bowl (bottom right), injection pump and the three fuel pipes which deliver fuel to the injectors.* (Massey-Ferguson)

neck if there is any doubt about the quality of the fuel being used.

Some fuel lift pumps have a water trap and filter. Make sure the filter is cleaned when changing the main filter element and drain off any trapped water.

Regular servicing of the main filter element (some engines have two) is essential because with use, the element will become clogged with dirt and cannot work efficiently. Most diesel fuel systems require that the main fuel filter be changed at 600 hour intervals. Check the engine instruction book to find the filter change periods for any engine you have to maintain. Always replace the rubber sealing rings when fitting a new element. Some filter units have a water trap at the base, below the element. This water trap should be emptied and thoroughly cleaned when the filter element is renewed.

Care must be taken when cleaning parts of the diesel fuel system during routine maintenance. The parts should be washed in clean fuel and left to drain. Fluffy rags should not be used because tiny pieces of material can cause problems if they find their way into the injection pump. The water trap should be drained from time to time between filter changes. After servicing the filters, it will be necessary to bleed the system to remove all traces of air from the fuel.

Very few diesel injection pumps require any maintenance between major engine overhauls. See the instruction manual to find out if any is required on your tractor. Some in-line pumps have an oil level plug which should be checked periodically.

The injectors will wear and should be checked by a mechanic with an injector tester at intervals of about 600 hours. Faulty injectors need to be replaced either with new ones or with cheaper exchange units which have been overhauled and tested.

Bleeding the Fuel System

There will be air in the fuel system after it has been serviced. The same problem will arise if the fuel tank runs dry. Before the engine can be restarted, the air must be removed by bleeding the fuel system. To carry out this task, first check that there is plenty of fuel in the tank and make sure the tap is turned on. Then:

- Slacken or remove the bleed screw from the top of the filter body. Operate the lift pump priming lever until the fuel flowing from the bleed point is free from air bubbles. Tighten the bleed screw. Some bleed screws have a vent hole drilled in them, and this type only needs to be slackened a turn or two to release the air when the priming lever is used. Fuel systems without a lift pump can be bled by gravity. After removing the bleed screw, wait for the fuel to flow free of air bubbles and then retighten the screw.
- Bleed the second filter, if there is one, in the same way.
- Next, slacken or remove the bleed screw on the injection pump and repeat the procedure to clear all air bubbles from the fuel. Some injection pumps have two bleed points.
- It may be difficult to start a diesel engine with a rotary pump after the fuel system has been serviced and bled. This problem is solved by bleeding the fuel injector

pipes. Slacken or remove at least two pipes from the injectors. The fuel stop button must be in the 'run' position. Turn the engine with the starter motor until fuel is seen coming from the pipe ends. Then replace and tighten the pipes. The engine should now start and soon run smoothly.

Cold Starting

In cold weather, some diesel engines, especially the indirect injection type, will be difficult to start. To overcome this problem, some form of cold starting aid is provided. The main types are:

- *Excess fuel device* Extra fuel can be supplied to the engine when starting from cold by pushing in the excess fuel button on the injection pump. The button will return to its normal running position once the engine has started.
- *Heaters* There are two types. Some engines have a small heater coil, or glow plug, in each cylinder. This warms the air inside the cylinder before the engine is started. A pilot light on the instrument panel shows the driver when the glow plugs are hot. The second, and more common type of heater, warms the air in the inlet manifold before it enters the cylinder. Extra fuel is automatically sprayed into the hot air through a thermostatic valve. The fuel burns in the manifold and this is the point at which the driver should use the starter.
- *Aerosols* Liquids such as ether which burn very rapidly are available in aerosol form and are another type of cold starting aid. The liquid is usually sprayed into the air cleaner immediately before operating the starter motor. Aerosols are usually successful but should be used as a last resort. However this may be the only way to start a worn engine on a cold day.

Diesel Fuel Storage

It is essential to use clean fuel to ensure long, trouble-free service from diesel injection equipment. The storage tank must be capable of keeping the diesel fuel as clean as when it was delivered by the fuel tanker.

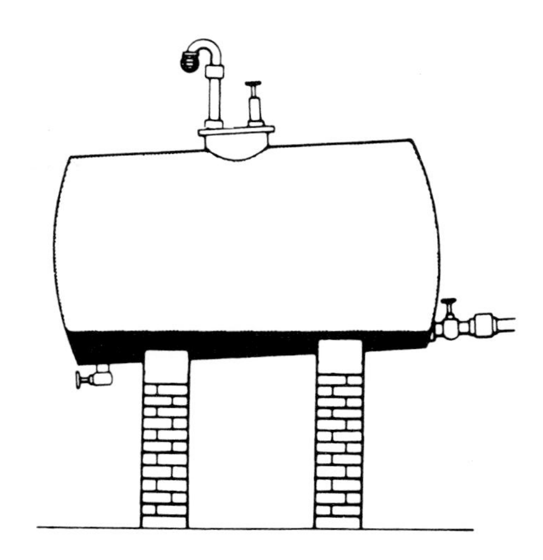

FIGURE 2.8 Diesel fuel storage tank.

The storage tank must not be made from galvanised iron because the fuel will attack the zinc coating used in the galvanising process. The tank should be high enough to allow gravity filling of the tractor tanks and it should be positioned to give easy access for tractors and the delivery tanker. The bottom of the tank should slope away from the outlet point with a sludge tap at the lower end so that water and sludge can be drained off. This should always be done before a new delivery of fuel arrives.

The cover on the tank filling point, usually a short intake pipe with a screw cap, should be airtight. A vent pipe is needed at the top of the tank to allow air in when fuel is drawn off; it should be a type which minimises the entry of dust and rainwater. A dipstick or sight gauge is required to check the fuel level.

Fuel should not be taken from the storage tank immediately after a delivery, as the new fuel will stir up the sludge in the bottom of the tank, leaving the fuel unfit for use. When a delivery is expected, fill the tanks of any equipment likely to be used that day and then drain off the sludge from the sludge tap. After delivery, leave the tank to settle for 24 hours before drawing off any fuel.

Turbochargers

The power developed by an engine is determined by the amount of fuel it can burn during the brief period of combustion. An engine needs plenty of air to burn the fuel completely, and the amount of air taken into the engine cylinders can be increased with a turbocharger. It has a fan, driven by the hot exhaust gases, which draws air from the air cleaner, compresses the air slightly and feeds it to the cylinders. Turbocharged engines must be allowed to idle for a short period after starting up to ensure the turbocharger is fully lubricated. Similarly, the engine should be left to idle for a while before stopping. This allows the turbocharger shaft, which can reach speeds of 80,000 rpm, to cool as it slows down. Few tractors used for horticultural work have turbocharged engines, which are more common on medium and high powered agricultural tractors and large irrigation pump power units.

PLATE 2.3 *The dry air cleaner on a compact tractor diesel engine.* (Massey-Ferguson)

Air Cleaners

Every engine from the smallest single cylinder to the largest multi-cylinder model has an air cleaner, which does the important job of removing impurities from the air supplied to the cylinders. Dust and grit will damage the cylinder walls, pistons and valves if not removed from the air.

Dry air cleaners
The engine air supply is drawn through a filter element which removes any dust it contains as it passes to the engine. The filter element may be made of various materials including thick paper, felt, foam, plastic mesh or fine plastic gauze. Small single cylinder engines have a simple dry air cleaner contained in a clip-on housing. The element must be cleaned at frequent intervals to ensure an unrestricted flow of air to the engine. Many manufacturers advise servicing at 25 hour intervals in normal working conditions and as frequently as 3 hours when operating in a very dusty atmosphere.

Felt, plastic and foam filter elements can be washed in soapy water, thoroughly dried and then refitted for further service. Some manufacturers advise that felt and foam elements should be soaked with oil after cleaning. It is important to squeeze surplus oil from the element before it is replaced.

Paper filter elements can be cleaned by tapping them on a firm surface to remove the dust or by blowing compressed air from the inside to the outside. After the elements have been cleaned many times, they will become unserviceable and need replacing. Some types of paper filter element can be washed; however, the instruction book should be consulted before washing one.

The type of dry air cleaner commonly fitted to multi-cylinder engines consists of either one or two filter elements made of felt or thick paper contained in a housing with a removable end cap. (See Fig. 2.11.) It has a pre-cleaner, usually dome shaped with angled air inlet slots, which separates large pieces of dirt from the air before it passes to the filter element(s) where the fine particles of dust are removed. A gauge or warning light on the instrument panel warns the driver if the air flow to the engine is restricted.

A dry air cleaner element is cleaned after removal from its housing, by tapping it gently against a rubber tyre or similar semi-

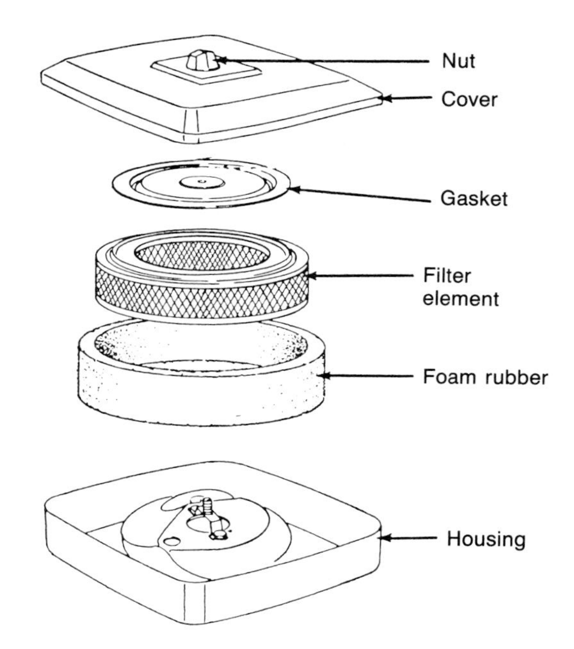

FIGURE 2.9 Dry air cleaner for small single cylinder engine with paper and foam rubber filters. *(Briggs and Stratton Corp.)*

FIGURE 2.10 Dry air cleaner for single cylinder engine. *(Briggs and Stratton Corp.)*

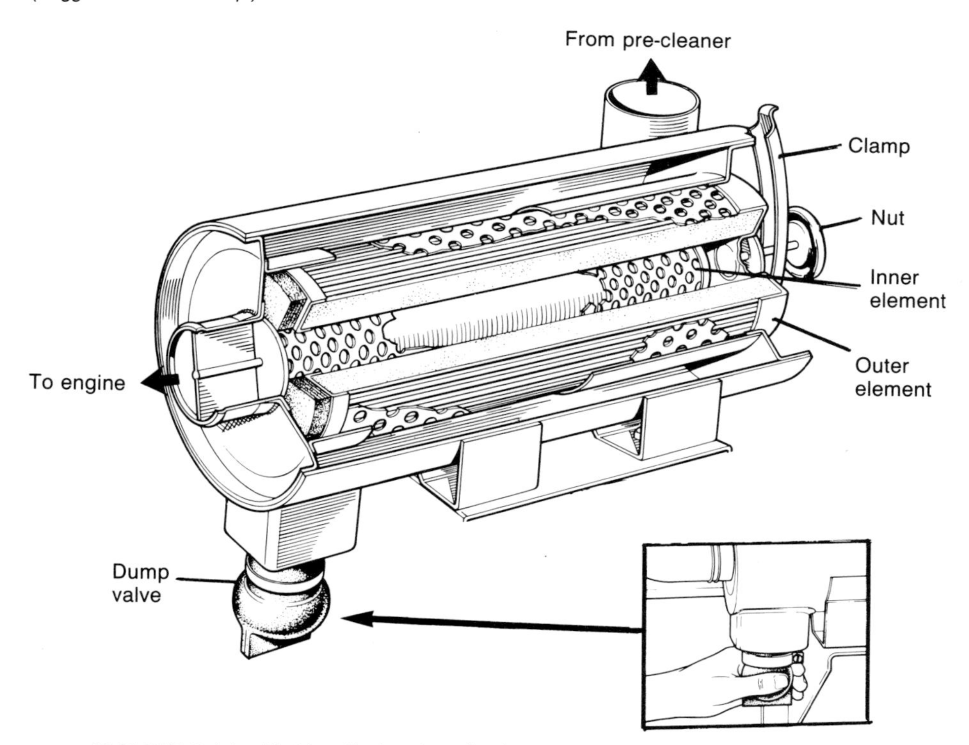

FIGURE 2.11 Multi-cylinder dry air cleaner. *(Massey-Ferguson)*

hard material to knock out the dust. The element should be rotated between each tap on the tyre. This method will not harm the element, which will be useless if damaged. Alternatively, the element can be cleaned with compressed air. The pressure should not exceed 2 bar (30 psi). Blow from the inside of the element outwards; do not hold the air nozzle close to the element as this may cause damage. Some instruction manuals advise washing felt outer elements in soapy water after 300 hours' service. Use cool water, shake the surplus away and leave it to dry naturally.

When servicing dry air cleaners with two filter elements, the inner element should not be removed. However, if the air cleaner warning light is still on after the outer element has been serviced, the inner element should be replaced—a job for a trained mechanic.

In normal conditions, the dry air cleaner element should be checked weekly and cleaned as necessary. In dusty working conditions it should be checked daily.

Dry air cleaners are efficient in most working conditions and simple to maintain. However this type of air cleaner is not ideal for engines which have to operate in very dry, dusty conditions. Tractors and engine driven machines which have to work in such adverse conditions are often equipped with the more efficient oil bath type of air cleaner.

Oil bath air cleaners

Oil bath air cleaners have the same type of pre-cleaner as dry air cleaners. With the larger pieces of dirt removed, the air passes through the oil bath and oil impregnated wire gauze filter. Dust is trapped by the oil in the gauze as the air passes into the engine. When the engine is stopped, the oil will drip back into the oil bath and the dirt settles out. As the dirt level builds up, the oil level will also rise. To prevent the oil level becoming too high, the oil bath must be emptied and the dirt cleaned out.

Maintenance is very important. A clogged air cleaner starves the engine of air, causing loss of power, waste of fuel and undue wear. In normal working conditions, check the oil bath every week; in dusty conditions it should be checked every working day. When about 12 mm of dirt has built up in the oil bath, it must be cleaned out and refilled with new engine oil to the level mark. The gauze filter can sometimes be removed from the housing for cleaning. When this is possible, it can be washed with petrol and left to drain or be cleaned with compressed air.

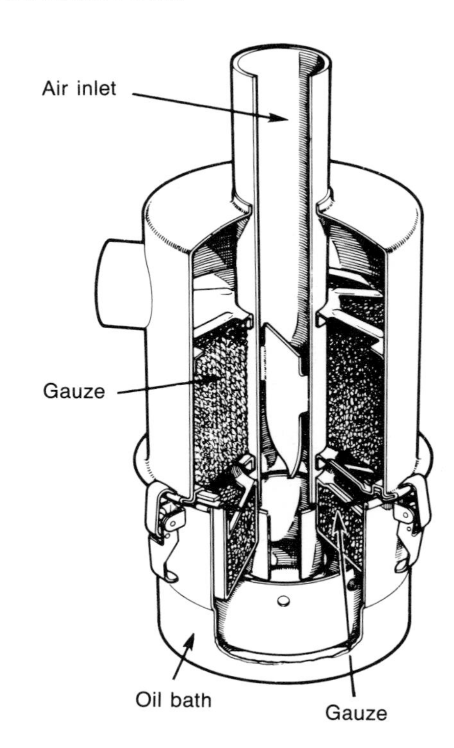

FIGURE 2.12 Oil bath air cleaner.

Petrol Engines

Petrol engines are in common use in horticulture. Air cooled single cylinder four stroke engines drive small ride-on mowers, pedestrian controlled machines and some hand held equipment. Petrol engines with two or more cylinders, either air or water cooled, are used to power many types of four wheel horticultural tractor and specialist grass machinery. Two stroke petrol engines are used to drive chain saws, strimmers, lightweight rotary mowers, etc. A carburettor

PLATE 2.4 *Single cylinder four stroke petrol engine with a power output of 2.5 kW (3.5 hp). It has a vertical crankshaft and horizontal piston. The recoil starter is at the top of the engine.* (Briggs & Stratton Corp.)

supplies and mixes the fuel with the air. The ignition system provides a spark to ignite the compressed mixture of fuel and air in the cylinder. The spark is produced by a magneto. Multi-cylinder engines have a coil ignition system.

The Fuel System

Fuel flows from the tank through a tap to the carburettor float chamber. A small needle valve, controlled by the float, maintains the correct fuel level in the float chamber. The petrol flows from the float chamber into a small diameter pipe (main jet) in the tube which carries the air from the air cleaner to the cylinder. The diameter of the air tube around the main jet is reduced, often with an insert called the venturi. Because the venturi reduces the diameter of the air tube, the speed of air is increased at this point. The increase in air speed around the main jet causes fuel to be drawn from the main jet and to mix with the air as it passes to the engine.

The *choke* is an adjustable valve situated between the air cleaner and the carburettor. It is used to start a cold engine by restricting the air flow to the engine and in this way supplies an air/fuel mixture with a higher petrol content. This is called a rich mixture.

The *throttle* controls the amount of air/fuel mixture supplied to the engine. The throttle valve, which is between the carburettor and the engine, is used to select engine operating speed by means of the throttle lever or pedal. The governor can override the throttle lever setting to maintain the required engine speed.

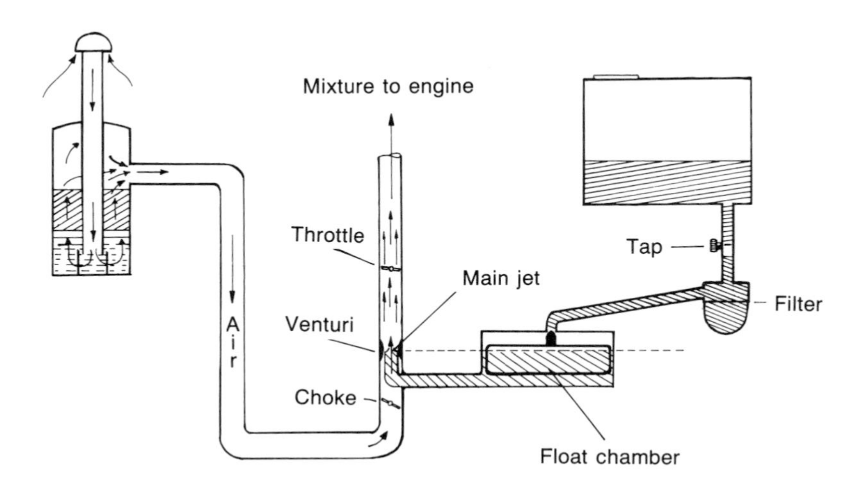

FIGURE 2.13 Petrol engine fuel system.

Air/fuel ratio
For normal engine running, the carburettor is set to mix one part of fuel with fifteen parts of air by weight. This means that 1 kg of petrol mixes with 15 kg of air to give an air/fuel ratio of 15 : 1. When the choke is used, the carburettor supplies a much richer mixture with an air/fuel ratio of 8 : 1. It is difficult to understand air/fuel ratio by weight, but in terms of volume, with a ratio of 15 : 1, approximately 9,000 litres of air will mix with just one litre of fuel. An engine will not develop full power when set to run on a weak mixture such as 17 : 1. A rich mixture, e.g. 12 : 1, wastes fuel and causes a build-up of carbon around the valves and combustion space. Excessive deposits of carbon reduce engine efficiency.

Smoke signals from the exhaust give clues about the way an engine is performing. Black exhaust smoke when the engine is warm is usually caused by a rich mixture; if the choke is out, this will also cause black smoke. It can also be caused by a clogged air cleaner which is restricting the air supply. Do not confuse black smoke with blue smoke, which is a sign of an engine using (burning) oil. This occurs when the engine is worn. A two stroke engine may also produce blue smoke if too much oil is mixed with the fuel (see page 43).

Diaphragm carburettor
Engines on chain saws and some garden machines have to run with the carburettor tilted sideways. A normal float chamber type carburettor is likely to spill petrol when it does not run in an upright position.

This difficulty is solved by using a diaphragm carburettor which is not affected by the angle at which the engine is required to run. One type has a single diaphragm instead of the usual float chamber and float. Suction created by the induction stroke of the piston draws fuel from the carburettor fuel chamber through the jet into the air as it flows through the venturi to the cylinder. The suction draws the diaphragm upwards against a spring; this movement opens the fuel inlet valve, allowing more fuel into the chamber. During the remainder of the cycle, there is no suction in the venturi and the spring returns the diaphragm to rest, where it will remain until the next induction stroke.

Another design has two diaphragms. One draws fuel into the carburettor chamber and the other supplies the required amount of fuel to the air flowing through the venturi to the cylinder.

Care of the fuel system
It is important to use clean fuel and handle it in clean containers. Many fuel systems have a filter and a sediment bowl, often combined with the fuel tap on the tank, to trap dirt and water mixed with the fuel. This needs to be cleaned periodically.

The carburettor float chamber will collect dirt after a time. The chamber should be removed, taking care not to damage the paper sealing gasket, and then cleaned with petrol. Avoid the use of a fluffy rag. Reassemble the carburettor making sure the gasket is fitted correctly.

The carburettor jets are very small so it is important to use clean fuel to reduce the risk of a blockage. Never attempt to clear a blocked jet with a piece of wire, as this may enlarge the diameter of the jet. Blockages should be cleared with compressed air.

Carburettor adjustments
The slow running or idling speed of the engine is usually adjusted with the throttle stop screw, which is turned by a small screwdriver until the engine runs at tick-over speed with the throttle lever fully closed. A fast idling speed wastes fuel and should be avoided. Some horticultural machines—for example lawn mowers with a centrifugal clutch to control the cutting cylinder and roller drives—need a low idling speed to ensure that drive is fully disengaged when the engine is at tick-over.

Many carburettors have another adjusting screw used to set the air/fuel mixture when the engine is at idling speed. The engine may misfire or stop if the mixture is too rich or too weak. Only a fraction of a turn of the carburettor adjusting screws is normally necessary to achieve smooth running at idling speed. When adjusting a carburettor, turn only one adjusting screw at a time and do so very gradually, taking note of changes in the way the engine runs.

Some carburettors, particularly on chain saws, have a main jet (or power jet) which can be adjusted to give maximum power at high engine speeds. Most engines develop maximum power output at less than full revs; this gives a reserve of engine power when the tractor is heavily loaded.

Ignition Systems

The spark which ignites the compressed mixture of air and fuel in the cylinder must be timed to occur just before TDC on the compression stroke of the four stroke cycle. Electricity, supplied at low voltage, is converted into high voltage current and supplied to the sparking plug at the correct time by means of a coil ignition system or a magneto. Single cylinder engines have a magneto, and many multi-cylinder engines have coil ignition. The magneto makes its own low voltage current, whereas a coil ignition system needs a battery.

The Magneto
The magneto generates its own low voltage electricity and converts it into high voltage current—between 10,000 and 15,000 volts—which is then passed to the sparking plug at the correct time in the four stroke cycle. The main parts of a flywheel magneto are the contact breaker points, the coil, the condenser and the flywheel itself. A switch is usually provided to enable the operator to earth out the current flow to the sparking plug, thus stopping the engine.

Electricity is produced when an electrical conductor is rotated in a magnetic field. Alternatively, a magnet can rotate around a static conductor. [This principle is explained on page 52.] Single cylinder engines have a magneto built into the flywheel. The magnets on the inside of the flywheel rim rotate in close proximity to the coil, a stationary conductor, which is attached to the engine housing inside the flywheel. The coil in an ignition system is a type of transformer which converts the low voltage current, produced by the magneto, to high voltage current, when flow in the low voltage circuit is interrupted.

The *coil* has two separate circuits. The low voltage or primary circuit, which includes the contact breaker points, has a coil with a small number of turns of insulated wire (windings) around a central iron core. The coil in the high voltage or secondary circuit, which feeds current to the sparking plug, has a large number of windings wrapped around the primary coil. There is a layer of electrical insulation between the primary and secondary windings, which are not connected in any way.

In a standard transformer used to increase or decrease voltage supplied to an electrical appliance, the output voltage is related to the number of windings on the secondary coil. To illustrate this, a transformer having a primary coil with 200 turns and a secondary coil with 400 turns will give an output voltage twice that of the input voltage. Thus, with an input of 12 volts, the output voltage would be 24. The secondary coil in a magneto coil will have many thousands of windings to give the very high voltage output needed at the sparking plug.

The *contact breaker points* are a form of high speed electric switch in the low voltage or primary circuit, and they are opened by a cam on the engine crankshaft and closed by spring pressure. Some flywheel magnetos have a breakerless system; this means that the low voltage circuit is controlled electronically.

Working principles of a flywheel magneto
Electricity generated when the flywheel rotates flows from the coil through the closed contact breaker points to earth. When the contact breaker points open, the low voltage current can no longer flow through the primary circuit. At this point, a brief pulse of high voltage current of between 10,000 and 15,000 volts is induced in the secondary circuit of the coil. This brief pulse of high voltage current flows through the plug lead to the sparking plug.

When the cam opens the contact breaker points, the sudden interruption of flow in the primary circuit would cause arcing (spark-

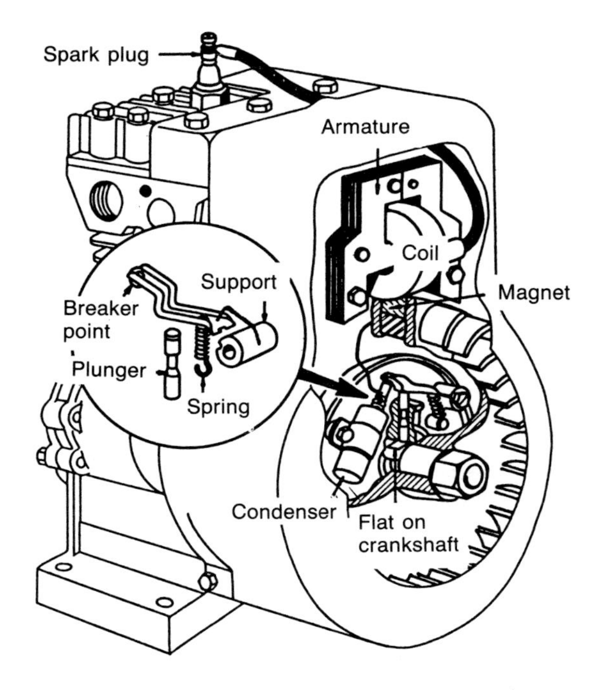

FIGURE 2.14 Flywheel magneto. Inset shows detail of the moving contact breaker point and its return spring. *(Briggs and Stratton Corp.)*

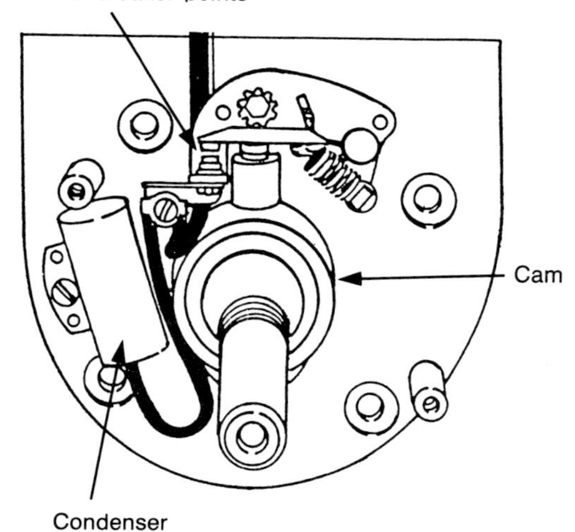

FIGURE 2.15 Contact breaker points and condenser. *(Briggs and Stratton Corp.)*

ing) across the faces of the points. This is undesirable because the tungsten faces of the points would rapidly burn away. The problem is overcome by the condenser, part of the primary circuit. It is a form of electrical shock absorber which reduces arcing when the contact breaker points open. The condenser absorbs the current and uses it to boost the strength of the current induced in the secondary winding (coil). The contact breaker points on a flywheel magneto should have a gap of between 0.3 and 0.5 mm when they are fully open. Timing of the spark is achieved by a cam which is usually an integral part of the crankshaft. Most flywheel magnetos can be fitted to the crankshaft in only one position, since a small key in the shaft must be aligned with a keyway in the flywheel hub before it is secured in position with a nut. A flywheel magneto without a locating key requires careful timing when it is fitted to the crankshaft.

Although no longer in commercial use, magnetos for multi-cylinder spark ignition engines can be seen at vintage tractor rallies. The working principle is similar to the flywheel magneto but, in addition, the magneto has a distributor cap which supplies the sparks to the plugs to suit the engine firing order.

Coil Ignition

The coil ignition system relies on a battery to provide the low voltage current for the primary circuit. Otherwise, the working principle is similar to the flywheel magneto. The main components of a coil ignition system are the coil and the distributor, which consists of the contact breaker points, condenser, rotor and distributor cap.

To understand how the coil ignition system works, careful reference should be made to Figure 2.16.

Low voltage current, usually 12 volts, is supplied by the battery through the ignition switch to the primary circuit terminal on the coil, usually marked 'sw'. The low voltage current flows from a second terminal on the coil, often marked 'cb', to the contact breaker points and condenser. The contact breaker

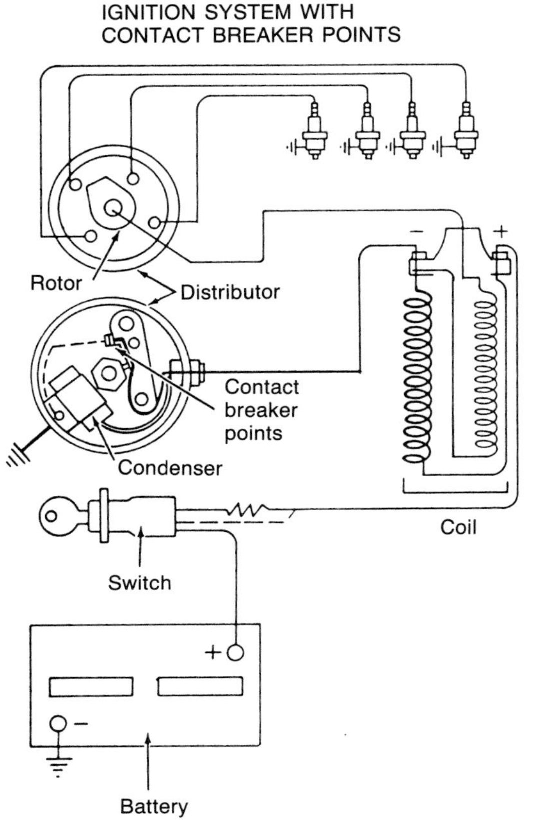

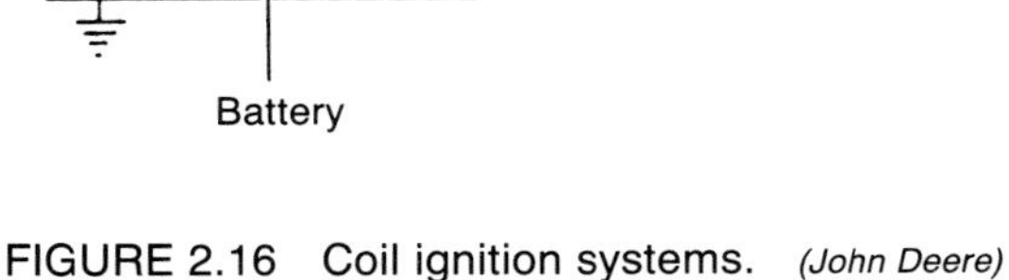

FIGURE 2.16 Coil ignition systems. *(John Deere)*

points are opened by a cam on the distributor drive shaft and closed by a spring.

When the cam opens the contact breaker points, a pulse of high voltage current, again between 10,000 and 15,000 volts, is induced in the secondary windings of the coil. The high voltage current flows from the larger centre cable on the coil, often called the king lead, and passes to the rotor arm inside the distributor cap. There are four equally spaced brass terminals inside the distributor cap on a four cylinder engine. Each terminal is connected by a plug lead to a sparking plug. The rotor arm is timed to pass close to one of the cap terminals every time the contact breaker points open.

The plug leads are connected to the distributor cap terminals in a sequence to match the engine firing order. In this way, a pulse of high voltage current reaches the sparking plug when the piston is a few degrees before TDC on the compression stroke. The current must jump the gap between the insulated central electrode to the metal body of the sparking plug. When the current jumps this gap a spark occurs, igniting the compressed mixture of air and fuel in the cylinder.

When an engine runs at 2,000 rpm, it requires 1,000 separate pulses of high voltage current at each sparking plug every minute. A four cylinder engine running at this speed will need 4,000 sparks every minute.

Electronic Ignition Systems

Contact breaker points suffer from mechanical wear and burning or pitting of the contact faces; this means frequent servicing and regular replacement is necessary. At high operating speeds contact breaker points tend to bounce, resulting in incorrect timing of the spark. Electronic ignition systems operate on the same electrical principles but by substituting an electronic unit for the usual contact breaker, it is possible to overcome most of the problems of pitting, bounce and wear.

The *breakerless system* has an electronic trigger switch and rotor unit which does the same job as the contact breaker points. It solves the problems of bounce and wear, requires no maintenance and, should it fail, is easily replaced.

The *transistor assisted coil ignition system* has contact breaker points but they are used only to operate a primary circuit transistor, which controls a second transistor that supplies low voltage current to the induction coil. This, in turn, produces the pulse of high voltage current which passes to the rotor arm and on to the sparking plugs. As very little current flows across the contact breaker points, the problem of pitted contact faces is overcome, but bounce and mechanical wear still occur.

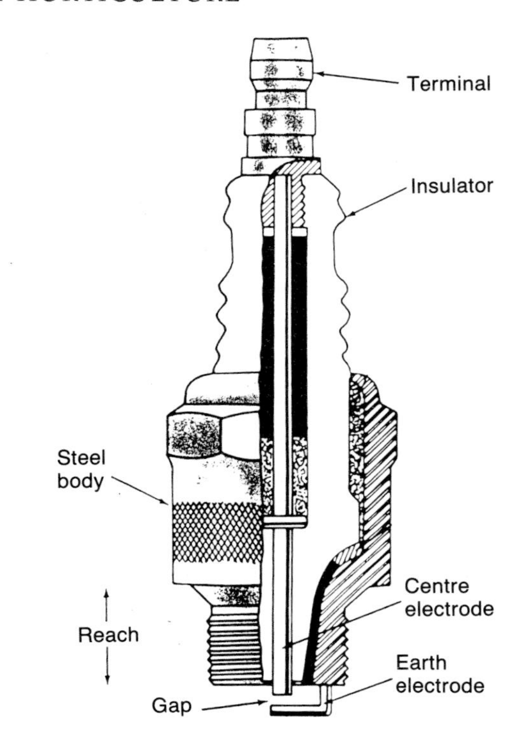

FIGURE 2.17 Section of a sparking plug. The gap is adjusted by carefully moving the earth electrode. *(John Deere)*

Sparking Plugs

A sparking plug has a metal body which is screwed into the cylinder head. The central part of the plug is ceramic, an electrical insulator. A thin metal rod, called an electrode, running through the centre of the ceramic material carries the high voltage current. A second or earth electrode is attached to the bottom of the sparking plug body. The spark occurs when the current jumps from the centre electrode to earth.

There are many types of sparking plug, made to suit different engine designs and working temperatures. Sparking plugs are made with three different lengths of thread (reach), most being 14 mm in diameter. The engine running temperature influences choice of plug. Some engines need cold running plugs which have the metal body designed to take the heat away from the electrode. Hot running plugs have a shorter body which keeps the electrode at a high temperature. The sparking plug heat range indicates whether the plug is designed for hot or cold running.

It is important to use the correct sparking plug as specified in the owner's handbook. A cold running plug will not be efficient if the engine needs a hot one. The reach must also be correct. For example, a long reach plug used where a short reach plug is required will reduce engine efficiency, especially on a small single cylinder engine. It is important to fit a matching set of plugs to a multi-cylinder engine, as an engine with odd plugs will waste fuel and lose power.

Care of the Ignition System

Contact breaker points The gap between the contact breaker points should be between 0.3

and 0.4 mm when open. The stationary point is adjusted after slackening the fixing screws. Make sure the points are fully opened by the cam before setting the gap. The points can be removed for cleaning with a fine oil stone but take great care to keep the faces parallel. It is best to fit a new set of points when the faces are pitted. Make sure the insulating washer is in place between the stationary and moving parts of the contact breaker set as, if the washer is missing, the unit will not spark.

Sparking plugs The gap between the centre and earth electrodes should be about 0.6 mm. Check it with a 0.6 mm feeler gauge, and adjust the gap by moving the earth electrode. Never try to move the centre electrode, as the ceramic material may crack and the plug will be useless.

Sparking plugs should be cleaned periodically to prevent a build-up of carbon around the electrodes. This can be removed with emery cloth or a wire brush. After cleaning, check and if necessary reset the gap.

Plugs have a limited life and should be renewed as directed by the manufacturer. The continued use of old and worn plugs is false economy. The settings given in this chapter for contact points and sparking plugs are purely a guide. Check the instruction book for the correct settings and renewal periods for any engine you wish to service.

Ignition Troubles

Starting and running problems are frequently caused by ignition or fuel system faults. Provided there is plenty of fuel in the tank and the carburettor is clean, look for the following minor faults in the ignition system before seeking professional advice.

- When there is no spark, it may be due to:
  (a) Dampness in the magneto or coil ignition system.
  (b) Contact breaker points stuck open or shut.
  (c) Very dirty or pitted contact breaker points.
  (d) Broken or loose wire.
  (e) Faulty sparking plug.
  (f) Damp or cracked distributor cap (coil ignition).
- When there is a very weak spark, look for:
  (a) Dirty spark plug or incorrect gap setting.
  (b) Dirty contact breaker points.
  (c) Loose wires or terminals.
  (d) Dampness.
  (e) Discharged battery (coil ignition).

THE TWO STROKE PETROL ENGINE

Because they are small, lightweight power units, two stroke engines are ideal for driving a variety of small horticultural machines. Some lawn mowers have two stroke engines but their main use is for hand held horticultural and estate equipment, including chain saws, grass trimmers, brush cutters, hedge trimmers, dusters, blowers, etc.

The main differences between two stroke and four stroke engines are:

- Every other stroke of the piston in a two stroke engine is a power stroke compared with one in every four in a four stroke engine.
- The two stroke petrol engine has no valves. It has openings, called ports, in the cylinder walls which serve the same purpose as the valves in the four stroke engine. There are three ports:
  1. The *inlet port* connects the cylinder to the air and fuel supply from the carburettor and air cleaner.
  2. The *transfer port* connects the crankcase to the combustion space above the piston.
  3. The *exhaust port* connects the combustion space above the piston to the exhaust pipe.
- The crankcase is airtight and contains no oil; lubrication is by oil mist (oil is mixed with the petrol). The instruction book states the proportions of oil and fuel to be mixed together, and this advice should be carefully followed.

Reference to Figure 2.18, noting the position of the components, will help the reader follow the working principles of the two stroke cycle.

First stroke As the piston nears TDC, a mixture of air and fuel (which contains some

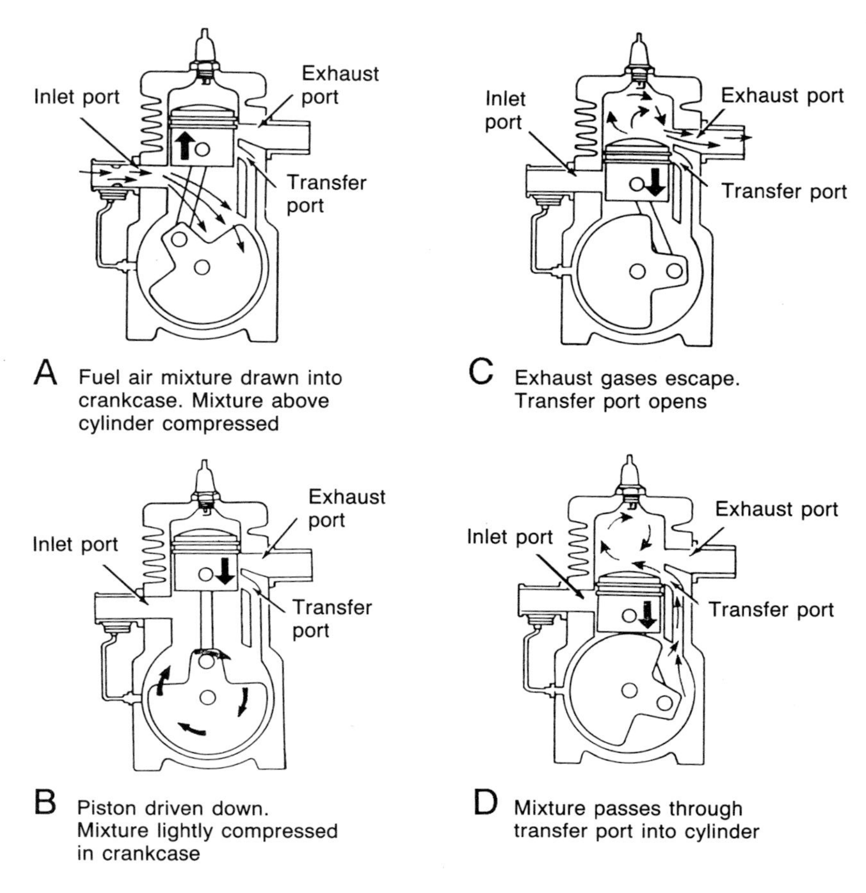

FIGURE 2.18 Principle of a two stroke engine. *(John Deere)*

oil) is drawn into the crankcase. The mixture of air and fuel already above the piston is compressed by the rising piston and then ignited by a spark from the flywheel magneto. Expansion from the heat produced drives the piston down, turning the crankshaft. Towards the end of this stroke, the piston closes the inlet port and the fresh mixture of air and fuel, trapped in the crankcase, is slightly compressed.

As the piston nears BDC, it uncovers the exhaust and transfer ports. The waste gases are released from the combustion space and to the exhaust system. The slightly compressed mixture in the crankcase, below the descending piston, passes through the open transfer port to the combustion space.

Some two stroke engines have a piston with a specially shaped top which causes the incoming air to swirl, helping to clear the exhaust gases from the cylinder.

Second stroke The piston starts its upward stroke and the remaining exhaust gases are forced out. The piston now covers the transfer and exhaust ports and the fresh mixture in the combustion space is compressed. By this time, the piston has uncovered the inlet port and more mixture is drawn into the crankcase because the rising piston creates a slight vacuum in the crankcase. The compressed mixture above the piston is ignited just before TDC and the cycle starts again.

Two stroke engines are air cooled. Fan

blades on the flywheel provide a constant flow of cold air to the cooling fins around the cylinder. The cooling fins need regular cleaning to prevent overheating as this type of engine runs at very high speeds.

Most two stroke engines run at much higher speeds than four stroke engines used in horticulture. Two stroke engines on chain saws run at 12,000 rpm or more. A typical single cylinder four stroke engine on a lawn mower will run at 3,000 to 4,000 rpm. A multi-cylinder diesel engine on a compact tractor runs at little more than 2,500 to 3,000 rpm.

Suggested Student Activities

1. Look for the various types of air cleaner and find out how to service them.
2. Find out how to bleed the fuel system on a diesel engine. Do not wait until you run out of fuel before learning how to do this task.
3. Study a coil ignition system on a car engine and identify the various parts described in this chapter.
4. Inspect the condition and setting of the sparking plug and contact breaker points, if fitted, on a single cylinder engine. Check that the gaps are set according to the instruction book.

Safety Checks

When filling a petrol engine fuel tank, make sure there are no naked flames nearby and do not smoke. Petrol fumes can be ignited by a tiny spark.

Ensure there is plenty of ventilation when running an engine inside a building. When servicing a lawn mower or garden tractor, it is best to take it outside before starting the engine.

Chapter 3

ENGINE LUBRICATION AND COOLING

It is not possible to produce a perfectly smooth finish on the surface of metal. When two dry metal surfaces run against each other at speed, there is a great deal of friction which generates heat and causes rapid wear. A thin film of lubricant between two metal surfaces will reduce wear and disperse much of the heat.

The basic principle of operation of an internal combustion engine is the creation of heat inside the cylinders by burning air and fuel. It is vital to remove much of this heat from the moving parts to ensure that the engine does not overheat or in extreme situations, especially with single cylinder air cooled engines, seize up completely. It is the job of the cooling system and to some extent the lubrication system to remove this unwanted engine heat.

ENGINE LUBRICATION

Engine oil has four main functions:

- To lubricate the moving parts to reduce friction.
- To assist in engine cooling by removing some of the heat caused by friction.
- To act as a sealant between the pistons and cylinder walls. Without this seal, there would be very little compression in the combustion space.
- To help keep the engine clean.

Oil Classification

Lubricating oil is classified according to its viscosity or fluidity using a range of viscosity numbers devised by the Society of Automotive Engineers (SAE). Thin oils have low SAE numbers and thick oils have high numbers. The viscosity rating refers only to the thickness of the oil.

The viscosity range includes:

SAE 20 and SAE 30	engine oil
SAE 40 and SAE 50	transmission oil
SAE 90	occasional use for transmission systems
SAE 120 and SAE 140	a very thick oil, used to lubricate some bearings

The viscosity of any lubricating oil alters as

PLATE 3.1 *Removing the dipstick to check engine oil level; the oil filter element is partly hidden by the operator's hand.*
(Massey-Ferguson)

its temperature changes. The oil will become thinner as the engine gets hot. Some oils have a special additive to help keep the viscosity stable as the temperature changes.

Multi-grade oils are now used for many engines. They span a range of viscosity ratings that make them suitable for engines and some transmission systems. Multi-grade oils also meet the need for an oil which is thin enough to ease starting the engine in cold weather, yet thick enough to be efficient when the engine is at its normal running temperature.

The cheapest oils, which are known as straight oils, contain no additives to improve their performance. Most engines, including high performance engines, need a better quality lubricant with various chemical additives to ensure long life of both the engine and its lubricating oil.

Additives

Most lubricating oils used in tractor and horticultural machines contain a variety of additives. A good quality engine oil should have the following additives:

Detergent An additive which stops serious build-up of the harmful deposits which occur when fuel is burned, causing piston rings to stick in their grooves and sludge to build up in oilways and bearings. When the oil is drained away, most of the harmful deposits leave the engine in the waste oil.

Dispersant Added to keep low temperature contaminants such as soot suspended in the oil. Again, this prevents a build-up of sludge deposits in oilways, bearings and filters.

Anti-oxidant Reduces the effects of oxidation (heat and chemical breakdown) which thickens the oil.

Corrosion inhibitor Protects the engine from rust and corrosion caused by acids and water formed when fuel is burned in an engine.

Anti-wear An additive which helps to sustain the strength of the oil film on the metal surfaces, especially when the engine runs under a heavy load.

Pour point depressant Lowers the temperature at which the oil starts to thicken to ensure the lubricant is free flowing on winter days.

Viscosity index improver Prevents the oil thinning when the engine runs at full operating temperature. It also keeps the oil sufficiently thin when cold to give easy starting. Thick oil puts an additional load on the starter motor.

Most oil companies offer a multi-purpose oil which is suitable for the engine, transmission and hydraulic systems of the larger horticultural tractors. However, these multi-purpose oils must not be used for some models of tractor since, without special additives, they could cause major damage. For example, tractors with oil immersed brakes need a special oil to ensure that the brakes operate efficiently. If in doubt about the use of multi-grade oil, consult the instruction book.

Horticultural oil, usually an SAE 30 oil with no special additives, is widely used to lubricate many single cylinder four stroke petrol engines on horticultural equipment. It is not suitable for diesel engines.

Oil Contamination

The efficiency of lubricants, and engine oil in particular, is reduced with use. A typical oil change period for a multi-cylinder four stroke petrol engine is 100 hours. Small single cylinder petrol engines need an oil change at 25 hour intervals, and many of the larger single and twin cylinder engines used in horticulture require an oil change at 50 hour intervals. Diesel engines need less frequent oil changes, 250 hours being a common service period. However, engine oil change periods vary, so consult the instruction book to find the correct oil change period for a particular engine.

It is necessary to change engine oil because with use it becomes contaminated with various substances including:

- *Sludge* is a mixture of carbon, water and other deposits formed when fuel is burned in an engine. If allowed to build up, sludge will partly block oilways and bearings and, in time, cause piston rings to stick. Oil with detergent properties will combat this problem but sludge will build up in time even with top quality oil if the engine manufacturer's advice on oil change periods is ignored.

- *Dust* will enter the engine through the air cleaner and crankcase breather. This can be minimised by regular servicing of the air cleaner and filter in the crankcase breather.
- *Metal particles* which break or chip off bearings, gears and other parts of the engine will find their way into the oil, especially when the engine is new.
- *Engine heat* causes a gradual breakdown of the structure of lubricating oil, which will also undergo chemical changes due to the effects of oxygen (oxidation).
- Small quantities of unburnt *fuel* will get past the pistons and mix with the oil in the sump, especially when the engine runs below its normal working temperature.
- *Water* is produced when fuel is burned, and still more water is formed through condensation inside the engine. A build-up of yellowish sludge on the underside of the oil filler cap is a sure sign of water in the oil.

Partly blocked oil filters, poorly stored new lubricants, as well as dirty oil jugs and funnels, will contaminate oil before it is put into the engine. These problems will be avoided by regular engine servicing combined with careful lubricant handling and storage.

Engine Lubrication Systems

Splash Feed Lubrication

This is used for small single cylinder engines. The bearings and other moving parts are lubricated by oil splashed over them by a small dipper scoop attached to the big end bearing cap as the crankshaft rotates. The dipper scoop collects the oil from a trough at

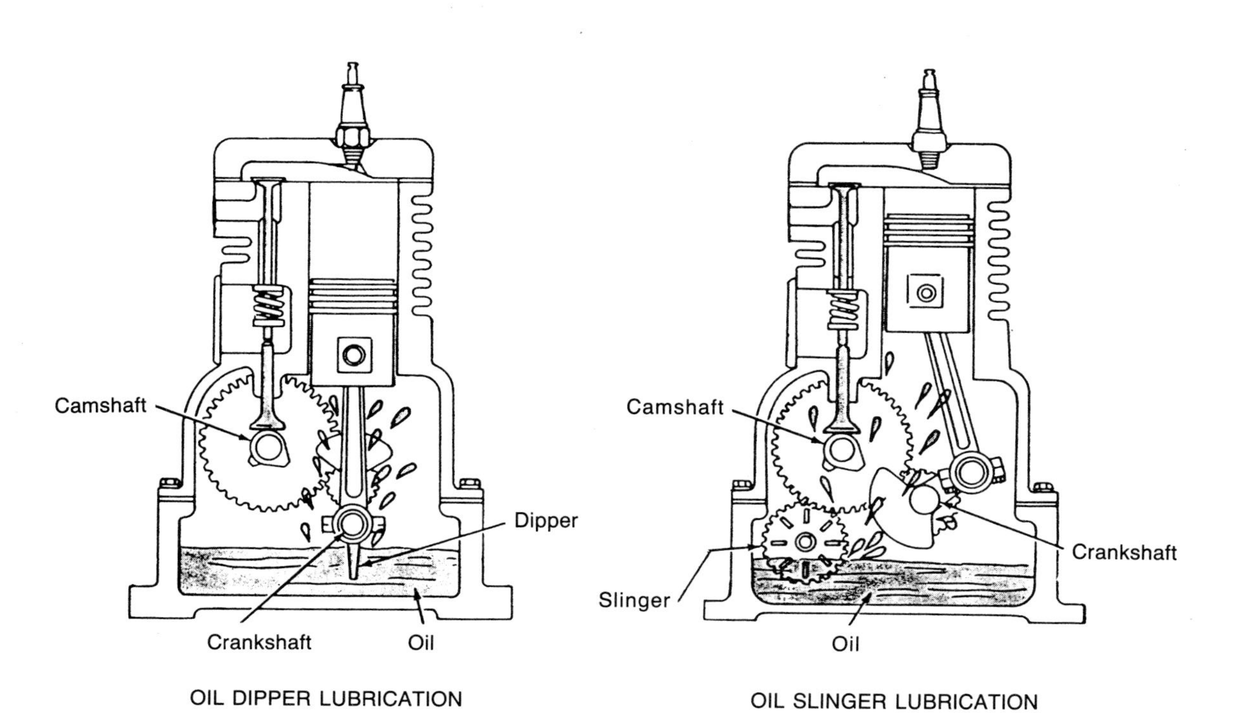

FIGURE 3.1 Splash feed lubrication systems. *(John Deere)*

the bottom of the crankcase. Some engines have an oil slinger or a small paddle wheel type of pump attached to the timing gears instead of a dipper scoop on the big end cap.

Oil on the cylinder walls is collected by the oil scraper ring on the piston. The gudgeon pin is lubricated by some of the oil which passes through holes in the piston. The oil also runs down the con rod through a hole to lubricate the big end bearing and then returns to the bottom of the crankcase. The main bearings and timing gears are lubricated by oil splashed around the crankshaft.

The oil in the crankcase must be maintained at the correct level to ensure that the splash feed system provides full lubrication to all moving parts. Splash feed lubrication engines are not suitable for applications which entail the engine being tipped or tilted sideways, as the bearings may run dry.

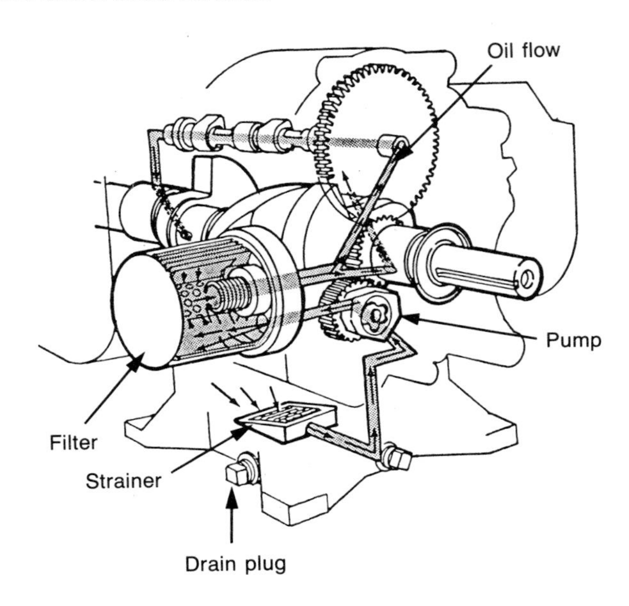

FIGURE 3.2 Force feed lubrication system for a single cylinder engine. The pump draws oil through the strainer which is pumped through the filter and then along the oilways to lubricate the moving parts. *(Briggs and Stratton Corp.)*

Force Feed Lubrication

Figure 3.3 shows the layout of a typical force feed lubrication system which pumps oil, under pressure, around a multi-cylinder engine.

The oil pump, driven by the timing gears or a gear on the camshaft, draws oil from the sump through a strainer and feeds it under pressure to the filter element. The filtered oil is pumped to the camshaft and crankshaft bearings. Drillings in the crankshaft carry the oil to the big end bearings. Oil is also pumped to the valve rocker shaft bearings. The push rods and cam followers are lubricated by the oil as it runs back to the sump. Oil splashing around the engine lubricates the timing gears and the cylinder walls. The oil scraper rings collect oil from the cylinder walls, and this oil passes through holes in the oil ring grooves to lubricate the gudgeon pin.

A pressure relief valve is included in the force feed lubrication system to protect the pump from damage through overloading, especially at high engine speeds. In some engines the relief valve is positioned in the system to allow oil to continue circulating if the filter becomes blocked through lack of maintenance.

Some horticultural tractors and self-propelled machines have an oil pressure gauge on the instrument panel which is connected to the main oilway and registers the working pressure of the force feed system. A typical working pressure is 2.7 bar (40 psi). All engines with a force feed lubrication system have a pressure switch in the main oilway connected to an indicator light on the instrument panel, and this warns the driver of low oil pressure. The light will come on when the ignition is switched on but should go out as soon as the engine runs. Always stop the engine immediately the warning lamp lights up, even when the engine appears to be running normally.

Oil pressure will fall away when the engine gets hot, but a very high or low oil pressure should be investigated.

Causes of low oil pressure include:

- Low oil level.
- Too thin oil, because the engine is overheating or the wrong grade of oil is being used.

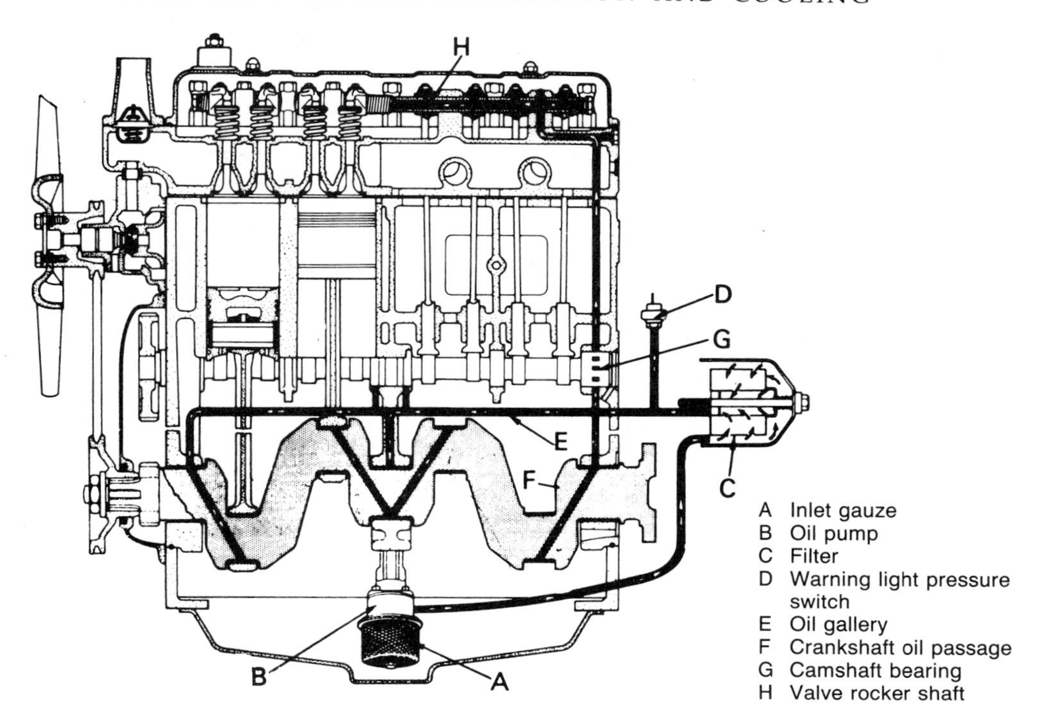

FIGURE 3.3 Force feed lubrication system for a multi-cylinder engine.

- Blocked oil filter—but only if the oil is filtered before it reaches the pressure gauge in the oil flow circuit.

Other causes of low oil pressure such as worn oil pump, worn bearings, blocked oilway or the relief valve remaining open will require the attention of a mechanic.

Causes of high oil pressure include:

- Blocked oil filter—but only if the pressure gauge is before the filter in the oil flow circuit.
- Too thick oil, because the engine runs too cold, the oil is very dirty or the wrong grade of oil is being used.

Other causes of high oil pressure, including the relief valve stuck shut or blocked oilways, are problems for a mechanic to solve.

Crankcase breather

This is a small air vent with a filter which prevents a pressure build-up in the sump. It may be part of the oil filler cap or a separate breather in the side of the cylinder block. Pressure is more likely to build up in older engines where worn piston rings allow gases to escape from the combustion chamber into the sump. The gases must be released to prevent condensation of water in the sump.

The crankcase breather must not become blocked. If it has a filter, it should be periodically washed in petrol.

Oil Mist Lubrication

This system is used to lubricate two stroke engines. As the crankcase must be dry and airtight, the moving parts are lubricated by oil mixed in the petrol. When the mixture of fuel, oil and air is drawn into the crankcase an oil mist is created, and this lubricates the moving parts. The mixture of petrol and oil may vary from about 16 parts of fuel and 1 part of oil up to as much as 50 parts of fuel to 1 part of oil. The amount of oil mixed with the petrol depends upon type of engine and quality of lubricant used. For example, a typical chain saw engine requires 1 part of oil to 40 parts of petrol when a high quality two stroke oil is used. The mix is reduced to 25 parts of fuel to 1 part of oil when a lower quality lubricant is used.

An advantage of engines with oil mist lubrication is that they can be used in any position, even upside down, since all moving parts will still receive full lubrication. Chain saw engines have oil mist lubrication for this reason.

Mixing two stroke fuel

Check with the instruction book to find the correct mix for any engine you may use. Measure the required quantity of oil and put it into a clean can. Add the correct amount of petrol and shake the contents to ensure thorough mixing. Only mix sufficient oil and fuel for about one week's supply because stale fuel is likely to cause starting problems. Lead free petrol can be used with two stroke engines, provided the oil mixed with it is a high quality lubricant with special additives needed for this type of fuel.

The crankcase of a two stroke engine may have a drain plug. This is provided for draining fuel and oil which may collect there, as some engines are difficult to start if there is any amount of liquid in the crankcase. Always follow the fuel mixture instructions given in the owner's manual. The addition of extra oil in the fuel will result in carbon deposits building up inside the engine and lots of blue smoke from the exhaust pipe.

Changing Engine Oil

All engines need the oil drained and renewed at regular intervals. This can vary from every 25 hours for a small single cylinder air cooled engine to 300 hours for a multi-cylinder diesel engine. Refer to the instruction book to find the oil change period for engines you have to maintain.

It is best if the oil is drained from the engine when it is hot. Make sure that the container for catching the old oil is big enough. Lawn mowers and other pedestrian machines with single cylinder engines should be tipped to ensure all the old oil is drained away. Replace the drain plug and refill to the correct level with the grade of oil stated in the instruction book. Some engines have a dipstick; others are filled to the top of the oil filler hole.

Multi-cylinder engines have an oil filter element, which must be replaced when the engine oil is changed. Many engines have a self contained screw-in type filter element which should be removed and replaced with a new one. Check that the sealing rings are in good condition. If in doubt, fit new ones. The filter element of some engines is in a metal bowl. Discard the old element and clean the bowl to remove all traces of sludge before fitting a new element.

After replacing the drain plug, refill the sump with the correct grade of oil to the level mark on the dipstick. Run the engine for a short while to fill the filter bowl, and then recheck the dipstick level. Finally check for leaks and record details of when the oil change was carried out and, where appropriate, the engine hourmeter reading, so you will know when the next change is due.

Engine Cooling

All engines create a vast amount of heat when fuel is burnt, and this heat must be removed to ensure the engine runs efficiently over long periods. An overheated engine will not operate at maximum efficiency, and in extreme situations, the piston may seize in the cylinder. The cooling system removes most of the surplus heat and more is taken away by the engine lubrication system. Multi-cylinder engines on some specialist equipment are required to run for long periods under a full load, and these engines may have an oil cooler with its own radiator which removes much of the surplus heat from the lubricating oil and helps to keep the engine oil at its correct working temperature.

There are two types of cooling system: air and water. Two stroke and most single cylinder four stroke engines are air cooled. Most multi-cylinder engines are water cooled. Air cooled engines may overheat if the cooling fins are not kept clean, but there is no chance of frost damage or leaks which can occur with water cooling.

Transfer of Heat

Engine cooling involves the transfer of heat from the cylinders to the atmosphere. Heat can be transferred from one point to another in three ways:

Conduction If you hold a metal rod in a fire, before long the end held in the hand becomes warm. Heat moving this way is transferred by conduction, and metals are good conductors of heat, especially copper, brass and aluminium. In an engine cooling system, heat is conducted from the combustion space through the cylinder walls.

Convection Hot air will rise above cold air because it is less dense (lighter), and hot water rises above cold water in the same way. Heat transfer occurs when hot water, or air, rises above cold water, or air. A domestic convector heater works in this way. In engine cooling, hot water rises up from the bottom to the top of the cylinder block, removing heat which has been transferred by conduction through the cylinder walls.

Radiation Rays of heat can be felt from, for example, an open fire. Heat rays always travel in straight lines from the heat source. This form of heat transfer is radiation. In engine cooling, heat is lost through radiation from, for example, the exhaust manifold and the outer surface of the engine cylinder block.

Air Cooling

The cylinder block and cylinder head have cooling fins which conduct the heat away from the engine. The engine flywheel has fan blades which blow air over the cooling fins. The fan blades and cooling fins are surrounded by a metal cover or cowling, which directs the air flow over the cooling fins and in this way removes the surplus engine heat. The cooling fins provide a large surface area to help disperse the surplus heat.

A few multi-cylinder engines are air cooled.

PLATE 3.2 *Air cooled petrol engine.*
(Ransomes, Sims and Jefferies)

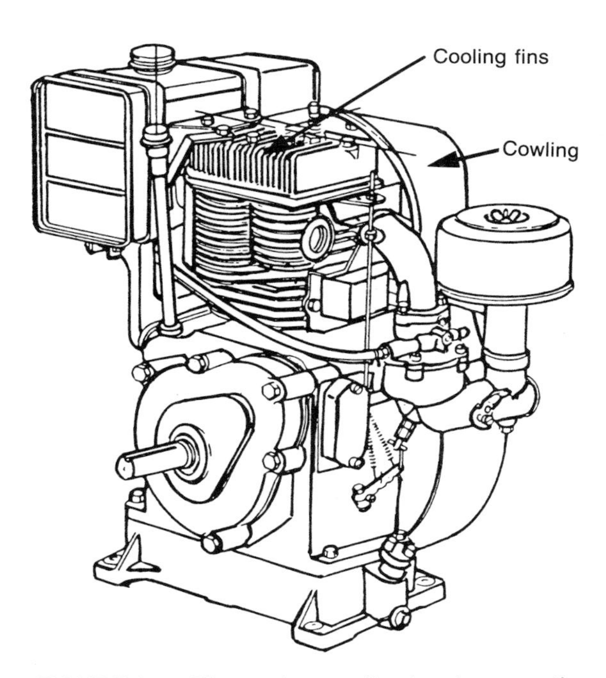

FIGURE 3.4 The engine cowling has been partly cut away to show the cooling fins. It is very important to keep the fins clean to prevent overheating. *(Briggs and Stratton Corp.)*

It is usual to have a separate fan inside a cowling extending around the cylinders to create a flow of air over the cooling fins.

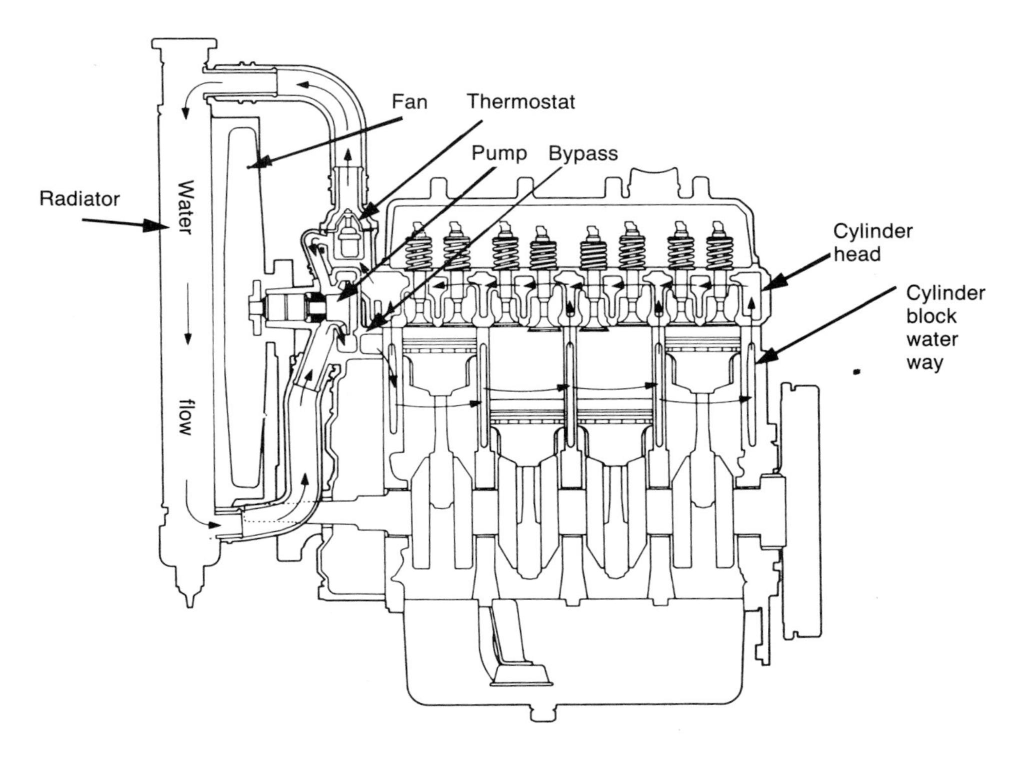

FIGURE 3.5 Impeller-assisted cooling system. Water moves up through the engine where it removes heat, then down through the radiator where it is cooled by the fan drawing air through the radiator core. *(Kubota)*

Care of air cooling systems
It is very important to keep the fins clean to ensure that the air cooling system works to maximum effect. It will be necessary to remove the cowling from time to time so that dirt can be brushed or blown from the spaces between the fins. Blocked fins restrict air flow and the engine will overheat. Oil on the fins, caused by a leak or careless filling, will collect dust, so always wipe surplus oil away immediately.

The air inlets to the fan should be cleaned frequently to ensure maximum air flow over the cooling fins. Lawn mower engines need special care and the air inlets and cooling fins frequently become blocked with grass. Check the belt tension on air cooling systems which have a belt driven fan.

Water Cooling

Thermo-syphon cooling
This system is only used for water cooled stationary engines. There is no fan or water pump, and cooling depends upon convection currents causing hot water from around the cylinder to rise to the top of an open water hopper. The surrounding air cools this hot water by radiation. The cooled water is then replaced by hotter water rising from the area of the cylinder. In this way, convection currents keep the water circulating to cool the engine.

Impeller-assisted cooling
The thermo-syphon principle still applies, but an impeller pump ensures full circulation

of the water. Figure 3.4 shows the arrangement of a water cooling system. The water rises up through the cylinder block as it takes up engine heat; it then cools as it flows down through the radiator which consists of a large number of small diameter pipes held together by thin metal fins. A fan helps to cool the water by drawing air through the honeycomb of radiator pipes. The impeller, often referred to as the water pump, assists with circulation of water around the system. The thermostat, a heat operated valve, controls the temperature of the water. When the engine is cold, the thermostat is closed and water cannot circulate through the radiator. The water can only rise up through the engine to the impeller, which returns it through the by-pass hose to the bottom of the cylinder block. When the cooling water reaches a temperature of about 75°C, the thermostat valve opens and the hot water is able to flow into the top of the radiator for cooling. Since the thermostat is closed when the engine is cold, full water circulation is prevented and the engine soon reaches its operating temperature.

The cooling system is pressurised by means of a special radiator cap with a pre-set relief valve, which raises the boiling point of the cooling water. This allows the engine to run at a higher operating temperature than would be possible if the water boiled at 100°C. As the cooling system has a higher boiling point, the engine can operate at higher working temperatures and therefore run more efficiently without the risk of boiling. A typical cooling system will have a radiator cap with a relief valve which opens at a pressure of 0.5 bar (7 psi). This will raise the boiling point to about 112°C.

Although the water does not appear to be boiling inside the pressurised system, removal of the radiator cap will release the pressure and the water may well be scalding hot. So, when removing the radiator cap from a hot engine, great care must be taken to turn it very slowly to release the pressure. To avoid the risk of scalding your hand, it is a good idea to cover the radiator cap with a piece of cloth before it is removed.

Some cooling systems have two drain taps, one on the side of the cylinder block and one at the bottom of the radiator. Other systems have only a single tap. If there are two taps, both must be used to ensure all the water is drained from the cooling system.

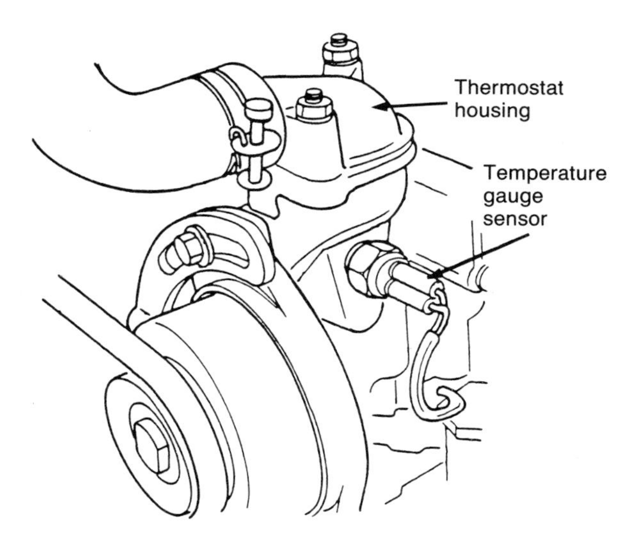

FIGURE 3.6 The thermostat housing is at the front of the cylinder head. The temperature gauge sensor is also placed here as this is the hottest part of the engine cooling system.
(Massey-Ferguson)

A small expansion tank, connected by a rubber tube to the neck of the radiator, is included in the cooling system of some engines. The expansion tank maintains the water level in the radiator. Other engine cooling systems have an overflow pipe. If excessive pressure builds up in the cooling system, the relief valve in the radiator cap opens to release the pressure, and sometimes some water, through the overflow pipe.

Frost Protection

An anti-freeze solution, ethylene glycol, is used to protect the cooling system from frost damage. A 25 per cent mixture of anti-freeze and water will protect the system from frost down to a temperature of about −18°C. A 25 per cent solution consists of 2 parts of anti-freeze to 6 parts of water (for example, 2 litres of anti-freeze and 6 litres of water). Some compact tractor engines have an aluminium alloy cylinder head on a cast iron cylinder block, and a special type of anti-freeze is necessary for these. Check the instruction book to find the recommended type and mix of anti-freeze and water for any water cooled engine you have to maintain.

Anti-freeze does not prevent freezing; it only lowers the freezing point of the coolant. It also lowers the boiling point. It is best to top up the cooling system with a ready mixed solution of anti-freeze and water.

Many anti-freeze solutions have a two year life. After this period, the coolant should be drained, the radiator flushed with clean water and then refilled.

Engine operating temperature
Neither a cold nor an overheated engine can work efficiently. The cooling water should be at a temperature of approximately 90°C to ensure efficient engine performance. Keep an eye on the temperature gauge on the instrument panel and if the temperature is very high or very low, investigate the cause.

Care of the Cooling System

- Check the water level in the radiator. This should be done at weekly intervals, and more frequently in very hot weather. The water level should be about 25 mm below the radiator neck to allow for expansion. Never add large quantities of cold water to the cooling system of a very hot engine, as this action could result in a cracked cylinder head or cylinder block.
- Flush the cooling system with plenty of clean water to clear any sediment when it is drained in preparation for a new solution of anti-freeze.
- Keep the outside of the radiator clear of dirt, chaff, etc. Wash or blow the dirt out the way it came in by using a hosepipe or airline from the fan side of the radiator. A dirty radiator will cause the engine to overheat.
- Maintain the correct fan belt tension. The adjustment procedure is described in Chapter 4.
- Check for cooling system leaks. Make sure hose connections are tight, and watch out for perished hoses, especially on the more aged engines.
- Few engines have an impeller pump which requires lubrication, but those which do should be greased weekly.

Causes of Overheating

- Low water level.
- Radiator blocked, sometimes internally but more often on the outside.
- Slack fan belt.
- Thermostat stuck in the closed position.

Suggested Student Activities

1. Locate the engine oil filler and the dipstick on a single cylinder engine.
2. Check a multi-cylinder engine instruction book to find:
   (a) the recommended grade of engine oil.
   (b) the sump capacity.
   (c) how often the oil and the filter element should be changed.
3. Read any literature you can find about oil refineries and how different types of fuel and lubricant are made.
4. Study the instruction manual for a number of two stoke engines to find how recommendations vary for the amount of oil which should be mixed with the fuel.
5. Find out the capacity of a multi-cylinder engine cooling system and calculate how much anti-freeze is needed to make up a 25 per cent solution.

Safety Check

Remember that the water in a pressurised engine cooling system will boil at a temperature higher than 100°C. If you need to remove the radiator cap when the engine is hot, do so slowly and protect your hand with a cloth or glove to avoid the risk of scalding yourself.

Chapter 4

STARTING AND CHARGING SYSTEMS

The battery supplies electric current to the starter motor, lights, etc., on horticultural tractors, ride-on mowers and some pedestrian controlled equipment. Chain saws and many pedestrian controlled cultivators and lawn mowers have a recoil starter which does not require a battery. This has obvious advantages.

Four wheel tractors, including compact models, ride-on mowers and specialist self-propelled machines, have other electrical equipment such as driving lights, warning lamps and gauges as well as an electric starting system. Petrol engines with a coil ignition system rely on a battery for the supply of low voltage current. Some of the more complex tractor mounted implements and machines have battery operated remote control systems.

Most engines with electric starting have a generator, usually an alternator, but sometimes a dynamo, to generate the electricity stored in the battery. Some pedestrian equipment has an electric starting system which is operated by a rechargeable dry battery. The engine does not have a generator and when discharged, the battery is recharged by a mains operated trickle charger, normally supplied with the machine.

The Battery

A battery stores direct current (DC). Mains electricity is alternating current (AC) which cannot be stored in a battery unless it is first converted (rectified) to DC current. A mains operated battery charger rectifies AC current to DC current and, at the same time, transforms it from mains voltage to battery voltage, which is usually 12 volts but sometimes only 6 volts on pedestrian operated machines. Most batteries used on horticultural machines are the lead acid type. Some equipment has rechargeable dry cell batteries.

Lead Acid Batteries

Batteries store electricity generated by an alternator or a dynamo. A battery consists of a hard rubber or polypropylene box containing a number of lead plates immersed in sulphuric acid (electrolyte). The plates are in sets called cells, each of which has a nominal value of 2 volts; a 12 volt battery thus has six cells. Electrical energy from the charging system is converted into chemical energy by the battery and stored in that form. During

PLATE 4.1 *The battery can be swung outwards for easy maintenance after slackening the wing nut near the air cleaner hose.* (Ford New Holland)

the charging process, hydrogen and oxygen gases are given off and the electrolyte becomes stronger. The acid strength will weaken as the battery discharges.

When the driver uses the starter or other electrical equipment, the chemical energy is converted back to electricity. The charging system provides most, if not all, of the electricity required for lights, horn, screen wiper, etc., when the engine is running at normal operating speed. The battery provides back-up when the engine idles.

The battery has two terminals, one the positive connection, the other the negative. On all but very old equipment, the negative terminal is the *earth* connection.

Battery Maintenance

Batteries with a screw cap to each cell need the regular maintenance described below. Most polypropylene cased batteries are maintenance-free or have very low maintenance requirements. The casing is translucent and the level of electrolyte can be seen. If it is below the minimum level mark on the casing, the cell caps can be pierced with a screwdriver and then unscrewed so that the electrolyte can be topped up.

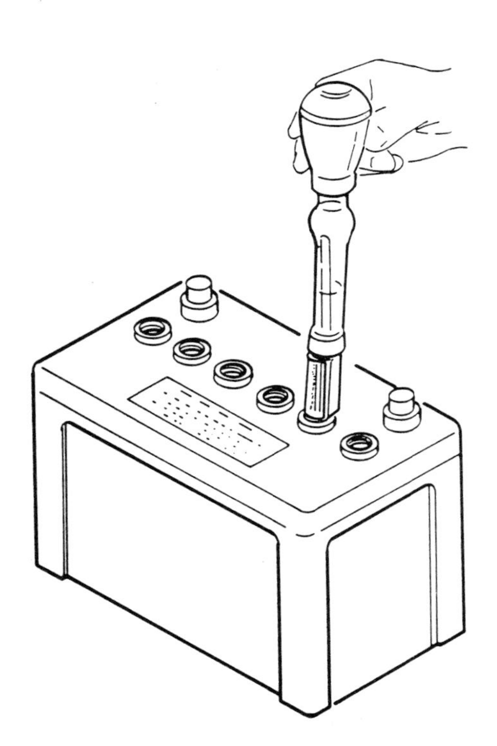

FIGURE 4.1 Testing a battery with a hydrometer.

Many batteries are ruined through neglect. A little tender, loving care should ensure long, trouble-free service. For maximum battery life:

- Clean any dirt from the top of the battery, then remove the cell caps and check the electrolyte level, which should be just above the plates. If low, top up to the correct level with distilled water. Replace the caps and wipe away any spilt water. Tap water should not be used as it contains chemical impurities.
- Keep the vent holes in the cell caps clear.
- Check that the terminal connections are clean and tight. A light coating of petroleum jelly will reduce corrosion.
- The negative terminal is the earth connection on all but the rather aged machines. Make sure that the earth connection on the chassis or bodywork is tight and corrosion-free.
- Keep the top of the battery clean and dry. A battery can self discharge if the top is damp and dirty.
- The battery must be secure in its mounting to prevent damage to the casing.
- Remove the battery from machines, such as grass care equipment, which are out of use for long periods. Store the battery in a cool, frost-free building and keep it well charged. Remember that a discharged battery will freeze.

Battery Testing

A hydrometer is used to check the state of charge of a battery. It measures the specific gravity of the acid, which indicates the strength of the electrolyte and therefore the state of charge—the stronger the acid, the higher the charge. A float in the hydrometer body has either a graduated scale which indicates the strength of the electrolyte or three coloured bands which show full charge, half charge or discharged.

The specific gravity readings are:

Specific gravity	*State of charge*
1.280	Full charge
1.200	Half charge
1.110	Discharged

To use a hydrometer, remove the cell cap and put the rubber tube into the electrolyte. Squeeze the rubber bulb, then gently release it to draw some electrolyte into the hydrometer body. Note the specific gravity reading or state of charge and return the liquid to the cell. Repeat the process for at least two more cells. Remember that the electrolyte is acid so keep it away from your clothes and skin. Rinse the hydrometer in clean water after use.

Recharging Lead Acid Batteries

A discharged battery can be removed from a tractor or machine and recharged with a mains charger unit. The most simple type of battery charger is connected to a 13 amp socket, and is usually called a trickle charger. It converts the mains supply from 240 volt AC to 12 volt DC current, which is stored in the battery. Some trickle chargers have an adjustment which offers the choice of 12, 6 or 2 volt DC output. The charging rate is very low, typically 2 or 3 amps. A trickle charger will take from 12 to 24 hours to recharge a battery. Large workshops often have a heavy duty charger which will recharge a battery very quickly.

Use the following procedure when connecting a battery to a trickle charger:

1. Clean the top of the battery where possible, remove the cell caps and top up the electrolyte with distilled water if necessary. Leave the caps loose in the holes to prevent gas pressure building up in the battery.
2. Connect the charger clips to the battery terminals, making sure the positive and negative clips are connected to the correct terminals. Check that the charger is set to suit the battery voltage.
3. Connect the charger to the mains supply and switch on.
4. Check the electrolyte every few hours with a hydrometer. When the battery is fully charged, switch off the mains supply before removing the terminal clips. Replace and tighten the cell caps.

Be warned: Batteries give off hydrogen and oxygen when being charged. Never use a lighted match or other naked flame to inspect a battery being charged in a badly lit area, as the result could be explosive.

Recharging Dry Cell Batteries

Dry cell batteries, which are mainly used on lawn mowers, will not be charged when the engine is running, so periodic recharging will be necessary during the grass cutting season. The battery should be removed from the machine and, after the top is cleaned, it can be connected to a mains charger unit. The charger leads must be connected to the correct terminals and only then should the charger be switched on. Most charger units will indicate when the battery is recharged and ready for use. Always switch off the mains supply before removing the charger terminals from the battery.

Using jump leads

Tractor and vehicle batteries are sometimes found to be discharged when they are needed to start an engine, perhaps because the lights have been left on all night. Jump leads—very heavy cables with a connecting clip at each end—can be used to overcome this problem in an emergency. If the battery is faulty, it will have to be replaced.

Be careful when using jump leads as failure to follow the correct procedure may result in personal injury or damage to the vehicle electrical system. For a negative earth system use jump leads as follows:

1. Connect the positive terminals of both batteries with one of the jump leads.
2. Attach the second lead to the negative terminal on the booster battery.
3. Select a good earth point on the tractor, at least 500 mm from the battery, and connect the other end of the second lead to that point.
4. Check that the leads are clear of any moving parts before using the starter. Once the engine has started, remove the

jump lead clip from the earth point on the engine and then remove the other clips in reverse order.

Be warned: It is a dangerous practice to connect the jump lead to the negative terminal on the discharged battery because severe arcing may occur. This will cause sparks, which, like a naked flame, may have explosive results. Never allow the jump lead clips to touch after they have been connected to the booster battery.

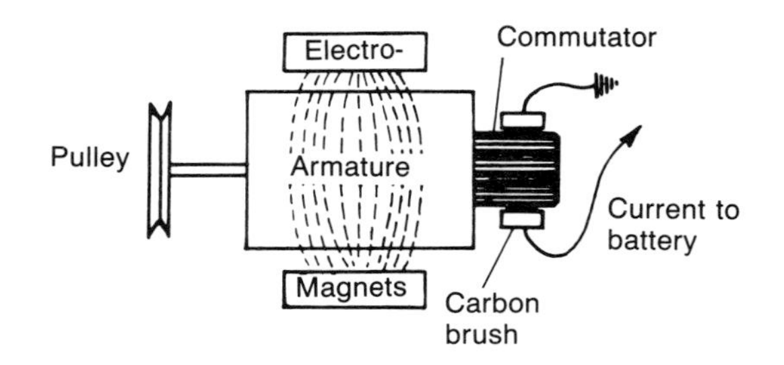

FIGURE 4.2 Principles of a dynamo.

The Charging System

An alternator or a dynamo is used to generate electricity for storage in the battery. Most horticultural tractors with a charging system will have an alternator, but some older models rely on a dynamo.

The advantages of an alternator are that it has a high output at low engine speed and a higher maximum output than a dynamo of similar size.

When a conductor rotates in a magnetic field, current is created in that conductor. A dynamo has stationary magnets and a rotating conductor (armature). An alternator differs in that it has rotating magnets (rotor) and stationary conductor windings (stator).

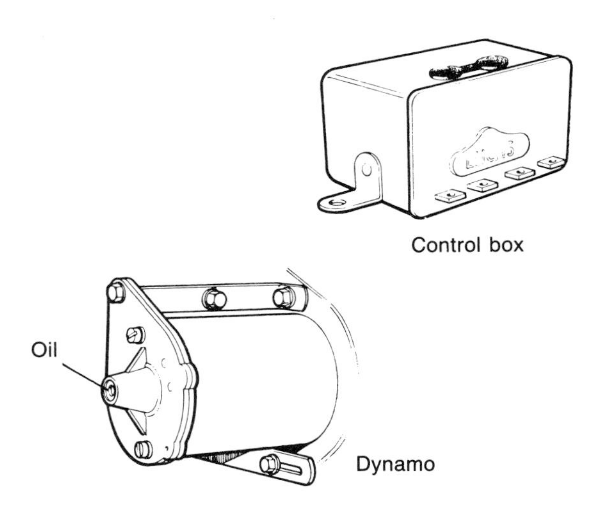

FIGURE 4.3 Dynamo and control box.

The Dynamo

Although the dynamo is less common, its working principles may be easier to follow than those of an alternator, so this type of generator will be dealt with first.

The armature consists of a large number of conductors, each insulated from the next. It rotates in a magnetic field created by two diametrically opposite electro-magnets attached to the dynamo housing. The magnets are energised by current from the battery, which is called the field current. There is some residual magnetism in the magnets but the main strength of the magnetic field is created and controlled by the field current.

When the armature is rotated—driven by the fan belt from the engine crankshaft pulley—the electric current generated passes to the commutator. Here, it is collected by a carbon brush which feeds it to the battery. Carbon is a very good material for conducting electricity. There are two carbon brushes, both held against the commutator by spring pressure. The brush which collects the current is connected to the dynamo output terminal, the larger of the two terminals at one end of the dynamo. The second brush is connected to earth.

It is necessary to limit dynamo output when the battery is fully charged and to prevent it discharging when the engine is stopped or running at very low speed. The control box, part of the charging circuit, carries out these functions.

The *voltage regulator* prevents overcharging by regulating the strength of the field current to the magnets. When the battery is partly charged, the voltage regulator provides a strong magnetic field to give a high rate of charge from the dynamo. When the battery is fully charged, the field current is reduced, thereby cutting charging rate to a minimum level.

When the engine is stopped or running at very low revs, current must be prevented from flowing back from the battery to the dynamo, as reverse flow of current would discharge the battery and damage the dynamo.

The *cut-out,* also in the control box, is an automatic switch which prevents back flow of current to the dynamo when it is not charging the battery.

The Alternator

The alternator works in a similar way to a dynamo but it has a stationary conductor (stator) and rotating magnets (rotor). When the engine runs at 2,500 rpm, the alternator will rotate at speeds of 6,000 rpm or more, depending on the diameter of the alternator pulley.

Current is supplied to the rotor windings by two carbon brushes to create the magnetic field. Alternating current is induced (created) in the stator windings when the rotor is driven by the fan belt. This current flows to a rectifier in the alternator housing which converts it to direct current before it flows to the battery.

A voltage regulating mechanism is built into the circuit. Instead of a cut-out, an automatic relay switch prevents backflow from the battery to the alternator.

PLATE 4.2 *A twelve volt alternator. Fan blades attached to the alternator pulley help keep it cool. The pump for the tractor power steering system and the radiator for the oil cooler can be seen below the alternator.* (Ford New Holland)

Simple Charging Faults

The ignition warning light on the instrument panel warns the driver that the alternator or dynamo is not charging the battery. The warning lamp will always light up when the ignition switch is on and the engine is stopped. It may also light up if the engine is running at very low speed but it should go out as soon as engine revs are increased. If the ignition warning light comes on when the engine is running normally, there is no need to stop the engine immediately. However, the fault should be investigated as the battery will soon become discharged, especially if electrical equipment such as the radio or headlights are being used.

Some simple reasons why an alternator or dynamo does not charge the battery include:

- Loose terminal connections or a broken wire.
- Broken fan belt or one with tension too slack.
- Worn, dirty or broken carbon brushes.

Maintenance of the Charging System

Correct fan belt tension is very important. The alternator or dynamo pulley is usually driven by the fan belt from the engine crankshaft pulley, which also drives the radiator cooling fan. Some engines have a separate generator drive belt. A slack fan belt will cause engine overheating as well as a low rate

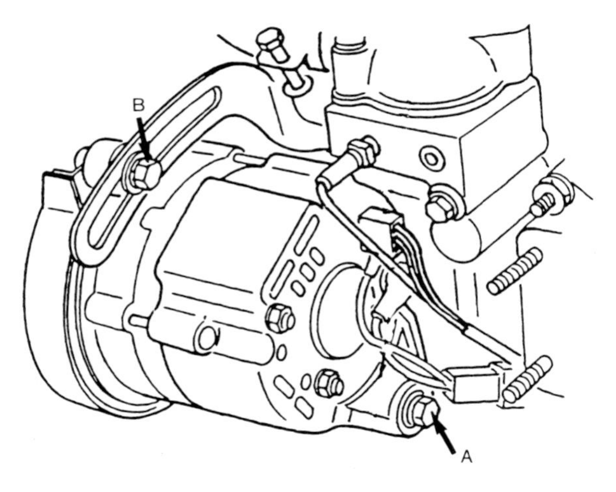

FIGURE 4.4 After slackening pivot bolts A and adjuster bolt B, the alternator is moved outwards to tighten the fan belt.

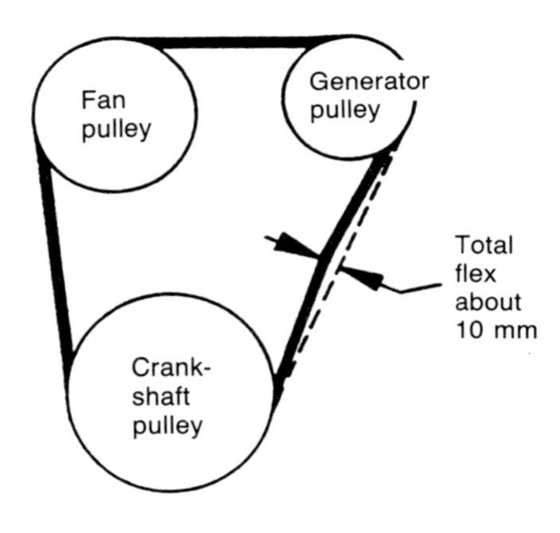

FIGURE 4.5 The total flex on the belt when correctly tensioned should be about 10 mm. *(Massey-Ferguson)*

of charge. An over-tightened fan belt may cause damage to the generator bearings.

The fan belt is adjusted by slackening the bolts which secure the alternator (or dynamo) to its mounting bracket on the side of the engine. Move the alternator on its pivot bolts until the correct tension is achieved. Finally retighten the bolts and check that the belt tension is correct.

Alternators and dynamos have a small fan, often on the drive pulley, to help keep them cool. The air vents at the ends of the generator must be kept clear to ensure maximum air flow. Some dynamos need occasional lubrication with a drop of light oil in the bearing end cap. No other attention is required except regular checks to ensure that the terminal connections are clean and tight.

Starting Systems

Multi-cylinder engines and some single cylinder four stroke engines have an electric starting system. Two stroke and many of the smaller single cylinder four stroke engines have a recoil starter.

Recoil Starters

Cord operated recoil starters have a coil spring which is tensioned when the cord is pulled to start the engine. The spring tension is used to rewind the starter cord. The engine is turned by two retractable pegs on the recoil starter drum when the cord is pulled. These pegs engage with slots in the engine crankshaft pulley.

The recoil starter cord and spring will give long service provided that the operator follows these simple rules:

1. Pull the cord handle slowly until the starter pegs engage; then pull rapidly to avoid kickback on the cord.
2. Do not pull the starter cord in rapid succession. After each pull, allow the cord to rewind in a controlled way before pulling it again.
3. The starter cord should not be pulled out to its fullest extent.

There are few problems with recoil starters. The whole unit can be removed from the engine cowling for cleaning by undoing three or four nuts or studs. A broken starter rope is not difficult to renew: take the starter off the engine, remove the broken pieces and fit a new rope. A broken recoil spring, however, is more difficult to replace. Some engine instruction books give advice on repairing recoil starters.

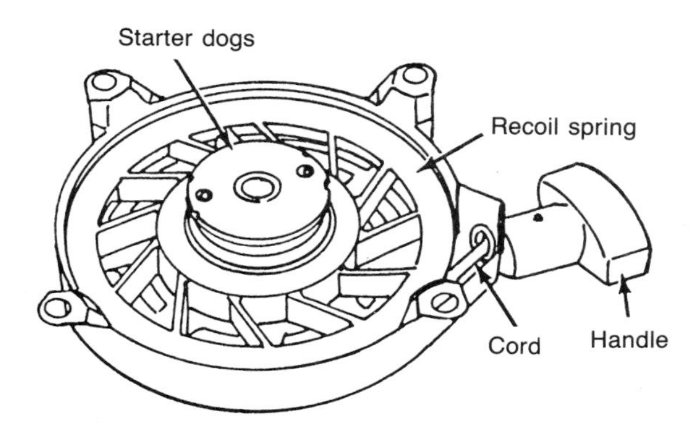

FIGURE 4.6 Recoil starter.
(Briggs and Stratton Corp.)

Electric Starting Systems

There are two main types of starter motor used on horticultural engines.

The *inertia starter* has its pinion (gear) turning at full speed before it engages with the flywheel ring gear.

The *pre-engaged starter* has its pinion engaged with the flywheel ring gear before it begins to turn the engine. Starter motors are controlled by the key operated switch on the instrument panel usually known as the ignition switch.

Many multi-cylinder spark ignition engines have an inertia starter. When the starter motor turns, the starter pinion moves along the shaft on a helix (a very coarse thread) against spring pressure. The pinion continues to turn as it moves towards the engine flywheel until it meshes with the ring gear. At this point, the starter motor turns the engine. When the engine starts and the starter switch is released, the starter motor slows down and the pinion is thrown out of mesh with the flywheel ring gear. The spring on the pinion shaft helps to return the pinion to the rest position.

Some lawn mowers have an inertia type electric starter, the current being provided by either a lead acid battery or a dry cell unit.

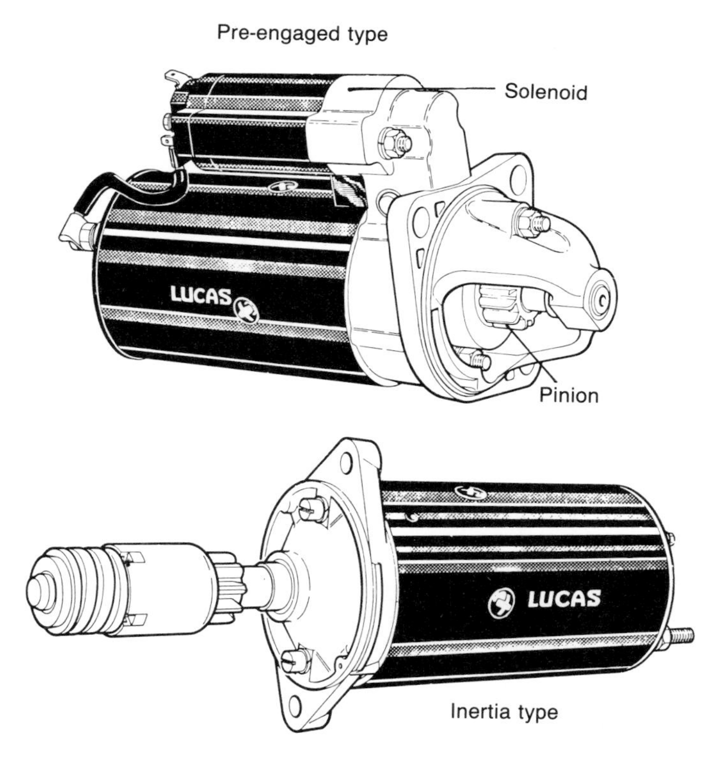

FIGURE 4.7 Inertia and pre-engaged starters.

It is common for lawn mowers with electric starting to have a recoil starter also. In many cases, the battery is recharged by either a small alternator or a generator built into the flywheel magneto, but with some models a mains operated charger must be used.

As diesel engines have a high compression ratio, there is a much higher load on the starter motor than is the case with a petrol engine. A pre-engaged starter motor is more suitable for starting diesel engines, and is also often used on multi-cylinder petrol engines. The pre-engaged starter has its pinion fully engaged with the flywheel ring gear before the pinion shaft is turned. A solenoid, operated by the ignition switch, is used to move the pinion along its shaft into mesh with the flywheel ring gear. After the pinion and ring gear are meshed, the same solenoid operates the switch which completes the electric circuit from the battery to the starter motor. The engine turns at a cranking speed of about 200 rpm. When the engine starts, the ignition switch is released and the solenoid spring returns the starter pinion to its rest position.

A hand lever is used to engage the starter pinion with the flywheel ring gear on some of the older diesel engines. Further movement of the lever operates a switch which completes the circuit from the battery to the starter motor.

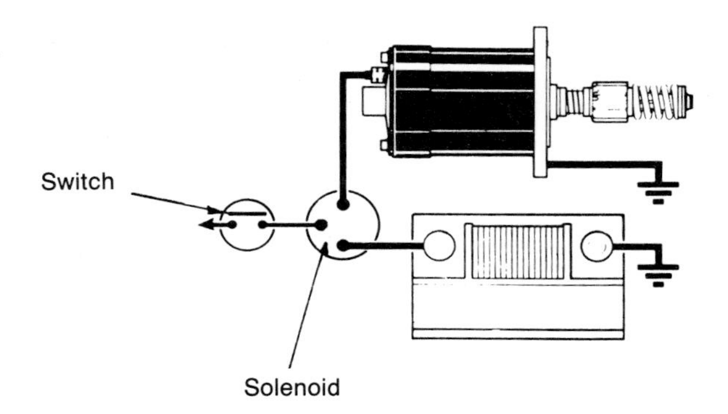

FIGURE 4.8 Starter circuit.

A solenoid consists of a coil in a casing with a soft iron plunger in the centre. When the solenoid coil is energised by the battery, a magnetic field is created. This pulls the plunger through the centre of the coil against a return spring. There are two terminals at one end of the solenoid casing, one con-

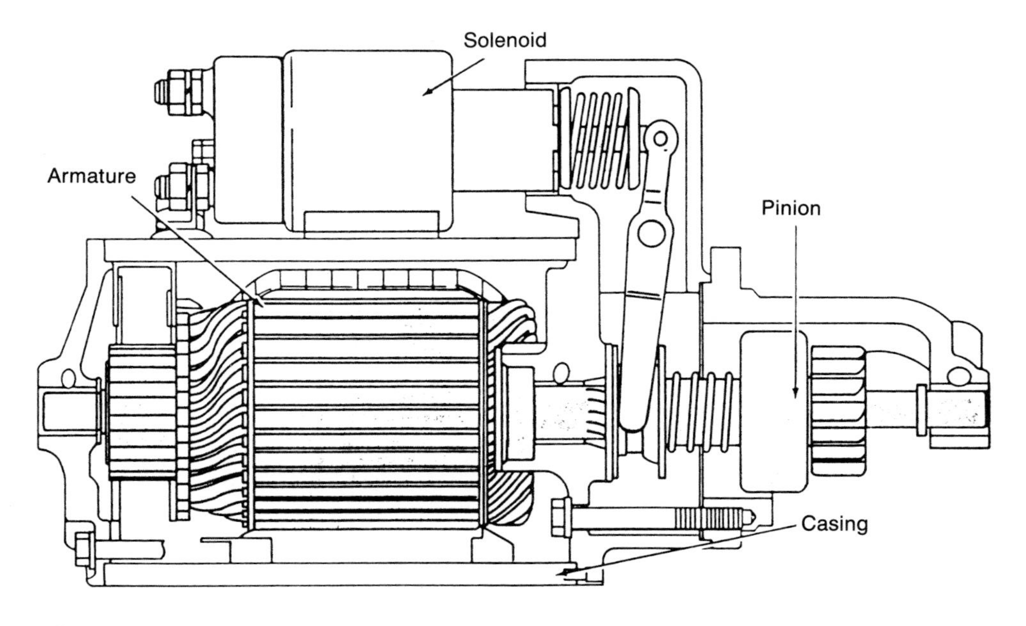

FIGURE 4.9 Section of a pre-engaged starter motor.

PLATE 4.3 *Compact tractor with a full set of lights.* (Ford New Holland)

nected to the battery and the other to the starter motor by heavy cables. Movement of the solenoid plunger on a pre-engaged starter moves the starter pinion into mesh with the flywheel ring gear. At the end of its travel, the plunger completes the circuit and current flows from the battery to the starter motor.

When the engine starts and the switch is released, the coil is no longer energised. The solenoid return spring now pushes the plunger back to the rest position and in so doing, the plunger stops the flow of current to the motor and withdraws the starter pinion from the ring gear.

The only driver maintenance required is to keep the terminals on the battery, solenoid and starter clean and tight.

Lights and Fuses

Fuses protect electric circuits from overload. Electrical systems on horticultural tractors are no exception. The fine wire in the fuse cartridge will melt (blow) if too much current flows through the circuit. This prevents damage to the circuit and the various electrical components such as lights, indicators and horn.

The flow of an electric current is measured in amps (see Chapter 21). A fuse is rated according to how many amps the fuse wire will carry without blowing.

The tractor instruction book should give information on the correct fuse size for the various electrical circuits. Try to find why a fuse has blown before replacing it, and always use a fuse of the same size as the original. Do not use a bigger fuse as it will not give adequate protection to the circuit.

Most four wheel horticultural tractors have a full set of lights which comply with the Road Traffic Regulations. Additional lights may be fitted on the rear of the tractor for night working. It is a legal requirement that if lights are fitted they must be in working order when the vehicle is used on the highway. To keep within the law, all lights should be checked at regular intervals and faulty bulbs replaced.

Suggested Student Activities

1. Check your tractor and lawn mower batteries at regular intervals, topping up the electrolyte with distilled water as necessary.
2. Find out how to connect a battery to a mains charger unit and learn how to use and read a battery hydrometer.
3. Check with the tractor instruction book to find the correct setting for the fan belt on a tractor or ride-on mower you use.
4. Locate the starter, solenoid, alternator or dynamo, fuse box and light switches, where appropriate, on different models of garden tractor and self-propelled grass mowers.
5. Investigate a recoil starter unit to find out how to replace a broken starter cord.

Safety Checks

Always follow the procedure described in this chapter when it is necessary to use jump leads to start a tractor or mower which has a discharged battery.

Keep your feet well clear of the blades when using a recoil starter to start a rotary lawn mower engine.

Chapter 5

TRANSMISSION SYSTEMS

Most tractor engines have a top speed of at least 2,500 rpm; the engines on many pedestrian operated horticultural machines have a top speed in excess of 3,000 rpm. In low gear the driving wheels turn at little more than 20 or 30 rpm.

An important function of the transmission system is to reduce the high rpm at the engine flywheel to provide a forward or reverse speed suitable for the work in hand. This will vary from low speed field operations such as drilling and planting up to high speed road haulage.

A tractor in low gear can pull a heavy load, but the forward speed will be low. Light loads can be pulled much faster because a high gear can be used. Fast speeds can also be achieved when towing heavy loads on hard surfaces such as the public highway. This is because the resistance to wheel rotation (rolling resistance) on the road is much less than when working on loose ground.

It can be seen that light loads can be pulled at high speed and, except when driving on hard surfaces, heavy loads can be pulled only at a much lower speed. In very simple terms, the power available from the engine equals the load multiplied by the forward speed. In the transmission system, the gearbox offers a range of forward and reverse speeds which allow the driver to select a suitable speed for efficient handling of heavy or light loads on soft or hard surfaces.

Four wheel drive improves the pulling performance of a tractor, and many compact and rowcrop tractors used in horticulture have four wheel drive. The front wheels are usually smaller than those at the rear, but some models have equal sized wheels.

The Transmission of Power

The power developed by internal combustion engines and electric motors can be transmitted in various ways.

Vee-belts are used to transmit drive on some pedestrian controlled machines, especially small rotary cultivators. Power is transmitted by the sides of the vee-belt which grip against the side faces of the pulley. The belt must be tensioned so that it does not run in the bottom of the pulleys.

A vee-belt drive provides protection from mechanical damage caused by overload or blockages because under these circumstances the belt will slip.

Chain drives provide a positive drive system which cannot slip. Where damage due to overload can occur, a safety overload clutch is built into the drive train. Chain drives are used on some pedestrian controlled machines including rotary cultivators and lawn mowers. They are also ideal for the transmission of power in situations where two or more components must be timed to give a definite sequence of operations. A common example is the chain drive used on some multi-cylinder engines for timing the crankshaft to the camshaft and ignition or injection system.

Gears are another positive drive system where slip cannot occur. Spur gears with either straight teeth or angled (helical) teeth and bevel gears are used in mechanical transmission systems. Spur gears transmit drive from one parallel shaft to another. Bevel gears are used to transfer drive through an angle from one shaft to another, a common example of which is turning the drive

PLATE 5.1 *Four wheel drive compact tractor.* (Kubota)

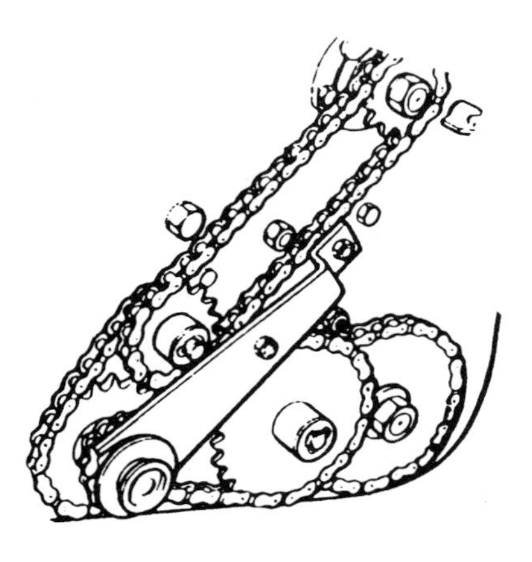

FIGURE 5.1 Cylinder mower chain drives to the cutting cylinder and rear roller.

through 90 degrees by means of the crown wheel and pinion in a tractor transmission system. The relationship between the speed of two meshed gears is directly related to the number of teeth on each gear. For example, a small gear on the driving shaft and a large gear on the driven shaft will result in a speed reduction on the driven shaft. (See Chapter 21.)

Line of Drive

Pedestrian controlled rotary cultivators have a simple drive system with a clutch, gear or chain reduction drive and a differential unit. Most self-propelled lawn mowers have a clutch and chain or belt drive.

Four wheel tractors have a more complicated transmission system. The engine flywheel is connected to the gearbox by

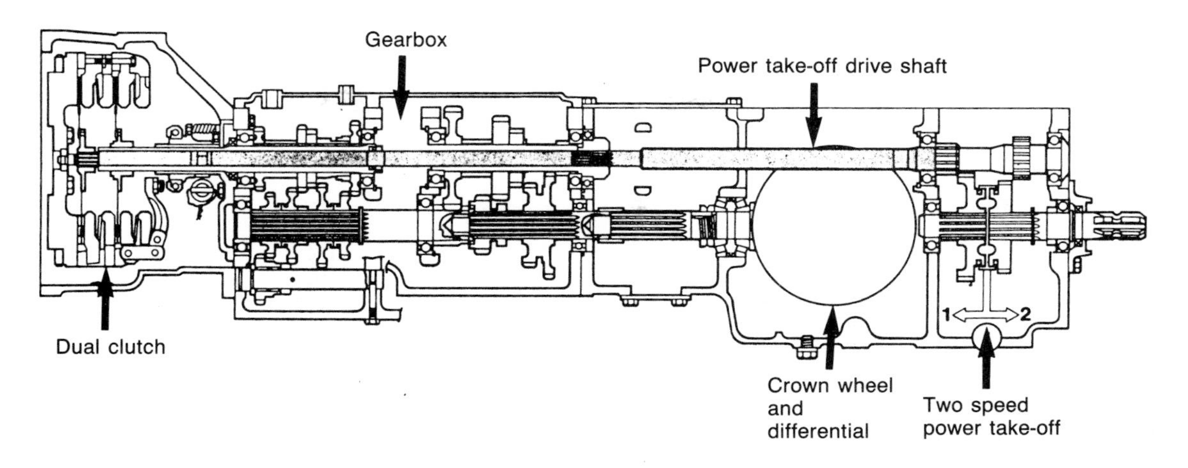

FIGURE 5.2 A compact tractor transmission system showing the line of drive to the differential and power take-off shaft. *(Kubota)*

the clutch which engages and disengages the drive to the driving wheels. Gear levers are used to select the required forward or reverse gears. The crown wheel and pinion puts the power from the gear box through a right angle. This is necessary because a tractor engine is, with very few exceptions, at right angles to the rear axle. The differential is carried on the crown wheel; it allows the driving wheels to turn at different speeds when the tractor turns a corner. The drive from the differential is transmitted by the half-shafts to the rear wheels. Many transmission systems have a final reduction unit which further reduces shaft speed before the power arrives at the rear wheels.

Tractors with four wheel drive have a shaft running forward from the gearbox to a differential and reduction gears in a housing on the front axle, where the drive is put through 90 degrees and the shaft speed is reduced before the power reaches the front wheels.

Many tractors have the power take-off shaft and hydraulic lift pump driven from the gearbox. Others have the hydraulic pump mounted on and driven by the engine; this arrangement provides a live hydraulic system completely independent of the tractor transmission.

THE CLUTCH

As well as engaging and disengaging the drive from the engine to the gearbox, the clutch enables the driver or operator to move off smoothly from rest.

The basic principle of a clutch is that a friction disc on the gearbox shaft, or drive shaft, is held against the flywheel, usually by spring pressure, so that power can be transmitted from the engine to the transmission system. Four wheel tractors have a friction disc clutch mounted on the engine flywheel. Pedestrian controlled machines may have a small multiple plate friction clutch or a centrifugal clutch or a dog clutch.

The main clutch assembly is attached to the engine flywheel. All parts of the clutch, with the exception of the clutch plate, rotate with the flywheel. The clutch plate, which has a friction disc on both faces, is carried on a splined output shaft connected to the gearbox. The flywheel and the pressure plate have smooth steel faces. When the clutch pedal is up, the clutch springs and pressure plate force the clutch plate against the flywheel face and power passes from the engine to the gearbox. The clutch spring pressure is released from the clutch disc when the pedal

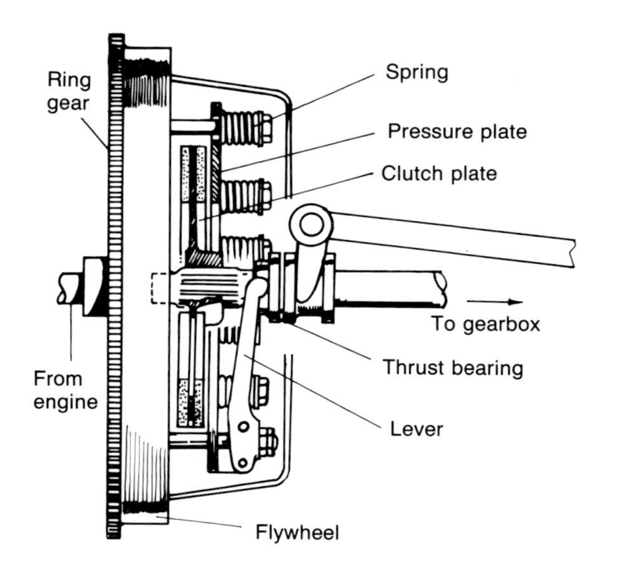

FIGURE 5.3 Single plate clutch with coil springs.

is pushed down; this breaks the drive to the gearbox. The pressure plate, springs and clutch cover continue to rotate with the flywheel, but the clutch disc is stationary.

The clutch pedal linkage operates a thrust bearing on the shaft which carries the clutch disc. When the pedal is pushed down, the thrust bearing moves the release levers towards the flywheel, and they compress the clutch springs to release the pressure on the clutch disc and disconnect the drive.

Many modern tractors have a hydraulically operated clutch similar to that used on a motor car. The clutch pedal operates a master cylinder, a simple pump, which supplies a special fluid under pressure to a small hydraulic cylinder and piston. Movement of the piston operates the clutch linkage which in turn disengages the clutch. When the pedal is released, the clutch springs re-engage the drive and the fluid returns to the master cylinder. This is similar to the hydraulic brake system shown in Figure 5.19.

Some tractor clutches have a diaphragm spring instead of the more common coil springs. The clutch operates on the same principle as a clutch with coil springs. The diaphragm spring (see Figure 5.5) is made of high quality spring steel and shaped like the concave rim of a dinner plate. When the clutch pedal is pushed down, the spring is flattened and the pressure is released from the friction disc to disengage the drive to the gearbox.

Many horticultural four wheeled tractors have a live power take-off. This means that the drive to the power shaft is not disconnected when the clutch is used to disengage the drive to the rear wheels. Some low horsepower compact tractors do not have the advantage of a live power take-off.

There are two ways of achieving live power take-off drive:

Independent clutch A single stage clutch on the engine flywheel makes and breaks the drive to the transmission but does not control the power take-off shaft. A second independent clutch in the transmission housing is used to engage and disengage the power take-off drive. A common design of independent clutch consists of a number of driving and driven plates which are operated hydraulically and controlled with a hand lever.

Dual clutch Although common on older tractors, the dual clutch has been replaced by the independent clutch system on most current models. The first stage of a dual clutch controls the transmission and the second stage stops and starts the power take-off shaft. The dual clutch works in the same way as a single plate clutch but has two discs, two pressure plates and two sets of springs. When the clutch pedal is pushed halfway down, the spring pressure on the transmission disc is released and the drive to the wheels is disconnected but the power take-off shaft continues to turn. When the pedal is pushed fully down, the second stage of the clutch breaks the drive to the power take-off shaft.

Clutch Pedal Free Play

The clutch pedal must have some free downward movement before any clutch spring pressure is felt. This free movement (clutch pedal free play) ensures complete clutch

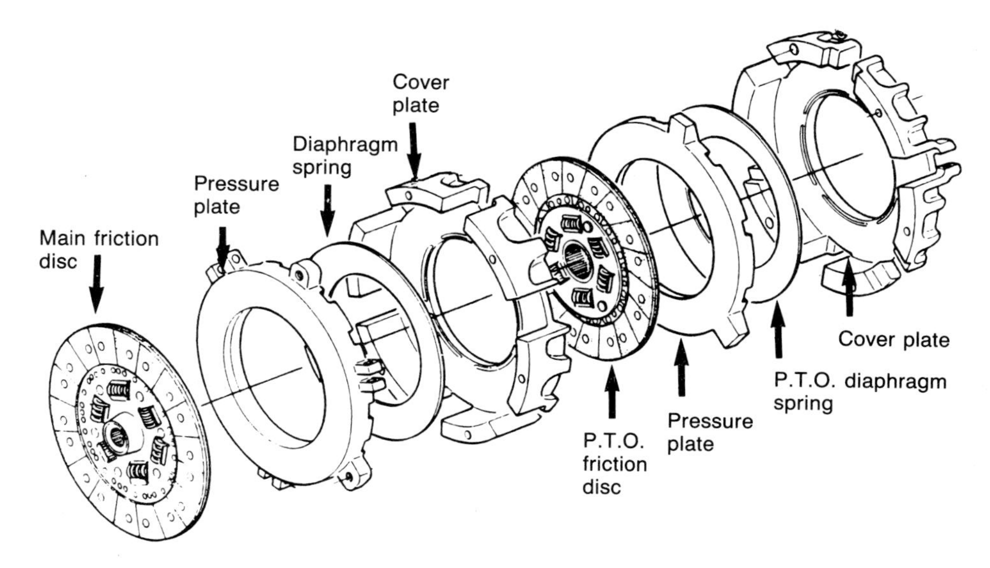

FIGURE 5.4 The parts of a dual clutch with diaphragm springs. *(Massey-Ferguson)*

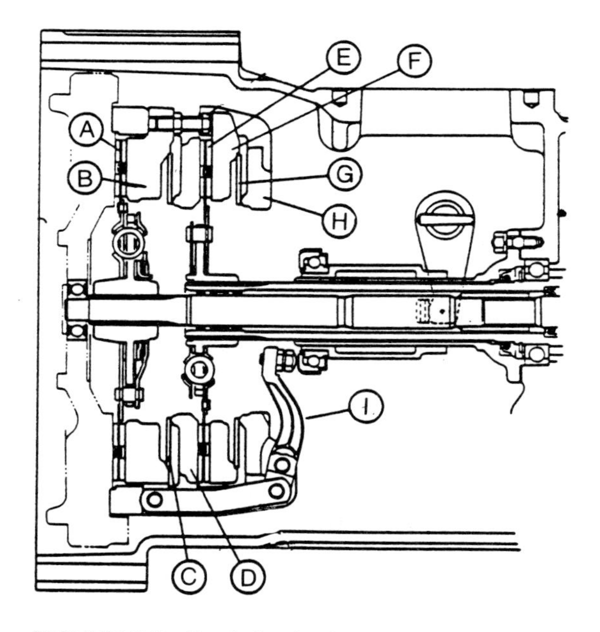

FIGURE 5.5 Dual clutch showing the arrangement of the main friction plate (A), pressure plate (B) and diaphragm spring (C) on the central shaft with the power take-off disc (E), pressure plate (F) and diaphragm spring (G) on the hollow outer shaft. The release levers (I) are operated by the thrust bearing, (D) is the main clutch cover and (H) is the p.t.o. clutch cover. *(Massey-Ferguson)*

engagement. The pedal linkage can be adjusted to obtain the correct amount of free play.

The amount of free play will vary with different models of tractor, and the setting and method of adjustment are given in the instruction book. A typical setting for free play is 20 mm; it is checked by measuring how far the clutch pedal can be pushed down by hand before the spring pressure is felt.

Too little free play may cause:

- Wear on the clutch thrust bearing.
- The clutch to slip because the pressure plate does not hold the clutch disc firmly against the flywheel.

Too much free play will make gear engagement difficult because the thrust bearing does not completely disengage the drive from the engine.

Never use the clutch pedal as a foot rest, because the weight of your foot will take up the free play and the thrust bearing will suffer excessive wear.

Other Types of Clutch

Centrifugal clutch Used on pedestrian controlled equipment, a centrifugal clutch engages the drive from the flywheel to the transmission

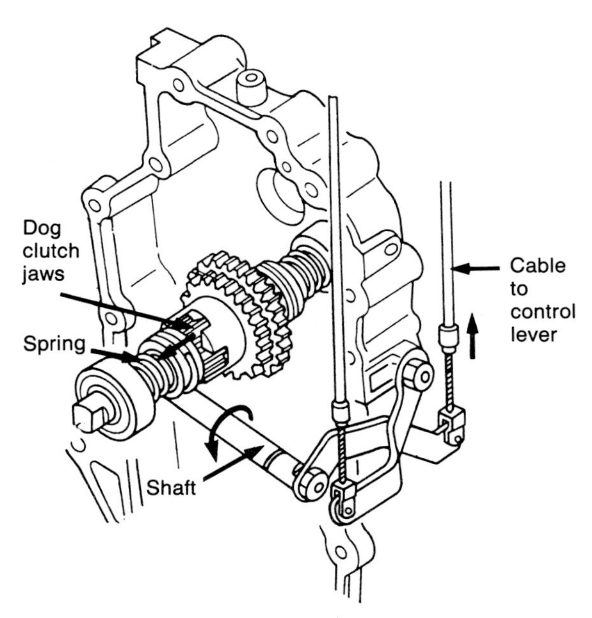

FIGURE 5.6a A dog clutch. The clutch cable rotates the actuating shaft to disengage the clutch jaws against the spring. When the clutch lever is released, the spring re-engages the clutch. *(Kubota)*

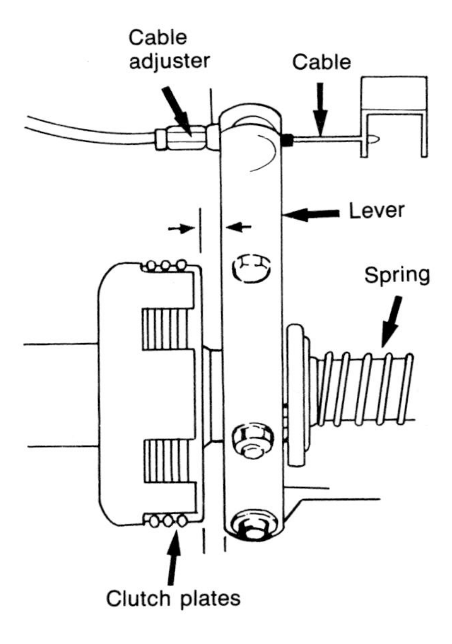

FIGURE 5.6b A small multi-plate lawn mower clutch. When the clutch lever is used, the cable pulls the lever against the spring and the clutch plates no longer transmit drive to the rear roller.

when the engine reaches a pre-set speed. The clutch consists of a pair of shoes with friction linings attached to the flywheel which rotate inside a hub on the transmission output shaft. The shoes are attached to a spring loaded linkage, which allows them to move outwards due to centrifugal force created when the flywheel rotates at speed. When the throttle is used to increase engine speed, the clutch shoes are thrown outwards towards the hub. The clutch shoe friction faces grip the clutch hub and drive is transmitted. (See Figure 5.7.)

The clutch disengages the drive when the engine returns to idling speed because the clutch spring pulls the friction faces away from the hub.

Chain saws, as well as some small lawn mowers and garden cultivators, have a centrifugal clutch.

Dog clutch This type of clutch can only be used to engage and disengage the drive when the shaft is stationary. The dog clutch consists of two hubs with square teeth on each face which are meshed together when drive is transmitted. The clutch is engaged and disengaged with a hand lever and held in the engaged position by a spring. The dog clutch is often used to connect secondary drives after the main drive from the engine has been broken with a friction type clutch. For example, the drive to the main roller on a lawn mower can be disengaged with a dog clutch so that the user can push the mower in confined spaces with the engine still driving the cutting cylinder.

Belt drive Engaging and disengaging a belt drive is achieved by means of a jockey or tensioner pulley operated by a hand lever. Some vee-belt drives have an over-centre device which locks the jockey pulley in position after it has tensioned the belt and engaged the drive.

The Gearbox

Self-propelled horticultural machines, including four wheeled tractors, ride-on lawnmowers, turf care machines and pedestrian controlled cultivators, have a gearbox which

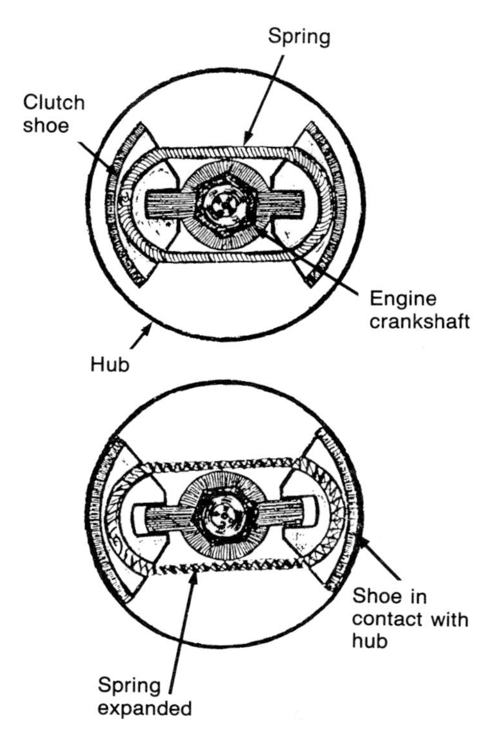

FIGURE 5.7 Centrifugal clutch. The coil spring holds the clutch shoes away from the hub until the engine gains sufficient speed so that the centrifugal force can overcome the spring pressure. The shoes grip the hub and drive is transmitted.

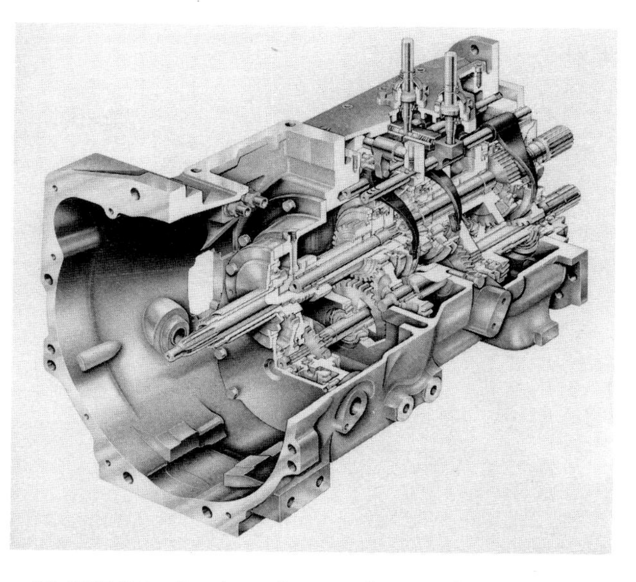

PLATE 5.2 *Section of a gearbox with straight tooth gears. The selector forks and gear levers can be seen at the top of the gearbox.* (Ford New Holland)

provides a selection of forward and reverse speeds. This gives the driver a range of forward speeds from less than 1.6 km/h (1 mph) to about 32 km/h (20 mph). Choice of gear will also determine how much the tractor can pull. Pedestrian controlled machines have a narrow range of gears with top speed limited to walking pace.

Many four wheeled tractors and ride-ons have a manual gearbox, which has a series of gears on two main shafts which give a fixed number of forward and reverse gears. Compact tractors usually have a gearbox with either six forward and two reverse gears or eight forward and two reverse gears. A typical example of a ride-on grass mower has five forward and one reverse gears.

Most compact tractors have two gear levers: the range lever engages either the high or low speed range, and the main gear lever selects the forward or reverse speed best suited to the work in hand.

Some four wheeled tractors have a hydrostatic transmission system rather than the more usual manual change gearbox. It is operated hydraulically and provides a stepless range from a very low creep speed to a fast road speed. Hydrostatic gearboxes are widely used on high specification compact tractors and ride-on grass mowers.

The simple manual gearbox used on most horticultural tractors will have either six or eight forward speeds and two reverse. The most basic type of manual gearbox is sometimes called a crash box because if the driver attempts to change gear with the tractor in motion, the gear teeth will grate against each other making a loud noise; excessive gear crashing will damage the teeth.

Large tractors with limited horticultural application may have as many as sixteen forward and eight reverse speeds, and some have twice this number of gears. The range lever has three positions giving low, medium and high ratios and there is an electro-

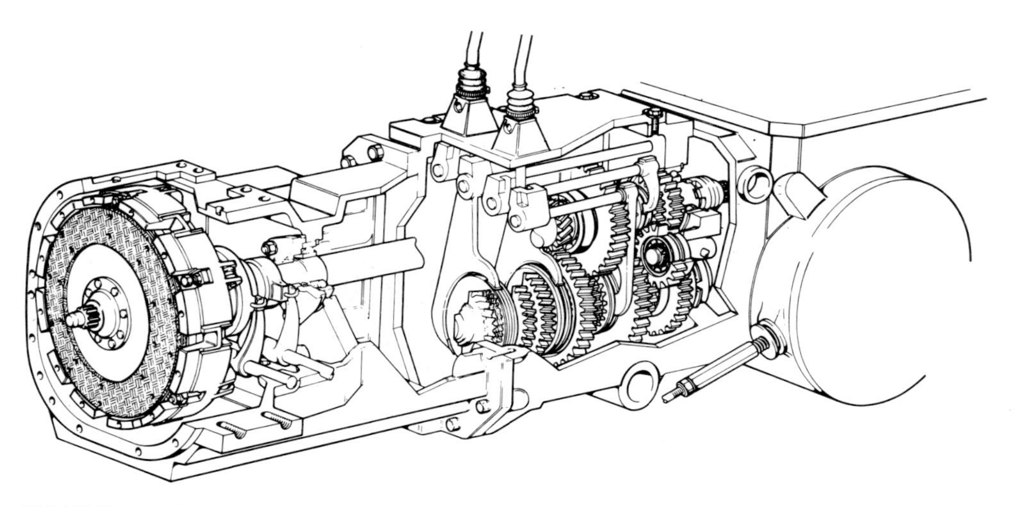

FIGURE 5.8 Sliding mesh gearbox, with straight tooth spur gears which are moved into and out of mesh with the two gear levers. The gear selector forks can be seen. *(Fiatagri)*

hydraulic control unit to provide additional gear changing at the flick of a switch.

Manual gearboxes on horticultural tractors usually have two gear levers. The range of speed is obtained by selecting high or low ratio with the range lever, and the required gear with the main lever. The different gearbox output shaft speeds are obtained by varying the gear combinations. A small gear on the input shaft driving a larger gear on the output shaft will give a big reduction in speed. A large gear driving a smaller gear will increase the output shaft speed (see page 285).

A typical gearbox with eight forward speeds will have four different gear trains which are engaged by using the main gear lever. The four speeds can be doubled up by selecting either high or low ratio with the range lever. Reverse is achieved by introducing a reverse idler gear on a third shaft in the gearbox.

The sliding mesh gearbox is one type of manual gearbox used on tractors. It has straight toothed (spur) gears which can be moved into and out of mesh with each other. When the driver changes gear, the selector fork operated by the gear lever moves the appropriate gear along its shaft until it meshes with another gear to complete the required gear train. Spur gears are rather noisy in operation.

The constant mesh gearbox is another type of manual gearbox. It has either spur or helical gears (see Figure 5.9), and the gears, which are carried in small groups on short shafts in the gearbox, are always in mesh. The different output speeds are obtained by connecting groups of gears in various sequences between the top and bottom shafts by means of special couplings controlled with the main gear lever.

Helical gears are quieter running than spur gears. Their angled teeth make them stronger than spur gears but they are more expensive to manufacture.

A change-on-the-move mechanism is fitted to the manual change gearbox on some tractors. A normal two lever gearbox is used to provide, for example, eight forward and two reverse speeds selected by using the clutch and gear levers. In addition a switch or a lever on the instrument panel can be used without using the clutch to select a high or low ratio drive within the chosen gear while the tractor is moving. This allows the driver to change speed within limits while on the move. It is a useful aid when, for example, a rough spot is encountered with plough or cultivator.

Synchromesh gearboxes are fitted to many models of tractors. Used on motor cars for many years, this type enables the driver to

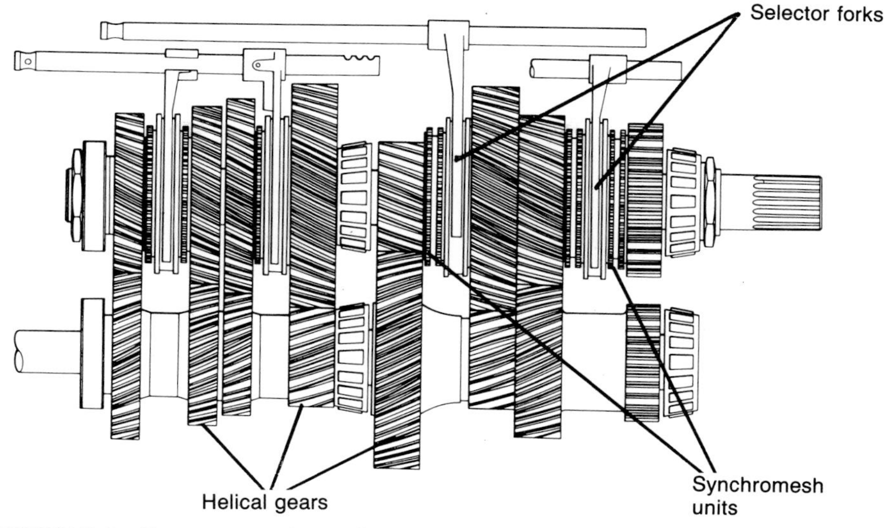

FIGURE 5.9 Constant mesh gearbox. *(Fiatagri)*

change up and down through the gearbox while on the move. It has special couplings which brings pairs of gears, about to be meshed together, to the same speed before the teeth are fully engaged. Many tractors have synchromesh on all forward gears. Others have synchromesh units on the higher gears only so the tractor must be stationary when, for example, changing up or down between first and second gear.

Hydrostatic gearboxes provide an infinite range of speeds from a very low creep speed up to a high road sped. This type of transmission is widely used for ride-on mowers and compact tractors because it allows the driver to vary forward speed in a very simple way without using the throttle lever. This is a great advantage when cutting grass or doing other tractor work where frequent changes in forward speed are necessary.

The hydrostatic gearbox has a range lever for selecting high, low or reverse ratio. Either a hand lever or, more commonly, a foot pedal is used to control forward and reverse speed. Some models have two pedals, one for forward movement and one for reverse. Others have a single rocker pedal with a central pivot. Foot pressure at one end of the pedal gives forward travel, and pressure at the opposite end gives reverse. As the pedal is moved downwards, the speed increases until it reaches the maximum setting in that ratio. Gradual release of the foot pressure will reduce the speed until the stop position is reached. If a lever is used to control speed, moving it forwards increases speed and backwards reduces it.

The hydrostatic drive system consists of an engine driven pump which supplies oil under pressure to a hydraulic motor in the transmission housing. An oil cooler is built into the system to prevent the oil overheating. The speed control lever or pedal governs the rate of oil flow to the motor, and in that way controls forward or reverse speed because, as the oil flow to the motor increases, the tractor will travel faster.

Tractors and ride-ons with hydrostatic drive do not have a clutch, as there are no gears to change. However, a pedal is often provided to assist in controlling the machine in confined spaces and when hitching an implement.

Lubrication

Gearboxes are usually lubricated with SAE 40 or 50 gear oil. The oil level should be checked weekly. Some tractors have a combined gearbox and transmission housing with one supply of oil to lubricate the entire system.

PLATE 5.3 *Some of the driving controls of a compact tractor. The independent brake pedals (right) are locked together. The hydrostatic transmission is controlled by the rocker at the left of the brake pedals.*
(Massey-Ferguson)

Belt drives
Some pedestrian controlled machines have a small range of forward speeds provided by means of a vee-belt drive with multiple pulleys. Both the driving and driven shafts may, for example, have a multiple pulley with three different diameter pulleys side by side. By placing the belt on the largest pulley on the driving shaft and the smallest pulley on the driven shaft, a speed increase is achieved. When the position is reversed with the smallest pulley driving the largest one, a speed reduction is achieved. The belt is tensioned by a lever-operated jockey pulley. Tension must of course be released before the belts can be moved to a different pair of pulleys.

The Crown Wheel And Pinion

Drive from the gearbox must pass through a right angle because the wheels run at 90 degrees to the engine crankshaft. The crown wheel and pinion—a pair of bevel gears—change the direction of drive. The smaller bevel pinion on the gearbox output shaft drives the much larger crown wheel. When a small gear drives a larger gear there will be a speed reduction. This means that the crown wheel and pinion not only put the drive through a right angle but also give a considerable reduction in shaft speed.

The crown wheel and pinion bevel gears may have either spur or helical teeth; many tractors have the stronger and quieter running helical gears.

The Differential

It would be impossible to turn a corner with a tractor if the axle was a single shaft with a wheel at each end. The differential allows one rear wheel to turn faster than the other when cornering, but when moving straight ahead, the differential passes equal drive to both wheels.

Figure 5.10 shows the differential unit fixed to the crown wheel. A bevel gear on each reduction gearshaft meshes with the smaller differential pinions running in journals (bearings) attached to the crown wheel.

When the tractor travels in a straight line, the crown wheel, reduction shaft gears and differential gears all turn as one unit. When the tractor turns a corner, the inner wheel slows down. As a result the differential gears rotate around the slower moving reduction shaft gear. This action increases the speed of the other reduction gear connected to the outer wheel. The higher speed of this gear means that the outer tractor wheel must also turn faster, which is necessary for the tractor to actually get round the corner.

As soon as the tractor is back on a straight course, the reduction shaft gears and differential gears stop turning independently and both wheels run at the same speed.

When one tractor rear wheel spins at high speed and the other is stationary, the differential is in operation with one reduction shaft gear turning at twice the normal speed and the other not turning at all. This degree of wheel spin will occur when the tractor is

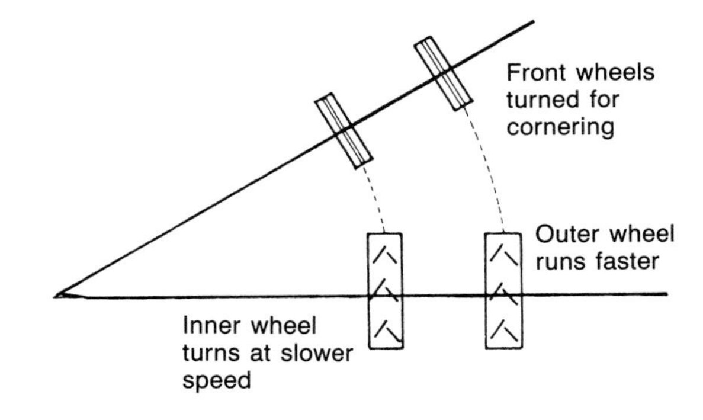

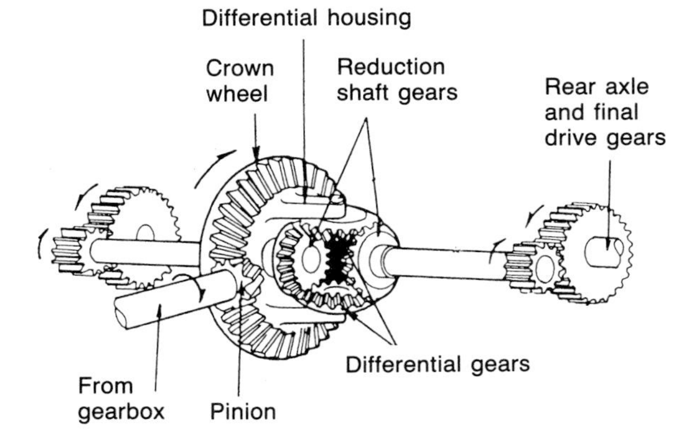

FIGURE 5.10 Principle of the differential.
(Kubota)

working in very wet conditions or is seriously overloaded.

The Differential Lock

Wheel spin can be overcome to a large extent by using the differential lock (diff-lock), which cuts out the action of the differential when the diff-lock pedal is pushed down. The two reduction gear shafts are prevented from turning independently because one of the reduction gears is locked to the differential housing by a pedal operated peg. This forms a single shaft between the rear wheels, making it impossible for only one wheel to spin when the tractor is working in difficult conditions.

When the diff-lock is engaged, the tractor cannot be driven round a corner. The lock mechanism is designed to disengage automatically when the pedal is released. If the diff-lock is still engaged after releasing the pedal, the problem can be overcome by pushing down the clutch pedal or one of the independent brake pedals. On some tractors, the pedal is pushed down to engage the diff-lock and the pedal will remain down after the driver removes his foot. When the pedal is pushed down again, the diff-lock is disengaged and the pedal returns to its upper position.

Some tractors have an electro-hydraulic control switch on the instrument panel for engaging the diff-lock. Disengagement is automatic. A warning light tells the driver when the diff-lock is engaged.

Final Reduction Gears

The speed of the output shafts from the differential is much lower than the engine crankshaft speed. However, many tractors have an additional speed reduction unit between the differential and the rear wheels. (See Figure 5.13, page 72.)

Epicyclic reduction units (see Figure 5.12) are common in the final drive of tractors designed for agriculture. An epicyclic unit in each axle housing gives a further speed reduction before the drive reaches the rear wheels. A small pinion on the half shaft transmits the drive through idler pinions to a large internal ring gear in the final drive hub. Since a small gear is driving a larger gear, a reduction in speed is achieved.

Most compact tractors have the power from the differential transmitted to the wheels through a bull gear and pinion in the transmission housing. A small pinion on each shaft from the differential drives a larger (bull) gear on the shafts which carry the rear wheels. Here again, a small gear drives a larger gear to provide a speed reduction.

Some models of self-propelled pedestrian operated rotary cultivators have bull wheel and pinion final drive.

Torque

It should be noted that as well as reducing the shaft speed, the reduction units give an increase in power output. The power or turning force in the half shafts is called torque. At each point in the transmission system where

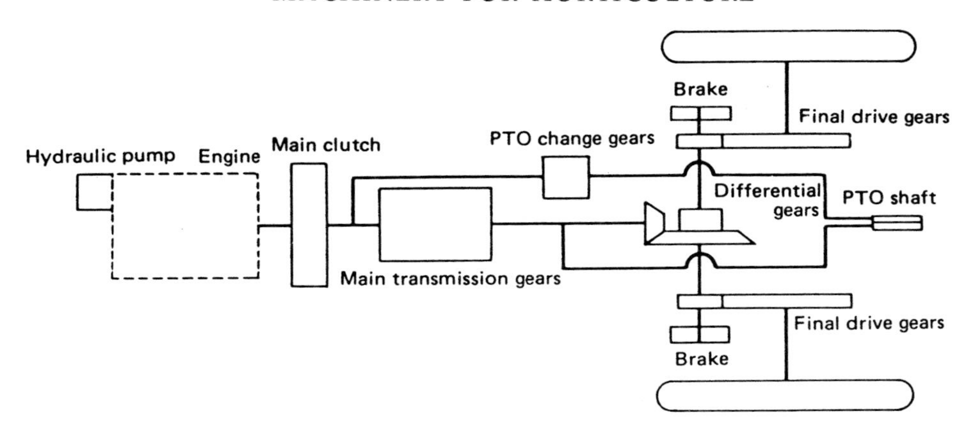

a. Final drive gears located near differential gears

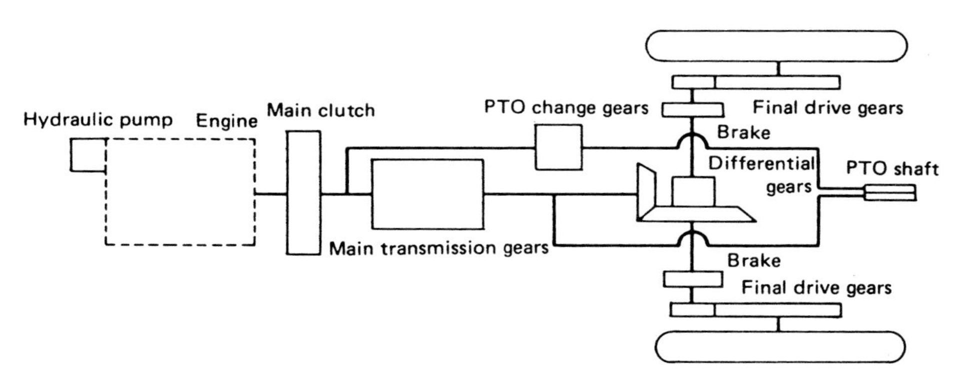

b. Final drive gears located near driving wheels

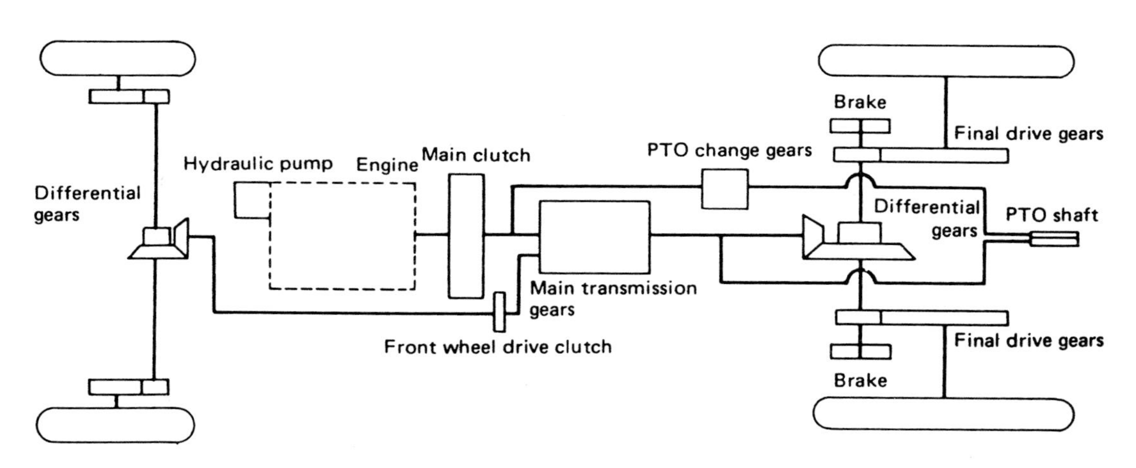

c. Four wheel drive tractor

FIGURE 5.11 Examples of final drive layouts. *(Kubota)*

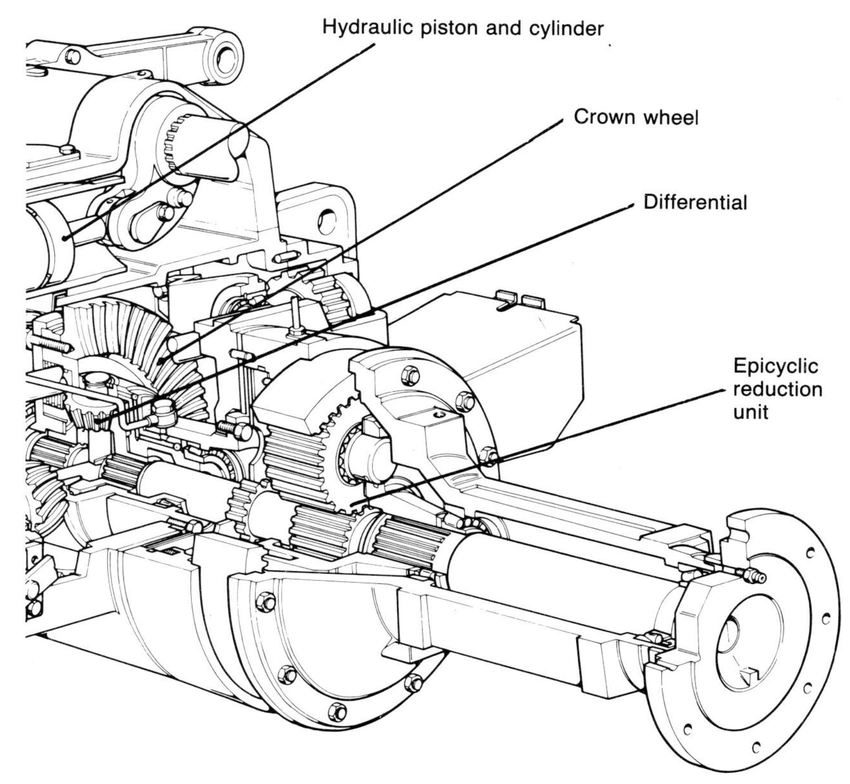

FIGURE 5.12 Epicyclic reduction unit. *(Fiatagri)*

a speed reduction occurs there will also be an increase in the available torque.

Power Take-off

The power take-off shaft is an important power source on four wheeled tractors. There is a British Standard which requires the power take-off shaft to run at 540 rpm (plus or minus 10 rpm). A mark on the engine rev counter indicates the engine speed at which the power shaft runs at 540 rpm. Another mark shows the engine setting which gives the second standard speed of 1000 rpm provided on some models of tractor. The power shaft must also meet the standard specification of 35 mm diameter with six splines.

Many agricultural tractors above approximately 45 kW (60 hp) have a 1,000 rpm power shaft for driving implements and machines which require a lot of power at the power take-off. On these more powerful tractors, the 1,000 rpm power shaft is 44 mm diameter with 21 serrations.

The standard dimensions for shaft size and speed are set to ensure that implement makers can produce power driven equipment which will fit and operate at the correct speed on any make or model of tractor within the power range for which the machine was designed.

Tractors with a two speed shaft have a high/low ratio gearbox with a lever to select 540 or 1,000 rpm. Some agricultural tractors also have a power take-off shaft at the front which is used to operate power driven implements attached to the front-mounted hydraulic linkage.

Compact tractors have a one, two or three

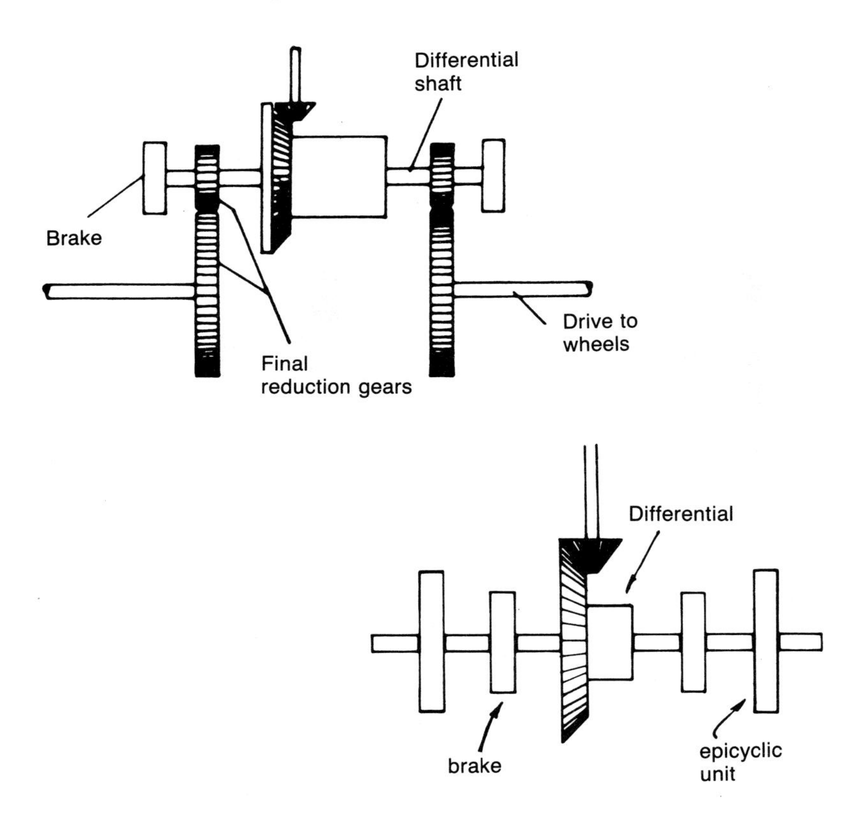

FIGURE 5.13 Final drive arrangements.

speed power take-off shaft at the rear. All have a standard 540 rpm shaft and many have the 540/1,000 rpm option. Some tractors have other specified power shaft speeds. For example, one model has a rear power shaft speed of 790 rpm at an engine speed of 2,370 rpm.

Some designs of compact tractor have a mid (central) and/or a front power take-off shaft which turns at approximately 2,500 rpm when the engine runs at a stated speed. However, front and mid power shaft speeds on different models of compact tractor vary from 2,100 to 2,600 rpm. The higher power shaft speeds—790, 1,000 and 2,100 plus rpm—are mainly used to drive mid- or rear-mounted grass mowers. Since grass cutting equipment is usually designed and supplied by the tractor manufacturer for a particular model, a standard power shaft speed is not so important.

Ground speed power take-off on some tractors has a shaft speed related to forward speed. When the tractor is reversed with the power take-off in gear, the shaft turns in reverse. Ground speed power take-off is used where the speed of the power shaft must increase and decrease in direct relationship to the forward speed. A trailer with a power driven axle and some power driven planters are examples of its use.

Nearly all tractors have a live power take-off. Tractors with a dual clutch must have the clutch pedal pushed to the floor before using the drive engagement lever. Tractors with an independent power take-off clutch do not require use of the transmission clutch pedal before using a lever or a switch, depending on the model, to engage the drive to the power shaft.

A few tractors have neither a dual nor an independent clutch to control the power

PLATE 5.4 *Power take-off shaft and guard.* (Massey-Ferguson)

take-off shaft. In such cases, the transmission clutch must be disengaged before operating the power shaft lever. A major disadvantage of this arrangement is that the power shaft is not able to reach full working speed before the tractor moves off.

It is important to ensure that the outer section of the power drive shaft from the tractor to the implement has a minimum overlap of at least 150 mm with the inner section when the shaft is fully extended, i.e. when the implement is at its lowest position. To prevent damage to the drive shaft when the implement is fully raised, it is important to have the inner shaft slightly shorter than the outer one. Shaft lengths should be checked

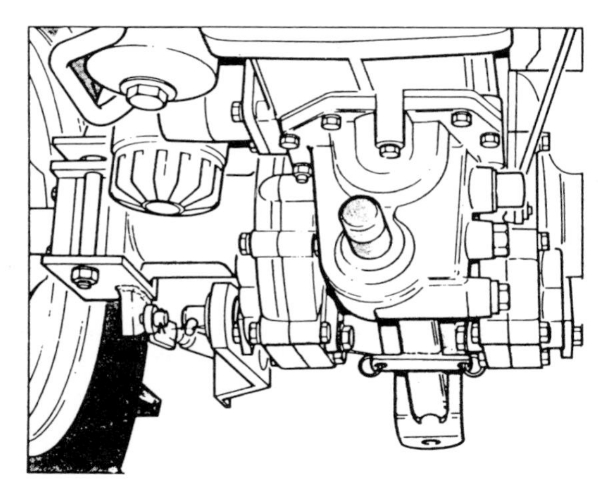

FIGURE 5.14 Mid power take-off shaft on compact tractor. *(Ford New Holland)*

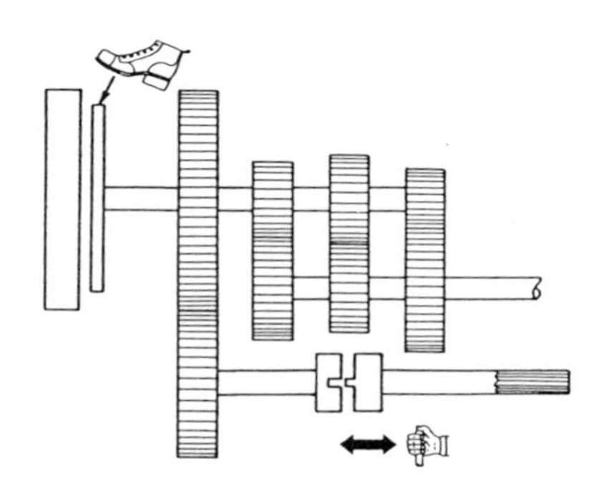

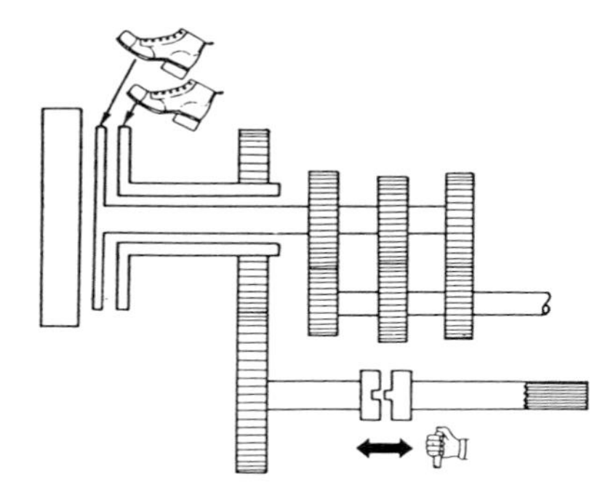

FIGURE 5.15 Drive engagement to the power take-off. Top: single plate clutch requiring main clutch to be disengaged before using the p.t.o. lever. Bottom: dual clutch arrangement with transmission clutch only disengaged. *(Ford New Holland)*

when fitting a new power take-off driven implement to a tractor for the first time. The same check is needed when fitting old implements to a different model of tractor.

Guarding the Power Take-off

There are a number of legal requirements concerning guards for power take-off shafts. A shield must cover the top and sides of the splined shaft on the tractor. It must be capable of supporting a weight of at least 114 kg (250 lb). When the power shaft is not in use, the shield may be removed provided that the splined shaft is protected by a dust cover of equivalent strength.

The drive shaft from the tractor up to and including the first fixed bearing on the machine must be covered on all sides. The guards must be maintained in good condition at all times.

Brakes

Four wheeled tractors and ride-on grass cutting equipment have either expanding shoe or disc brakes. Many models of ride-on mower and garden tractor and some compact tractors have expanding shoe brakes, whereas agricultural tractors and high specification horticultural tractors usually have disc brakes.

Some tractors are fitted with dry disc brakes which have friction lining on the disc faces. Other models have wet disc brakes—either plain steel or friction lined disc faces immersed in oil—to keep them cool and extend their working life. Wet brakes do not fade, as is the case with some dry brakes which become less effective when the friction linings are hot.

A brake is always more effective on a fast moving shaft because there is less torque in that shaft. Tractors with final reduction gears normally have the brakes situated between the differential and the reduction gears. Models without a final reduction unit have the brakes on the rear wheel hubs.

The rear wheel brakes have separate pedals which can be operated independently to aid

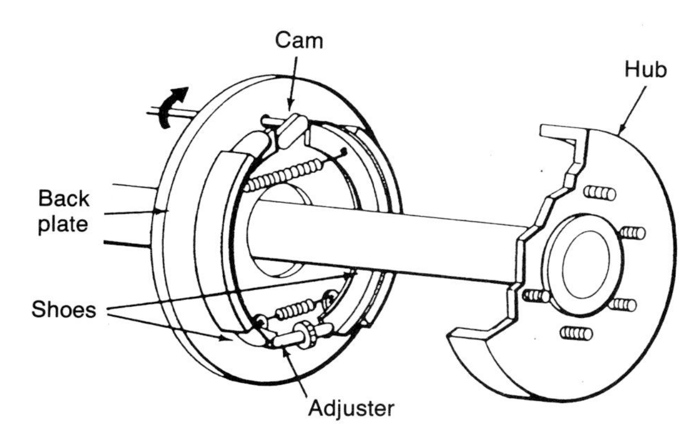

FIGURE 5.16 Expanding shoe brake. *(Kubota)*

turning. The pedals must be locked together before the tractor is taken on the road.

Expanding Shoe Brake

A pair of friction lined brake shoes are attached to the back plate. The shoes are held inwards, away from the brake hub, by springs. A pedal operated cam is situated between one end of each shoe.

The brake drum is secured to the shaft from the differential unit. When the brake is applied, the pedal linkage operates the cam and the brake shoes are forced outwards until the linings on both shoes grip the revolving brake drum. This action slows or stops the tractor. The brake springs pull the shoes away from the drum when the pedal is released.

Each brake has an adjuster unit which is used to set the shoes close to the drum. The length of the rods which connect the operating cams to the brake can also be adjusted.

When the independent pedals are locked together for road work, it is possible that only one brake will come on if there is uneven wear on the brake shoe linings. To overcome this problem, the brakes should be readjusted and then the pedals checked for correct balance. This means that when the unlatched brake pedals are pushed down as far as possible by hand, they should both be at the same height from the floor. Unbalanced pedals can be corrected by adjusting the length of the brake rod on the lower of the two pedals until both pedals are at the same height.

Unbalanced brakes are very dangerous, especially when driving at high speed.

Disc Brakes

Disc brakes have a greater friction surface area than expanding shoe brakes of equivalent size. A disc brake consists of two, sometimes three, discs which rotate between stationary plates. When the brake pedal is applied, a central expander plate forces the friction discs tightly against stationary outer faces. Depending on the amount of pedal pressure, the tractor will slow or stop.

Some disc brakes (wet brakes) run in oil which must have a special additive to stop the brakes juddering or 'squawking' when they are used. Wet brakes are more efficient because the oil washes and cools the brake assembly, making them less likely to stick or fade. Dry brakes tend to stick due to a build-up of dust from the disc linings as they wear.

The brake discs are attached to the rear wheel drive shafts. The stationary expander plates, which are held together by strong

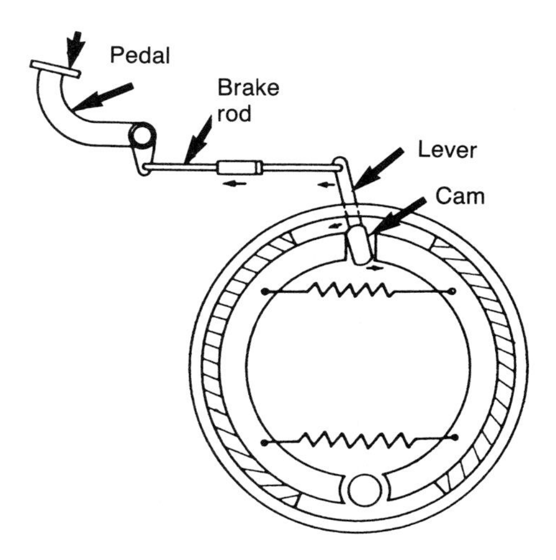

FIGURE 5.17 Operation of expanding shoe brakes. After the brake shoes have been adjusted, the pedals are balanced by shortening or lengthening the brake rod with the turnbuckle adjuster. *(Kubota)*

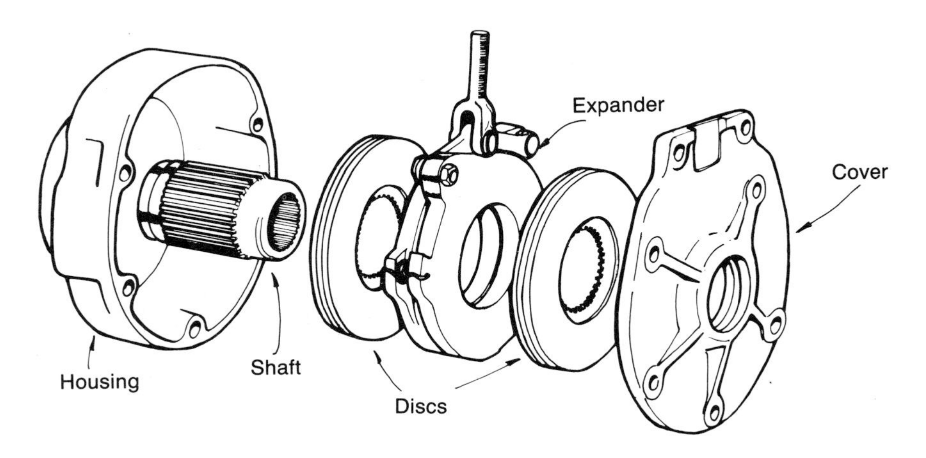

FIGURE 5.18 The parts of a disc brake.

springs, have two or three steel balls in tapered slots sandwiched between them. The smooth inner surfaces of the disc brake housing and the cover plate provide the outer faces on the type of disc brake shown in Figure 5.18.

When the brake pedal is pushed down, the linkage causes the pair of expander plates to contra-rotate through a few degrees. The resulting movement of the steel balls in their tapered slots forces the two expander plates apart. This movement forces the rotating brake discs against the outer faces to slow or stop the tractor. The springs pull the expander plates away from the friction discs when the pedal is released and the discs are free to rotate.

Some disc brakes have an adjustment for pedal balance and the running clearance between the discs and the expander unit. Others are self adjusting. Refer to the tractor instruction book before attempting to adjust disc brakes for the first time because there are numerous methods of completing this task.

In the same way that quiet cabs and modern design have brought about the use of hydraulic clutch control, many tractors now have hydraulically operated brakes. Each brake pedal operates a master cylinder which in turn applies one brake with a small ram cylinder built into the brake assembly. A balancing valve ensures that both brake cylinders work at the same pressure when the pedals are locked together.

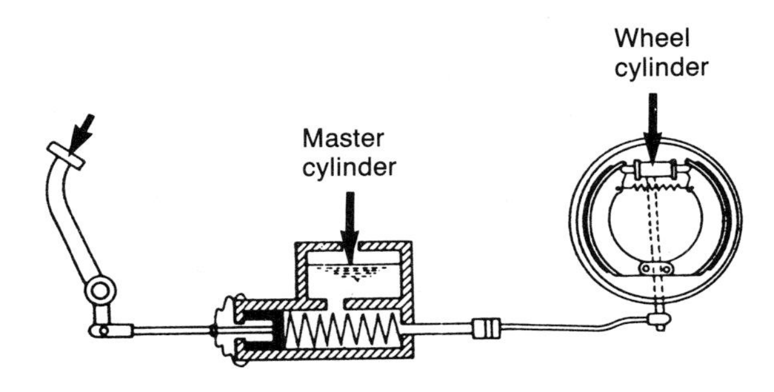

FIGURE 5.19 Hydraulic brake arrangement. The pedal operates a small master cylinder which sends pressurised brake fluid to the wheel cylinders. *(Kubota)*

Parking Brake

The parking brake or handbrake, which overrides the pedals, is used when the tractor is parked. The handbrake lever has a ratchet device which holds it in position. It is good practice to hold the ratchet away from its teeth by using the ratchet release lever when applying the brake. This action will extend the life of the ratchet teeth.

Some tractors have a locking device on

the pedals to hold on the footbrake when parked.

Lubrication

The type of lubricant used in the transmission housing will depend to a large extent on the design of the system. Special oils are needed for wet brakes and the efficient operation of some hydraulic systems depends on use of the correct oil.

Transmission oils have additives to give them the properties necessary for long and efficient service. They include:

- *Anti-wear*—an additive which, by sustaining the strength of the oil film, protects the transmission components from the very high pressures which occur between the gear teeth.
- *Anti-oxidant*—this combats chemical breakdown (oxidation), which tends to thicken the oil.
- *Wet brake additive*—needed in lubricants used for transmission systems with oil immersed brakes. It reduces 'squawking' when the brakes are applied.
- *Anti-foam*—an additive for reducing the build-up of air bubbles in the oil (foaming) caused by the high speed rotation of gears, and by the oil returning to the transmission housing from hydraulic rams. Foaming reduces the efficiency of the oil, especially in hydraulic systems.
- *Corrosion inhibitor*—this gives the oil the ability to combat corrosion, especially from traces of moisture caused by condensation in the transmission housing.

Regular transmission oil level checks are important, especially if the tractor is used to operate tipping trailers and other equipment with external hydraulic rams.

Transmission oil changes are often skimped; this is false economy because oil becomes less efficient with prolonged use. In many tractor transmission systems, the oil must lubricate moving parts, operate the hydraulic system, supply the independent power take-off clutch and ensure long trouble-free service from wet brakes.

Regular oil changes therefore are important. A general guide to oil change periods is given on page 107 but you should read and follow the recommendations in your tractor instruction book.

Four Wheel Drive

Difficult working conditions such as sloping ground and wet or loose surfaces may result in two wheel drive tractors being unable to work efficiently. The advantages of four wheel drive include:

- Increased traction (pulling power).
- Less wheel slip.
- The ability to work in adverse conditions.
- Greater safety on sloping land.
- Better flotation resulting in less compaction of soil and turf.

There are two main types of four wheel drive. Horticultural tractors, with a few exceptions, have smaller front wheels than rear ones. Some large agricultural tractors have equal sized wheels. It should be noted that some self-propelled front-mounted mowers are driven by the front wheels and steered from the rear.

Drive to the front wheels is taken from the main gearbox output shaft, by means of a transfer gearbox and forward running shaft, to a differential housing on the front axle. The housing contains the crown wheel and pinion which carries the differential unit. The output shafts from the differential transmit power to the front wheels. These shafts have a flexible drive unit to facilitate steering. Final reduction gears in the drive train, usually an epicyclic unit, reduce the shaft speed in order for ground speed of the front and rear wheels to be matched. It is usual for the front wheels to turn marginally faster than the rear wheels, as this improves traction and makes the tractor easier to steer.

A front axle diff-lock is usually provided. It is designed to work simultaneously with the rear wheel diff-lock. Some front wheel drive systems have a limited slip differential which allows the front wheels to rotate at different speeds when cornering, while at other times it maintains equal drive to both wheels.

Four wheel drive is engaged either by a switch on the instrument panel or by a

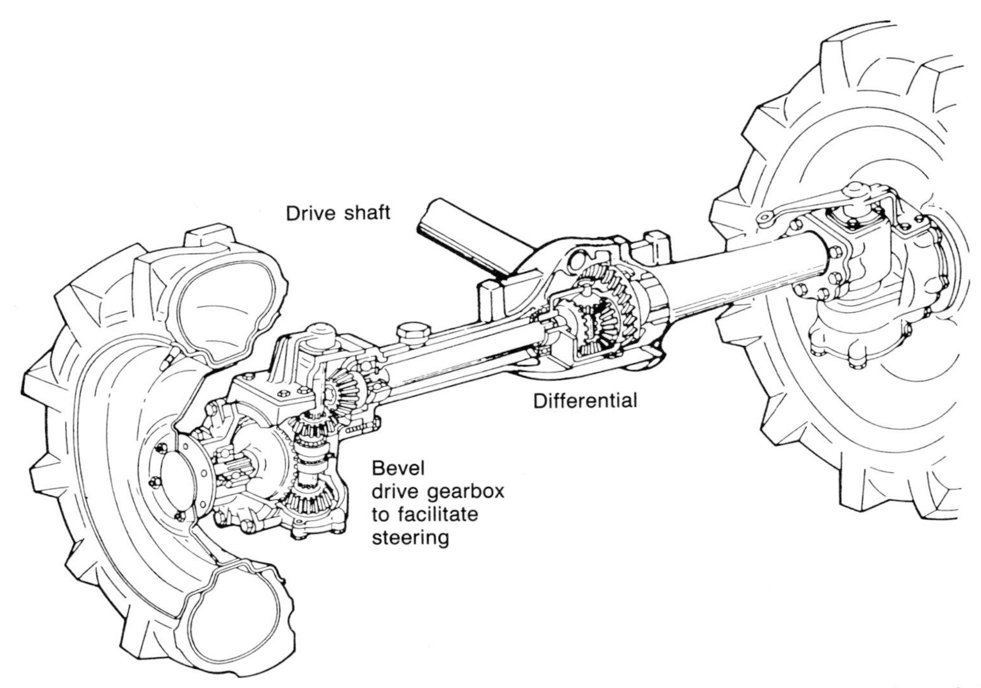

FIGURE 5.20 Front wheel drive system. The drive shaft from the main gearbox passes through the differential to both front wheel gearboxes. The bevel gear arrangement makes it possible to steer the tractor with the drive engaged. *(Kubota)*

hand lever, depending on the model. It can usually be engaged on the move, although it is necessary to stop some tractors first.

Maintenance

Oil levels in the transfer gearbox and front differential housing should be checked frequently. Gear oil is used and this needs to be changed at intervals of approximately 500 hours. Hubs and universal steering joints also require regular lubrication. Check your tractor instruction book for correct service intervals.

PLATE 5.5 *The front axle of a four wheel drive compact tractor. The gearing is totally enclosed and runs in oil.* (Massey-Ferguson)

Suggested Student Activities

1. Check the clutch pedal free play on your tractor. The correct setting and method of adjustment should be detailed in the instruction book.

2. Follow the line of drive on at least two different models of pedestrian controlled cultivator or lawn mower.
3. Get to know the position of the diff-lock pedal, the front wheel drive engagement lever or switch, the range and main gear levers and the power take-off lever on as many tractors as possible.
4. Locate the transmission oil level plug or dipstick, the gear box oil level plug and the front wheel drive oil level plugs on a range of tractors.
5. Check your tractor brakes often. When the brakes are applied, the tractor should pull up squarely with the independent pedals locked together; if it does not, make sure the fault is corrected.

Safety Checks

Never engage the drive to a power take-off shaft until you have checked that all guards are in place.

You will find it much easier on your knuckles when connecting a power shaft if the splines on the shaft and coupling are clean and lightly lubricated.

Always make sure that the independent brake pedals are locked together before taking a tractor on a private or public road.

Chapter 6

STEERING, WHEELS AND TYRES

Steering

Horticultural tractors, ride-on mowers and other self-propelled equipment may have manual, power assisted or hydrostatic steering. Tractors are front wheel steered but some models of ride-on mower and garden tractor are steered by either the rear wheels or by all four wheels. Four wheel steering allows the driver to make very tight turns, a considerable advantage when working in confined spaces.

Power assisted steering which has a hydraulic ram to help turn the wheels to left or right makes the use of front-end loaders and other front-mounted equipment much easier.

Hydrostatic steering has no mechanical linkage between the steering wheel and the front (and/or rear) wheels, and is ideal for air conditioned, quiet cabs.

Manual Steering Systems

The steering wheel moves the drop arm backwards and forwards by means of the gearbox at the bottom of the steering column. The drag link connects the drop arm to one of the steering arms on the front axle. When the drop arm pulls the drag link towards the back of the tractor, it turns the steering arm and in so doing changes the direction of travel. Both front wheel stub axles have a steering arm; they are connected by the track rod which gives the same degree of movement to both front wheels. The length of the track rod must be reset after changing the front wheel track width.

A second type of manual steering has two drop arms, both operated by the steering box. When the steering wheel is turned, one drop arm moves forwards and the other backwards. The drop arms are connected to the steering arms by drag links which turn both front wheels to the left or right. There is no track rod. Some tractors need an adjustment to the length of the drag link after altering wheel track widths; others do not.

Power Assisted Steering

Power assisted steering has a similar linkage to manual steering with a small hydraulic ram to help steer the tractor to left or right. The ram is operated by a hydraulic pump which has its own oil reservoir, control valve and pipework. The pump is usually driven by the engine fan belt and the control valve is operated by the steering gearbox.

The steering ram is double acting; this means that oil can be pumped to either side of the ram piston to give movement in two directions.

When the driver wishes to turn to the left, rotation of the steering wheel causes the control valve to direct oil, pumped under pressure, to one side of the hydraulic ram piston. As the front wheels turn to the left, oil from the other side of the piston returns to the reservoir.

Tractors can be manually steered if the power steering system develops a fault.

Hydrostatic Steering

There is no mechanical linkage from the steering wheel to the front wheels. The main components are an engine driven pump, a control valve operated by the steering wheel, a double acting ram, an oil reservoir and high pressure pipework.

PLATE 6.1 *Turf care tractor with articulated steering.* (Ransomes, Sims and Jefferies)

The tractor can only be steered when the engine is running. It cannot be steered manually if the hydrostatic system fails.

When the driver turns the steering wheel, the control valve directs a supply of oil under pressure from the pump to one side of the ram piston. The oil from the other side of the piston returns to the reservoir. The control valve, operated by the steering wheel, controls the direction and rate of flow of oil from the pump to the double acting ram piston which, in turn, determines how far the front (or rear) wheels turn towards the left or right.

Many garden tractors and ride-on mowers have hydrostatic steering. Some are front wheel steered; others are rear wheel steered, often by means of an articulated linkage. The front and back parts of the machine are connected by a central pivot point. The steering ram linkage alters the position of the rear half in relation to the front, causing the tractor or mower to change direction.

Steering Boxes

Tractors with single drop arm manual steering usually have a recirculating ball steering gearbox. This changes the rotary movement of the steering wheel into the linear movement needed for the steering linkage to turn the wheels in the required direction. Tractors with power assisted steering usually have a recirculating ball gearbox with a control unit for the steering ram. When the steering wheel is turned, the control valve directs oil to the assister ram.

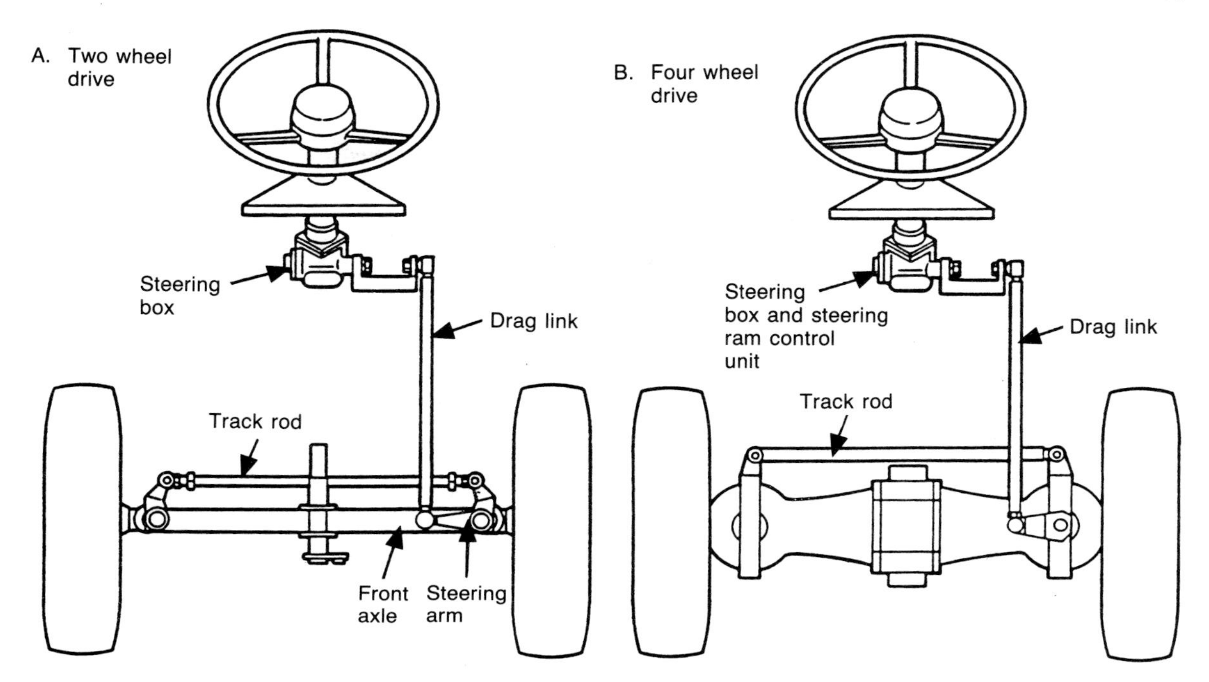

FIGURE 6.1 Single drop arm steering linkages for two wheel and four wheel drive tractors. The four wheel drive will have power assisted steering. *(Massey-Ferguson)*

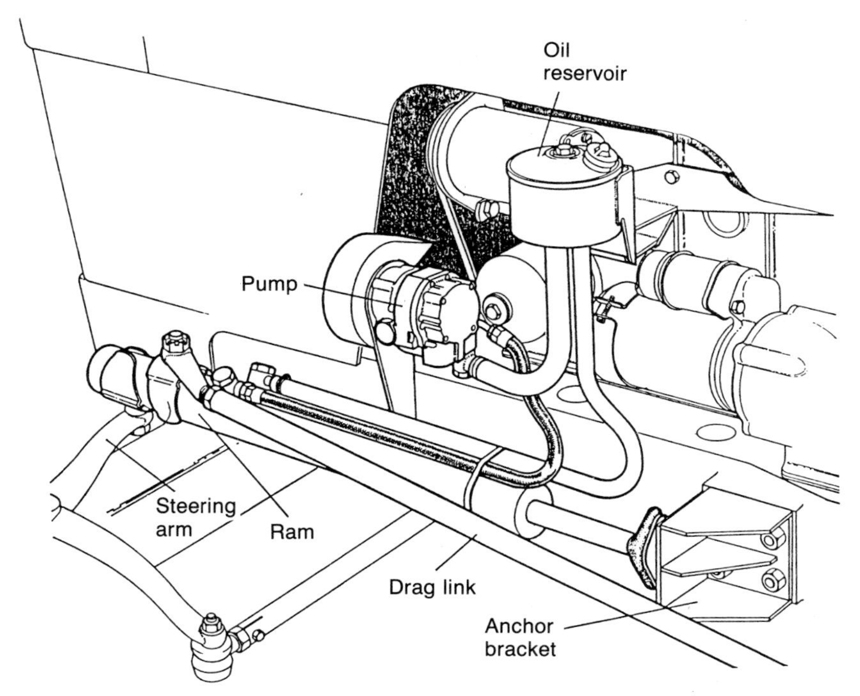

FIGURE 6.2 Power assisted steering. The control valve is part of the steering gearbox; it directs oil to the ram to help turn the front wheels in the required direction.

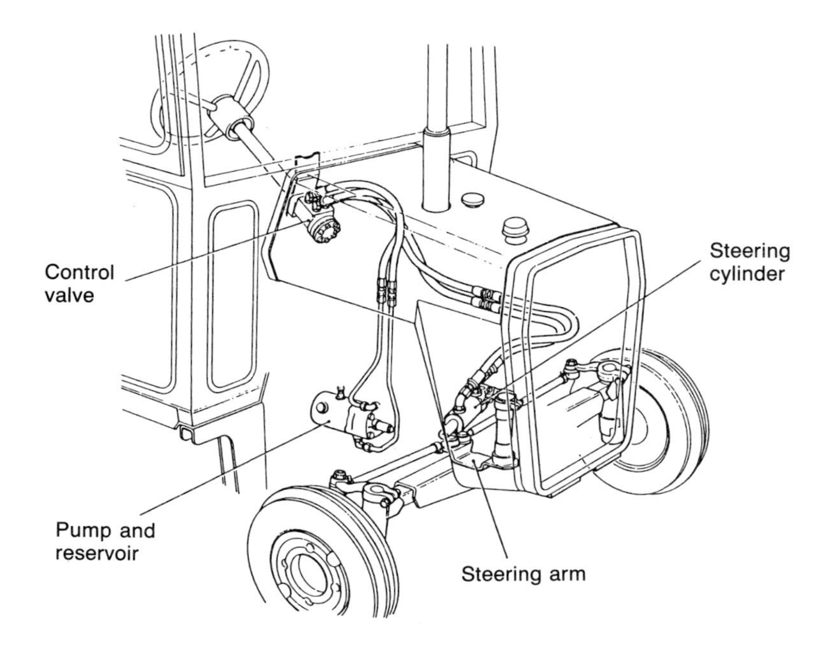

FIGURE 6.3 Hydrostatic steering. There is no mechanical linkage; the oil flow can be traced on this diagram from the pump to the steering ram.

The recirculating ball steering box (see Figure 6.4) has a thread (A) at the end of the steering column. A nut (B) moves up and down on the thread when the steering wheel is turned. The nut is connected to the rocker arm (D) by a peg (F). The steel recirculating balls (C) run in the groove formed by the nut and the thread on the steering column. When the nut (B) moves along the steering column thread, the rocker arm and splined shaft (E) rotate. The drop arm is attached to the splines and as the shaft rotates through an angle of about 30 degrees backwards or forwards, it pulls or pushes the drag link to change the direction of travel.

Double drop arm steering systems have either a gear and double sector plate gearbox or a modified recirculating ball gearbox. The gear and double sector gearbox has a gear at the end of the steering column which meshes with short sections of teeth on the two sector plates. One sector plate is connected to each of the two drop arms. The gear on the steering column rotates the drop arms, one forwards and the other backwards, turning the front wheels to the left or right.

Power assisted steering systems have a control valve unit built into the steering gearbox which controls the direction and rate of

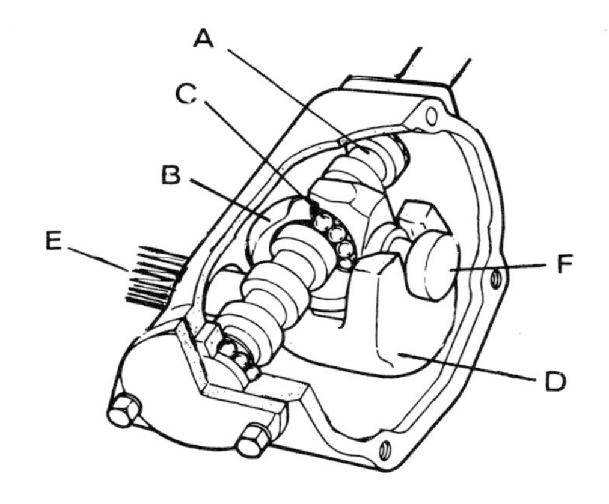

FIGURE 6.4 Recirculating ball steering box for single drop arm steering.

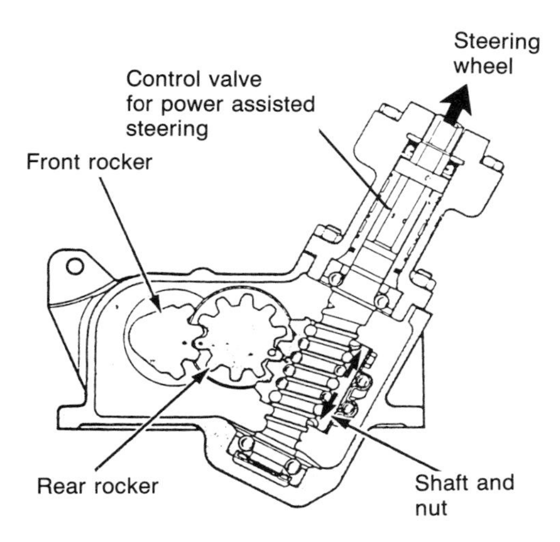

FIGURE 6.5 Recirculating ball steering box with built in control valve for power assisted steering. *(Ford New Holland)*

flow to the steering ram. Figure 6.5 shows an example of this type of steering gearbox with two rocker shafts for the double drop arm steering system. The two rocker shafts contra-rotate when the steering wheel is turned. This results in one drag link moving forwards and the other moving rearwards, and both front wheels turning to the left or right.

The illustration also shows the power assisted steering valve assembly which controls oil flow to the steering ram.

Maintenance

As a general rule, the steering linkage should be greased every 10 working hours. Some manufacturers suggest less frequent lubrication, so check your instruction book. Grease nipples are located on the track rod and drag link ends, steering arms and stub axles. Many steering linkage pivots have sealed bearings which require no lubrication. The steering gearbox may have an oil level plug which should be removed at 200 hour intervals in order to check the oil level. SAE 90 is commonly used for steering box lubrication.

The front wheel bearings should be greased at 50 hour intervals. Some wheel hubs have grease nipples; others require removal of the hub caps which are then packed with grease and replaced. From time to time, check for wear in the front wheel bearings. To do this, set the handbrake, chock the rear wheels and then jack the front wheel clear off the ground. With one hand at the top of the wheel and the other at the bottom, try to rock the wheel on its bearings. If there is movement, look to see if it is due to wear in the stub axle bushes or the wheel bearings. Replacement of worn stub axle bushes is a job for a mechanic but bearing adjustment is not difficult. Remove the hub cap and take out the split pin (if fitted) from the retaining nut. Next, tighten the bearing nut until the slackness has been removed but make sure that the wheel can turn freely. Fit a new split pin, grease the bearing and replace the hub cap.

Toe-in

Toe-in is necessary for efficient steering and to reduce front tyre wear on two wheel drive tractors. It is not usual for four wheel drive models to have toe-in as the front wheels are set parallel with each other.

The front wheels are said to have toe-in when the measurement between the wheel rims is less at the front than at the back. This measurement is taken at wheel bearing height. In Figure 6.6, measurement A should be between 3 and 6 mm less than measurement B. This is a typical toe-in setting, which will vary slightly with different models of tractor. Toe-in is altered by making a small adjustment to the length of the track rod on single drop arm steering, or by adjusting the length of the drag links on double drop arm steering.

The adjustable ends of the track rod, along with the drag links, are threaded into the tubular centre section and locked in position with clamp bolts. After slackening the clamp bolts, the track rod or drag link is adjusted in the same way as the top link on a tractor hydraulic system.

Four Wheel Drive Steering

Four wheel drive tractors have either power assisted or hydrostatic steering; heavy power driven front wheels are not suited to manual

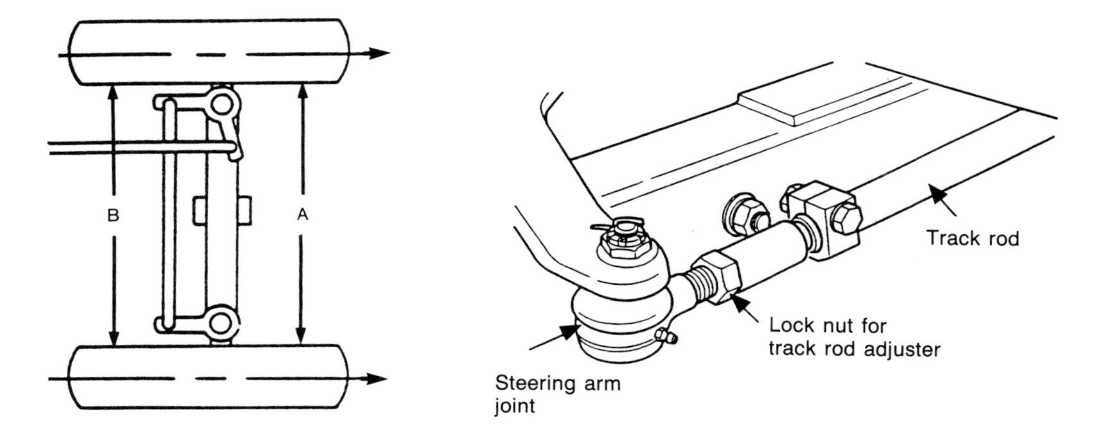

FIGURE 6.6 Setting toe-in. Measurement A will be between 3 and 6 mm less than B. It is adjusted by altering the length of the track rod. Lock nut B must be tightened after toe-in has been set.
(Massey-Ferguson)

steering. Universal couplings in the drive shafts make it possible to steer the tractor while full power is transmitted to the front wheels.

Four Wheel and Rear Wheel Steering

Tight turns can be made with four wheel (or all wheel) and rear wheel steered tractors and mowers. Ride-on mowers with a mid-mounted grass cutting deck and four wheel steering have a small turning circle which allows them to cut very close to flower beds and trees. Ride-on mowers with a front-mounted grass cutting deck usually have rear wheel steering, giving a high degree of manoeuvrability.

Rear wheel and four wheel steering may be manual or hydrostatic, with controls and linkages similar to front wheel steering.

Wheels

Pneumatic tyres are standard equipment on most horticultural tractors and machines. Exceptions include small rotary mowers and pedestrian controlled rotary cultivators, many of which have solid rubber tyres.

Compact and agricultural tractor rear wheels consist of a dished centre disc bolted to lugs welded to the rim. There are alternative positions for attaching the disc to the lugs; this provides the method of altering the wheel track setting to match the widths of various row crops or achieve the correct furrow width when ploughing.

Other horticultural machines, including some compact tractors, have split rim wheels. The two halves of the wheel are bolted together and attached to the tractor or machine axle. There is little or no provision for altering the track setting.

The front wheels of two wheel drive tractors are made in one piece, and track settings can be altered on agricultural models and many compact tractors. Ride-on mowers and garden tractors have either one piece or split rim wheels with no provision for track adjustment.

Tyres

Open centre tread tyres with bars or lugs in an open vee-shaped pattern are fitted to the driving wheels of most tractors used for land work, i.e. the rear wheels of two wheel drive models and all the wheels on four wheel drive tractors.

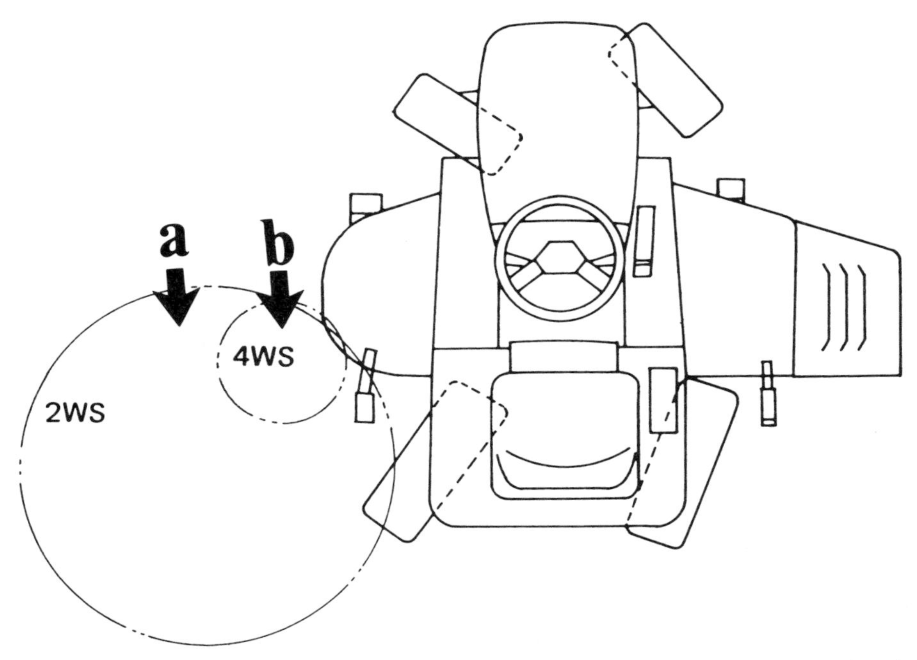

FIGURE 6.7 The advantage of four wheel steering with a garden tractor. Area (a) will be left uncut with two wheel steer compared with area (b) with four wheel steer. *(Kubota)*

PLATE 6.2 *Compact tractor with mid-mounted mower fitted with grassland tyres.* (John Deere)

Open centre tread tyres have a self-cleaning action. It will be noted that tyre walls bulge outwards at the 6 o'clock position of rotation and return to normal shape at the 12 o'clock position. The changing shape of the tyre walls as the wheel rotates flexes the tread bars, and soil trapped between the bars tends to fall out to give the self-cleaning action.

The tyres are fitted to the rims so that when viewed from the back of the tractor the arrow pattern of the lugs point upwards.

Narrow open centre tread tyres on purpose made wheel rims are preferred by some growers for rowcrop work. By using narrow wheels, crop damage is minimised.

Grassland tyres or turf tyres, with a wide diamond shaped or similar pattern tread, are fitted to tractors and ride-on mowers used on golf courses and sports grounds. High flotation tyres—very wide tyres with a ribbed tread—have a large footprint area, a desirable feature for grassland work. The various tread patterns on grassland tyres cause minimal damage to the turf, unlike open centre tread tyres which find little favour for sports ground and amenity turf maintenance work.

Ribbed tyres of various designs are used on the front wheels of two wheel drive tractors, implements and grass machinery. Tractor front tyres have a high centre rib which aids steering and helps to keep the tractor on the correct course.

Tyre sizes
The size of a tyre is marked on the tyre wall. There are three main systems of sizing:

- Two wheel drive tractor front tyres and those on trailers and many implements are sized according to wheel rim width

PLATE 6.3 *Rowcrop tractor with open centre tread rear tyres and ribbed front tyres.*
(Massey-Ferguson)

and diameter. For example, a 7.5—16 tyre fits a wheel rim 7.5 in wide and 16 in in diameter.

- Open centre tread and grassland tyres fitted to the driving wheels of two and four wheel drive tractors may, for example, be sized as either 14—28 or 16.9/14—28. This dual marking is explained as follows. The tyre was originally designed for a rim 14 in wide and 28 in diameter (14–28). Modern design, however, has altered the effective wheel rim to 16.9 in and the tyre size thus becomes 16.9/14–28.
- Very wide flotation tyres, which have limited horticultural use but are used on some trailers and implements where soil compaction is a problem, have a different sizing system. For example, a 9.0/75—18 tyre fits a rim 9 in wide and 18 in in diameter with a tyre aspect ratio of 75 per cent.

The aspect ratio is found when the tyre section height is divided by the tyre width. For example, a tyre with a height of 9 in and a width of 12 in will have an aspect ratio as follows:

$$\text{Aspect ratio} = \frac{9}{12} = 0.75 = 75 \text{ per cent}$$

The aspect ratio of 75 per cent indicates that the section height of the tyre is 75 per cent of its width. A very high aspect ratio indicates that the tyre has a very rounded section. The aspect ratio is also commonly found on car and lorry tyres. It will be noticed that radial tyres have a low aspect ratio.

Ply rating, normally marked on the tyre wall, refers to the number of layers of canvas or fabric used to make the tyre. Four, six and eight ply are common ratings. The heavier six or eight ply tyres are required on the front wheels of tractors with a front-end loader.

Radial and cross ply tyres differ in the construction of the tyre wall. The radial tyre is more expensive but has a longer

PLATE 6.4a *Ribbed tyre for implements and trailers. This type of tyre is also used for tractor front wheels, especially on turf.* (Michelin)

PLATE 6.4b *Narrow open centre tread tyre for working in rowcrops.* (Michelin)

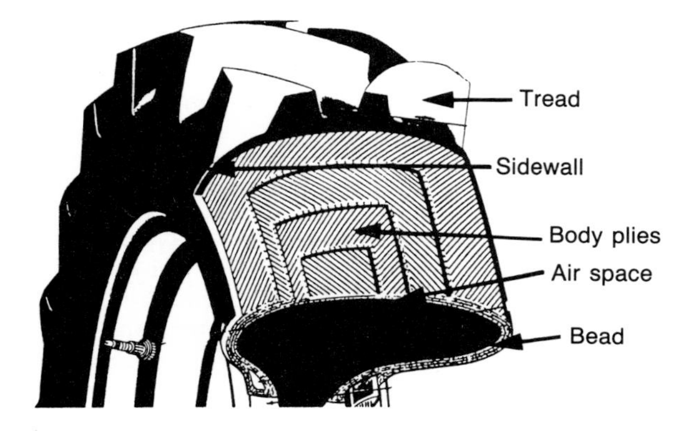

FIGURE 6.8 The construction of a rear tractor tyre.

life and offers improved grip with less soil compaction. It is usual for radial tyres to work at higher inflation pressures than cross ply tyres.

Inner tubes are fitted inside the tyres of most horticultural tractors and machines. Some radial tyres are tubeless.

Tyre care

As well as maintaining tyres at the correct working pressure, it is important to wipe any oil and grease from them and to remove stones or other sharp objects from the tread. Skidding and fierce use of the clutch will shorten tyre life.

Tyre pressure

Tyre pressure depends on type and size of tyre, tyre loading and the work being done. Radial tyres generally require higher pressures than cross ply tyres.

Tyre pressures may be stated in pounds per square inch (psi), bar or kilogrammes per square centimetre (kg/cm²). 1 bar equals 14.7 psi; so a tyre pressure gauge marked in psi and bar will show that a tyre at a pressure of 1 bar is also approximately 15 psi. (Note: 1 kg/cm² is approximately equal to 1 bar.) Many tyre pressure gauges measure the inflation pressure in bar and psi; some are calibrated in kg/cm². Always replace the dust cap on the valve after checking tyre pressure. A special gauge is needed when testing inflation pressures of water ballasted tyres.

Tractors A range of typical tyre pressures follows. It is most important to check the instruction book to find the correct pressures for the tyres on your tractor, taking into account the likely wheel loading.

Radial	Front	1.7–2.3 bar (25–35 psi)
	Rear	1.2–1.6 bar (18–24 psi)
Cross ply	Front	1.3–2.0 bar (20–30 psi)
	Rear	0.8–1.1 bar (12–16 psi)
Grassland	Front	0.8–1.6 bar (12–24 psi)
(turf)	Rear	0.7–1.2 bar (10–18 psi)

Increase these pressures by about 30 per cent for prolonged roadwork. Front tyre pressures should be increased by about 50 per cent on tractors with front-end loaders and heavy, front-mounted implements.

Mowers and garden tractors Reference to the instruction book is essential as there is a wide variation in tyre pressure settings for ride-on mowers and garden tractors. Some machines have very high pressures in excess of 2.6 bar (40 psi), while others work at no more than 0.4 to 0.7 bar (6–10 psi). Typical examples are:

Ride-on mower	Front 0.9 bar (14 psi)
	Rear 0.7 bar (10 psi)
Garden tractor	Front 0.8 bar (12 psi)
	Rear 0.5 bar (8 psi)
Rotary cultivator	Two wheel model 1.3 bar (20 psi)

Tyres transmit power partly by penetration of the tread bars or lugs and partly by adhesion (surface contact). Grassland tyres transmit drive by adhesion, as penetration by the tread bars would damage the turf. Tyre performance can be improved by adding weight, either water ballast or cast iron weights, to improve tyre adhesion and, where appropriate, penetration by the tread bars.

Water ballast is a cheap way of adding weight to the rear wheels. The water must have calcium chloride added to protect it from frost. It is usual to fill the inner tube about three-quarters full, a task which can be done by the driver without the need for special equipment. The calcium chloride must be added to the water—never add

PLATE 6.5 *Garden cultivator with open centre tyres.* (Dowdeswell)

water to the chemical—before it is put into the tube. A chemical reaction which generates heat occurs when the calcium chloride is added to the water so it must be left to cool before it is put into the tube. It is usual to mix 1 kg of calcium chloride with every 5 litres of water.

A special adaptor which releases the air to be displaced by the ballast water is needed to fill the tube to the correct level. The valve must be at the highest position when the tube is filled, either by gravity or with a special pump. When the tyre is three-quarters full (75 per cent ballasted), the water will be up to the level of the valve. The tyre is then inflated to its normal pressure, which must be checked with a pressure gauge designed for ballasted tyres. Rinse the gauge after use to remove all traces of chemical. A 13.6/12—36 tyre, 75 per cent ballasted will have 190 kg added weight.

Cast iron weights are a more expensive way of adding weight, but have the advantage that they can be removed quickly and mending punctures is less of a problem. Weights can be bolted to the rear wheel discs, the number and total weight added varying with model and size of tractor. A typical compact tractor can have up to three 27 kg (60 lb) weights bolted to each rear wheel; larger tractors may have four weights or more, each weighing 40 kg.

Weights at the front of a tractor, either bolted to the front wheels or carried on a front tray or frame, will improve stability and act as a counterbalance for heavy rear-mounted implements. A typical front wheel weight for a compact tractor will be 15 kg, while larger tractors have front wheel weights of 30 to 40 kg. A compact tractor front weight frame will usually take three detachable 19 kg weights; a 45 kW (60 hp) tractor can have about 200 kg of front end weight.

Weight transfer The hydraulic system on tractors with draft control can be used to transfer some of the weight of a mounted implement on to the tractor. This is most effective with implements which work in the soil and have no depth wheels.

Dual wheels are useful when extra pulling power is required without increasing soil compaction. Quick release fittings make it possible to add or remove a second pair of pneumatic tyred wheels in conditions where single rear wheels will cause unacceptable damage to the surface. Cage wheels which also reduce soil compaction and are very useful when preparing seedbeds are an alternative. A cage wheel consists of tubular or angle iron tread bars welded on to two large metal rings to form a wheel. A quick release linkage makes it a simple matter to attach

PLATE 6.6 *Compact tractor with front weights to give stability when ploughing and when the plough is raised on the hydraulic linkage.* (Ford New Holland)

PLATE 6.7 *Power adjusted rear wheel. The wheel disc is turned slowly, using a low gear so that the rim moves in or out on the rails to increase or decrease track width. The lock on the rail must be removed before adjustment can be carried out.* (Massey-Ferguson)

cage wheels to special lugs welded on the rear wheel rims.

Wheel Track Settings

Wheel track width can be adjusted on tractors and some pedestrian controlled cultivators suitable for working in rowcrops. Large models of rowcrop tractor have track settings from 1.23 or 1.32 to 1.93 m (48 or 52 to 72 in). Track adjustment is rather limited on compact tractors; one example has a range of 1.01 to 1.23 m (40 to 48 in) with agricultural tyres, and 1.14 to 1.36 (45 to 54 in) with turf tyres. Track adjustment, usually in 100 mm (4 in) steps, is a jack and spanner job, although some tractors have power adjusted rear wheels.

On pedestrian controlled cultivators where adjustment is possible, the track width is altered by sliding the wheels in or out on their axles after removing a locking pin or bolt in the wheel hub.

Rear wheel track is altered by fitting the

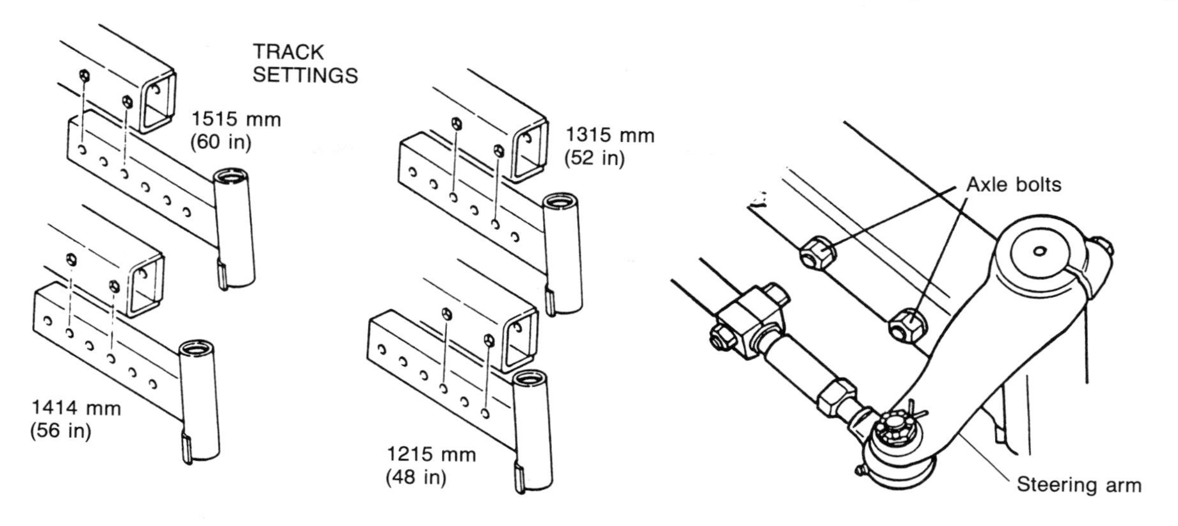

FIGURE 6.9 Front axle settings for various track widths. The axle bolts must be completely tight after adjusting the track; toe-in is restored by sliding the track rod out to its next notch. *(Massey-Ferguson)*

disc to any one of four positions on the wheel rim lugs. The wheel disc centre can be reversed on the hub to double the options up to eight possible settings. The front track setting on four wheel drive tractors is set in a similar way to the rear wheels.

Two wheel drive tractors have a centrally pivoted front axle with telescopic sections which can be moved in or out to obtain the required track width.

Suggested Student Activities

1. Locate the grease nipples on a tractor steering linkage. Check the instruction book to find the recommended lubrication periods.
2. Check the front wheel bearings on two wheel drive tractors periodically and adjust them if necessary.
3. Look for the different types of tyre mentioned in this chapter, study the markings on the tyre walls and interpret their meaning.
4. Check the pressure of your tractor and ride-on mower tyres at regular intervals.
5. Study the method of wheel track adjustment on at least one type of tractor and one pedestrian controlled cultivator.

Safety Checks

Never work on a jacked-up tractor or machine without first placing an axle stand or strong wooden block under it. Before removing a wheel, always support a tractor with a stand under the axle or some other suitable position. Apply the handbrake and for added safety chock the wheels on the ground. Slacken the wheel nuts slightly before jacking up the tractor or implement.

Chapter 7

TRACTOR HYDRAULIC SYSTEMS

The hydraulic system provides the power for lifting and lowering front-, mid- and rear-mounted implements and machines into and out of work. It also operates external (auxiliary) rams on front loaders, tipping trailers, etc., and supplies the oil required for hydraulic motors on implements such as hedge cutters and grass mowers.

The Principles of Hydraulics

Many liquids can be used to transmit power, but the lubricating and corrosion resisting properties of oil make it the most suitable one. In a simple hydraulic system, a pump supplies oil under pressure to a ram cylinder. The oil forces a piston (ram) along in its cylinder, and the movement of the piston is used through a linkage to lift a load.

A hand operated, hydraulic vehicle jack is an example of a simple hydraulic system. The hand lever operates a pump which forces oil into the ram cylinder. The oil pressure on the piston moves it upwards and the vehicle is lifted. The handle is also used to operate a control valve in the body of the jack. When the vehicle is lowered, the valve releases the oil from the ram cylinder and the piston moves downwards.

Figure 7.1 shows the working principles of a simple tractor hydraulic lift system. Oil is pumped from the reservoir—usually the transmission housing—through the control valve to the ram cylinder.

With the control valve set in the *lift* position, the oil pressure forces the piston along the cylinder. The piston is connected to the ram cross-shaft by a simple connecting rod. Piston movement turns the cross-shaft on its pivots, thus raising the lift rods and the lower lift arms which are connected to mounted implements.

Oil is released from the ram cylinder back to the reservoir when the control valve is moved to the *lower* position. The implement falls under its own weight, forcing the oil from the ram cylinder. A relief valve allows oil to return to the transmission housing or oil reservoir if the lift system is overloaded.

This simple type of hydraulic system requires a wheel to control the working depth or height of the implement. Most tractor hydraulic systems control working depth and height automatically, making a depth wheel unnecessary.

Pumps

Many hydraulic systems have either a gear or multi-piston pump located inside the transmission housing driven by the shaft which transmits drive to the power take-off. An alternative arrangement is to have the hydraulic pump mounted on and driven by the engine. A typical 22 kW (30 hp) tractor has a pump with an output of 30 litres of oil per minute; another with a 12 kW (16 hp) engine has a pump output of 14 litres per minute. The oil will become quite hot in the transmission housing or oil reservoir when the hydraulic system is used for long periods. Some tractors, especially those fitted with a digger, have an oil cooler in the hydraulic circuit.

Filters

The oil used in hydraulic systems must be clean to prevent damage to the pump and control system. Many models of tractor and ride-on mower have some form of hydraulic oil filtration, and these range from a simple

PLATE 7.1 *Compact tractor with front loader. Two pairs of auxiliary rams lift the loader and angle the loader bucket.* (Lewis Equipment)

magnetic plug to catch small particles of metal to a replaceable element type filter. Machines with a separate oil reservoir for the hydraulic system have an oil filter.

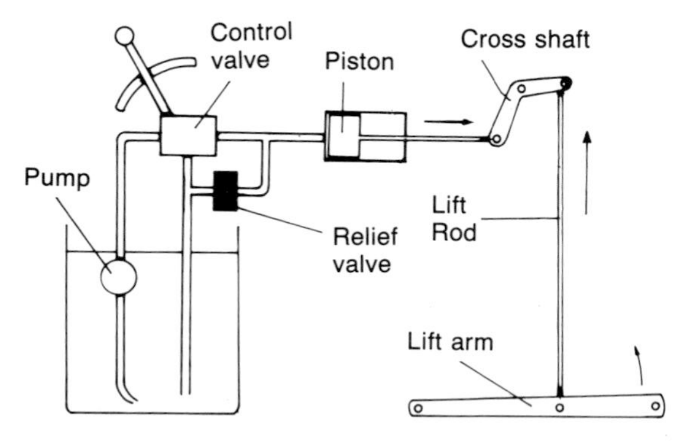

FIGURE 7.1 Layout of a simple hydraulic lift system.

Relief Valve

The relief valve protects the pump from damage caused by overloading by limiting the maximum working pressure of the pump. The valve will blow (and release the pressure) if, for example, the driver attempts to lift too much on the three point linkage or with a front-end loader. The relief valve also reduces the risk of damage to hydraulic motors which are supplied with oil by the tractor pump.

The delivery pressure of a hydraulic pump increases as the load on the system increases. When the circuit pressure equals the maximum allowed by the relief valve, it will 'blow', and a knocking sound can often be heard when this happens. Some examples of typical hydraulic system working pressures are a 22 kW tractor at 158 bar (2,250 psi) and a 12 kW tractor at 137 bar (1,990 psi).

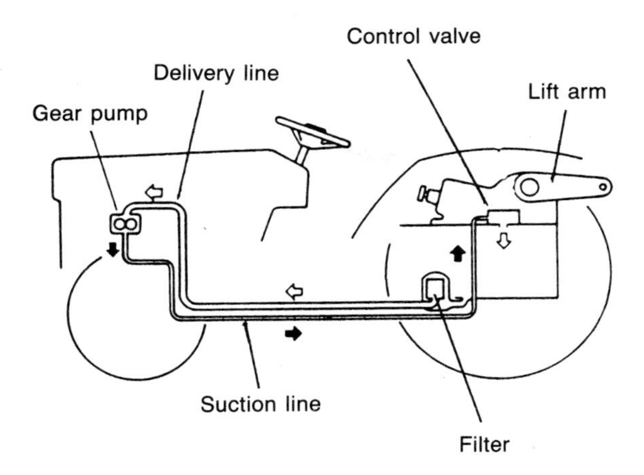

FIGURE 7.2 When the engine is running, the pump draws oil from the transmission housing through the filter. The oil then flows under pump pressure through the relief valve to the control valve. From here it either returns to the transmission housing or is directed to the ram cylinder when the control lever is in the lift position. *(Massey-Ferguson)*

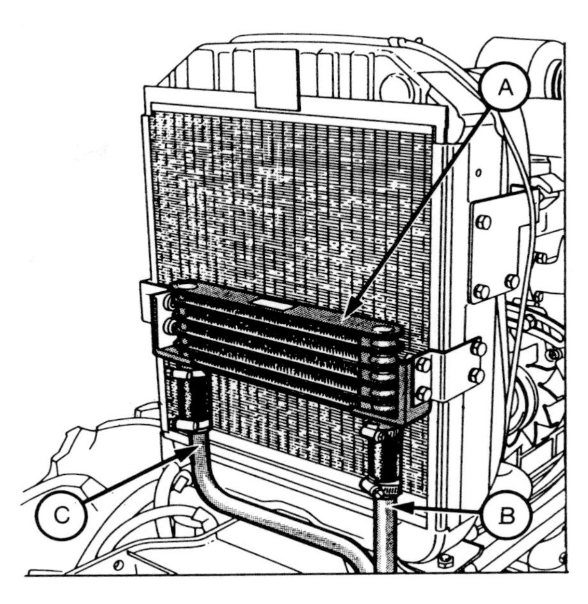

FIGURE 7.3 Hydraulic oil cooler. The oil is pumped through the cooler (A) through the flow and return pipes (B) and (C). *(Ford New Holland)*

Hydraulic Transmission Systems

Hydrostatic transmissions (see page 67) have a variable speed hydraulic motor instead of a manual gearbox. The rate of flow of oil to the motor determines the forward speed of the tractor or mower, and the direction of oil flow gives either forward or reverse travel.

To achieve this stepless speed control in forward and reverse, a hydrostatic transmission has an engine driven pump, a filter, a lever or pedal operated control valve unit and a variable speed hydraulic motor.

Many tractor mounted and trailed gang mowers have a hydraulic motor to drive each cutting cylinder. The motors are supplied with oil under pressure, from either the tractor hydraulic system or an independent hydraulic system with a power take-off shaft driven hydraulic pump on the mower.

An oil cooler, consisting of a small, air cooled radiator, is usually built into hydrostatic transmissions and hydraulic motor drive circuits on grass cutting equipment.

Control Systems

Tractor hydraulics have two basic control systems, draft control and position control, which meet the needs of the different types of three point linkage implements. The arrangement of the hydraulic controls varies from one model of tractor to another.

PLATE 7.2 *Draft control in use with a reversible plough.* (Ford New Holland)

It should be noted that many models of garden and compact tractors have a very basic hydraulic system which requires wheels on the implement to control the working depth. The tractor instruction book gives full information about the hydraulic controls, which should be studied to ensure maximum benefit is obtained from the system.

Draft Control

Draft control is used with implements which work in the soil, ploughs, subsoilers and cultivators being examples. Draft control maintains a constant load (draft) on the tractor. In a perfect working situation with soil type and condition the same all over the field, the load on the tractor, and therefore the working depth, would be constant. In practice, of course, this does not happen very often.

The tractor driver can use draft control to set the plough at a chosen depth. This setting will result in a certain draft being maintained by the hydraulic system. The draft control mechanism keeps the tractor load constant by lifting the plough slightly if the draft increases, and dropping it if the draft decreases. In this way, draft control will give an acceptable standard of depth control in most field conditions. The driver can use the control lever to override the system if changes in soil condition cause the implement to run too deep or too shallow.

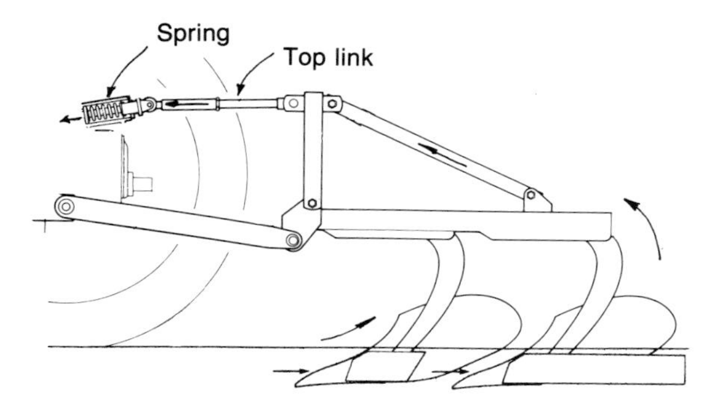

FIGURE 7.4 Top link sensing depends on the compression force acting towards the tractor. Increased force lifts the implement and reduced force allows the implement to drop.

Top link sensing
Mounted implements are carried on the lower hydraulic links, while the top link prevents them tipping backwards or forwards. When a plough or other soil engaging implement is used, the resultant compression force acts through the top link towards the tractor. The compression force will increase as the draft (working depth) increases.

The draft control system uses variations in the top link compression force to maintain a constant load on the tractor. This is achieved by a linkage connecting the top link to the hydraulic control valve. When the implement drops below the pre-set working depth, the increased force in the top link causes the control valve to lift the implement slightly, returning it to the required depth.

There is a heavy spring at the tractor end of the top link, sometimes visible but more often inside the transmission housing. This spring must be compressed by the forces acting along the top link before any correction can be made to the working depth of the implement. When the draft on the tractor is reduced, usually because the implement is not deep enough, there will be less pressure on the spring. When this occurs, the control valve returns the implement to the pre-set load (depth).

Very uneven soil conditions cause far too many corrections to the working depth. Most hydraulic systems have a damping or response control, with a range of settings from fast to slow, to overcome this problem. When draft control operation is sluggish, a fast response setting is needed; a slow response setting should be selected if the hydraulic system makes too many corrections to the working depth.

The draft control system described above is only affected by variations in the compression forces in the top link. An implement set to work at a shallow depth can create a tension (stretching) force in the top link rather than the compression force which occurs when it works at a greater depth. The tension force will make it very difficult to control implement depth.

Many tractors have a double acting top link. This transmits both tension and compression signals to the control valve to maintain a constant draft (or depth). The top link will be in tension when shallow ploughing. If the tractor front wheels drop into a furrow, the tension force in the top link will increase and the control valve will return the plough to its required setting. When deep ploughing, the compression forces in the top link will maintain the correct implement draft.

To prevent damage to the hydraulic system when moving over rough ground, a mechanism isolates the effect of the tension forces in the top link when the implement is fully raised.

Lower link sensing
The tension forces acting in the lower hydraulic links are used on some models of tractor to operate the draft control system. Lower link sensing relies on the fact that if the draft increases, the tension in the lower links will also increase. A linkage connects the lower link sensing mechanism to the hydraulic control valve. Increased draft on the lower links causes the control valve to lift the implement slightly. In the same way, when the draft on the lower links is reduced, the implement will be lowered.

Other draft control systems
Constant implement draft can also be maintained by using the changes which occur in the torque (twisting force) in the tractor transmission. Some larger models of tractor have a special coupling built into the gearbox output shaft which reacts to changes in torque. When the load on the tractor increases due to greater working depth or heavier soil conditions, torque also increases in the coupling. A linkage connects the torque coupling to the hydraulic control valve and when the draft exceeds a pre-set level, the implement will be lifted slightly until the correct draft setting is obtained. This system will also increase implement depth when the draft and torque falls below the pre-set level.

Electronic draft control is available on some high specification tractors. Sensors measure the draft (load) when the plough runs too shallow or too deep and automatically re-adjusts the depth of the implement. The

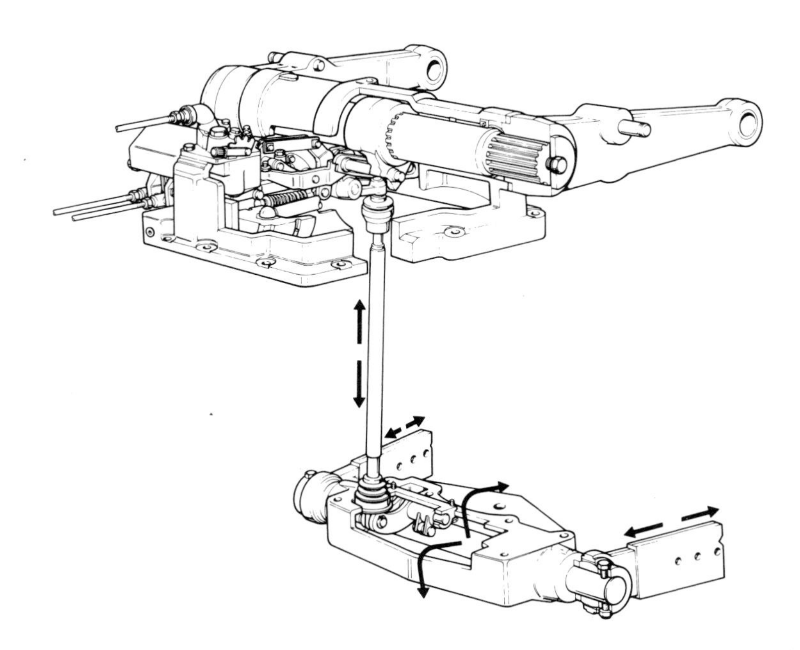

FIGURE 7.5 Lower link sensing depends on increasing or decreasing tension in the lower arms which results from changes in the load (draft) on the tractor. *(Fiatagri)*

PLATE 7.3 *Position control in use with a mounted fertiliser broadcaster on a golf course.* (Kubota)

system also controls the rate of implement drop and the maximum lift height, and locks the hydraulic system when the engine is stopped.

Position Control

Position control is used to set the working height of mounted implements such as sprayers, fertiliser spreaders, mowers and hedge trimmers which operate above ground level. Position control uses hydraulic power only for lifting; the implement is lowered under its own weight when oil is released from the cylinder. The lift arms are retained at the required working height, determined by the lever, when the engine is running. When the lever is moved, the control valve supplies oil to or releases oil from the cylinder until the implement attains its new working height. At this point, the control valve returns to neutral. Any variation in the forces acting through the top link has no effect on the control valve.

It is important to select position control for implements such as mounted sprayers and fertiliser spreaders. For example, if draft control is mistakenly used with a sprayer, the weight of liquid in the tank would set up both tension and compression forces in the top link and the result would be disastrous.

Flow Control

The rate of flow to and from the hydraulic ram cylinder can be adjusted with the flow control valve. It controls either the rate of lift or drop, depending on tractor model.

Mechanical Lock

Some models of tractor have a mechanical lock for the hydraulic linkage which, when engaged, prevents the implement being lowered. The manually operated locking mechanism removes the risk of damage to the hydraulic system when transporting an implement, especially when driving across uneven surfaces or transporting mounted implements at speed on the road.

Auxiliary Hydraulic Services

In addition to operating the three point linkage, tractor hydraulic systems provide other services. A number of horticultural machines have auxiliary or external rams and some have hydraulic motors.

The most common use of an auxiliary ram is on a tipping trailer. The trailer body is raised when oil is pumped from the tractor hydraulic system into a single acting ram. It is lowered by using the weight of the trailer to discharge the oil from the ram cylinder back to the tractor.

Double acting rams are used to give controlled movement under pressure in two directions. The ram has two hose pipes, one to supply oil to the ram and one to return it to the hydraulic system. Oil can flow in either direction in both pipes, depending on the position of the spool valve (control unit) lever. Examples of the use of double acting rams include hedge trimmers, ditch diggers and reversible ploughs.

Both single and double acting rams are connected to tractor hydraulic systems with heavy duty hoses by means of quick release couplings plugged into hydraulic sockets. Older tractors may have a screw type hose connector.

It is possible to control a single acting ram on, for example, a tipping trailer by using the normal hydraulic controls, provided that the tractor has an isolator valve and an auxiliary ram coupling. More complicated arrangements with two or more single acting rams or a double acting ram are controlled by spool valves. Some tractors have rocker switch operated control valves instead of hand levers to control oil flow to auxiliary rams and hydraulic motors.

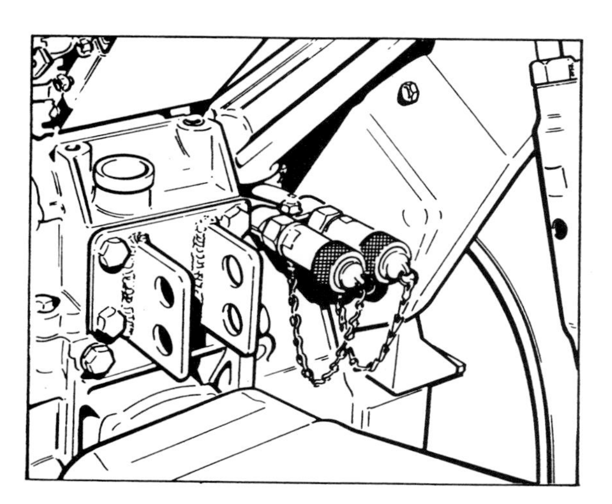

FIGURE 7.6 Auxiliary services sockets. The dust caps should be replaced when the sockets are not in use. *(Ford New Holland)*

PLATE 7.4 *Rocker switches on this compact tractor mudguard (or fender) are used to control oil flow to the mower lift ram. The tractor also has an auxiliary ram coupler socket at the back of the tractor.* (Massey-Ferguson)

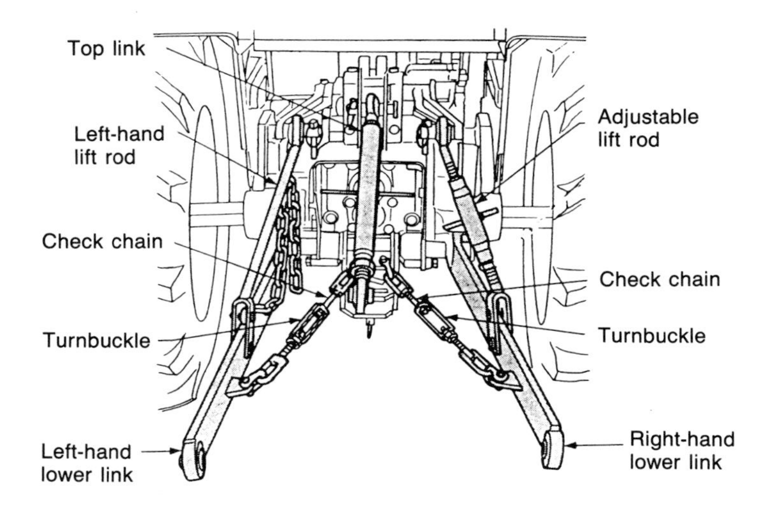

FIGURE 7.7 Hydraulic three point linkage. *(Kubota)*

The hydraulic system also supplies oil required to drive hydraulic motors, used for example to operate many models of gang mower and hedge trimmer. Two hose connections are required, one for flow and one for return. The advantages of hydraulic motor drive are that it eliminates the need for expensive drive shafts which must be guarded, and the oil flow can be regulated at the touch of a lever to give variable speeds in forward or reverse.

The Hydraulic Three Point Linkage

The *lower links* are connected to the ram cross-shaft by the lift rods. The right-hand lift rod has a levelling box which is used to adjust the height of the right-hand lower link. Some tractors have a levelling box on both lift rods.

The linkage hitch pins on mounted implements are made in various diameters and lengths. The dimensions are related to the size of tractor and implement. The measurement between the three hitch point positions also varies with size of implement.

Category '0' is the linkage with the smallest dimensions. It is used for implements attached to some garden tractors and ride-on mower power units.

Category '1' is the standard size of linkage found on compact tractors and the smaller models of farm tractor.

Category '2' linkage is used for many hydraulically mounted implements. The larger models of compact tractor often have a dual linkage with categories 1 and 2 achieved by changing the hitch balls at the implement ends of the top and lower links.

Category '3' linkage is used for very high power tractors but it has no application in horticulture.

The *top link*, usually adjustable in length, is used to adjust the pitch (angle) of implements such as a plough, cultivator, sprayer or fertiliser broadcaster.

Remember to attach mounted implements in the correct order:

First —attach the lower left-hand link.
Next —fit the adjustable lower right-hand link.
Then—connect the top link.

Stabiliser bars are used to stop the three point linkage implements swinging sideways. Some tractors have telescopic stabilisers which can be

PLATE 7.5 *Category '0' three point linkage.* (Ransomes, Sims and Jefferies)

set to hold the implement in a rigid position or adjusted to allow some sideways movement.

External check chains, which can be adjusted with a central turnbuckle, serve the same purpose as stabilisers.

Internal check chains prevent the lower links hitting the sides of the tyres when they swing sideways. They will not hold an implement rigidly on the linkage.

A *quick attach linkage system* is in common use with heavy mounted implements. The lower ball ends are replaced by telescopic hooked ends with self-locking latches. The hooked link arm ends connect with ball shaped couplings fitted to the implement hitching pins. To attach the implement, the tractor is reversed up to it with the link arms lowered and the telescopic ends extended. When the hook ends are aligned the link arms are lifted to engage with the couplers. The top link hook end is engaged with the implement and locked in position. When the implement is lifted, the telescopic ends are locked in the closed position.

The implement is uncoupled by lowering it and then releasing the hook end locks with a release cable from the driving seat. The top link is then lifted clear and the lower link arms are dropped to their lowest position.

Care of the Hydraulic System

- Grease the levelling box on the right-hand lift rod weekly.
- Check transmission oil level or the hydraulic oil reservoir every week. This is most important if a tipping trailer, or other implement with an external ram operated from the tractor hydraulic system, has been used and oil has been lost when coupling or uncoupling the hosepipe.
- *Do not lubricate* the balls in the lower links and the top link. The oil would collect abrasive particles of soil and cause rapid wear.
- Never turn a sharp corner with a mounted implement in the ground; this puts unnecessary strain on the hydraulic linkage.

Suggested Student Activities

1. Locate the hydraulic controls on the tractor after studying the instruction book.
2. Find out how to convert a tractor hydraulic lift linkage from category 1 to category 2.
3. Look for examples of stabilisers and check chains on different tractors.
4. Practise reversing up to a mounted implement until you can accurately line up the lower links with the hitch pins on the implement. Don't cheat by moving the implement manually.

Safety Checks

Never tow anything from the top link position on a tractor. This is a very dangerous practice because it can result in the tractor somersaulting backwards. At a forward speed of 3.25 km/h (2 mph) a tractor will rear up through 90 degrees in one second.

Always lower a mounted implement or loader to the ground when parking your tractor.

Chapter 8

TRACTOR MAINTENANCE

All tractors used in horticulture need regular maintenance, the frequency of which will vary with different models of tractor. This chapter deals with the routine maintenance for compact and other four wheeled horticultural tractors. Maintenance of self-propelled grass cutting equipment including engine servicing is dealt with in Chapter 14.

The operator's manual sets out the full service schedule, and you should refer to this before attending to the routine maintenance of your tractor. Remember that for most tractors the service hours refer to the *engine running hours* shown on the proof meter. Some tractor proof meters register the hours the engine has run, with no relationship to its speed, but in most cases, the meter registers the hours the engine has run at about two-thirds maximum speed. So, in a normal day's work, the proof meter will register at least 10 hours. Most proof meters state the engine speed at which engine hours are recorded, usually between 1,800 and 2,200 rpm.

A Typical Compact Tractor Service Schedule

The following maintenance schedule is typical of the recommendations for a compact tractor.

10 Hours (or daily)

- Check engine oil level and top up with the correct grade if necessary.
- Check coolant level in the radiator and top it up if low. Check that air flow through the radiator is not restricted by dust, grass mowings, etc.
- In dusty conditions, check the air cleaner oil bath and service it if necessary. For tractors with a dry air cleaner, check the filter element and clean it if required.
- Lubricate steering linkage. Some operator's manuals suggest this task should be carried out at 50 hour intervals.
- Check the fuel filter sediment bowl for dirt or water and clean if required.
- Make a visual check of the tyres, and test air pressure with a gauge if it is thought necessary.

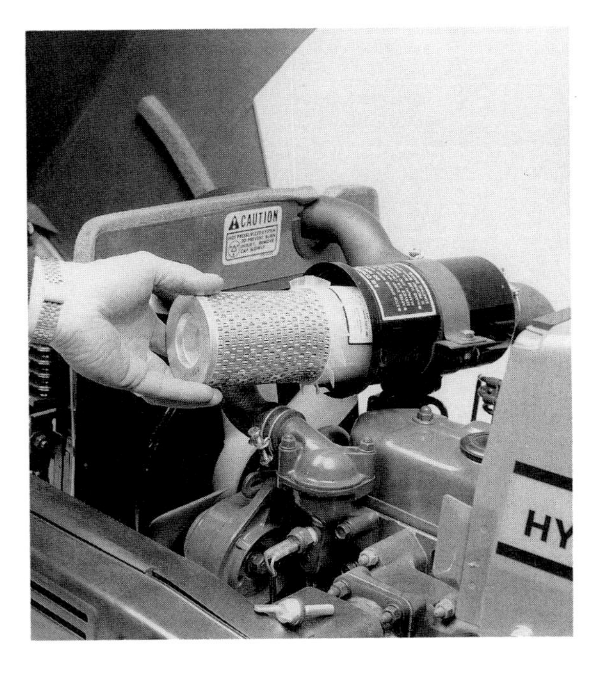

PLATE 8.1 *Removing the dry air cleaner element for routine service.* (Massey-Ferguson)

LUBRICATION POINTS

GREASE EVERY 10 HOURS, OR DAILY

1. Front hubs
2. Front axle kingpins
3. Front axle pivot pin
4. Main steering shaft
5. Power steering ram pivot
6. Lift rods
7. Levelling box
8. Hook links
9. Auto-hitch latches

FOUR WHEEL DRIVE

10. Front support pivot
11. Drive shaft couplers front and rear
12. Rear support input housing right-hand side
13. Drive shaft centre bearing
    Grease every 100 hours or in wet or muddy conditions every 10 hours or daily.

EVERY 100 HOURS

Apply a smear of light oil to the cab door hinges and locks, hood side panel hinges and latches, and front grille latch.

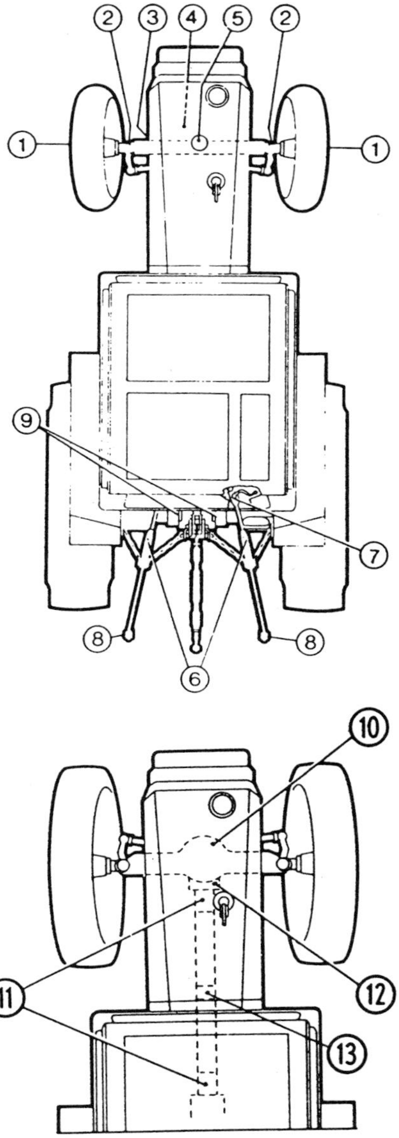

FIGURE 8.1 A page from a tractor maintenance schedule listing routine lubrication requirements. *(Massey-Ferguson)*

50 Hours (or weekly)

- Check oil levels in the gearbox and transmission, and top up as required.
- Lubricate wheel bearings, clutch and brake pedal linkages and the hydraulic linkage levelling box.
- Service the air cleaner.
- Check the battery and top up if necessary with distilled water. Clean off the top of the battery and check that the terminals are tight and corrosion-free.
- Inspect the tyres and test the air pressure with a gauge. Inflate any tyres with pressures below the minimum recommended level.
- Check for loose nuts and bolts, especially on the wheels and front axle.

200 Hours
Complete the 10 and 50 hour service tasks and in addition:

- Drain the oil from the engine, preferably when it is hot. Fit a new filter element and refill the sump with fresh oil. Oil change periods vary for different tractor models within a range of 100 to 300 hours.
- Clean the engine crankcase breather.
- Check clutch pedal free play and reset if required.
- Check fan belt tension and adjust if necessary.
- Make sure the radiator core is clean externally. Remove grass and dust with an airline if possible by blowing it out from the fan side of the radiator.
- Check brakes for adjustment and, where appropriate, balance the pedals.

500 Hours
In addition to the tasks listed above:

- Renew the diesel fuel filter element and bleed the fuel system.
- Arrange for a trained mechanic to test the fuel injectors and replace them if worn.
- Change transmission and hydraulic oil and renew the oil filter element.
- Repack front wheel bearings with grease.

The strength of the anti-freeze in the cooling system should be checked and topped up in readiness for the winter months. It is usual to drain and flush the cooling system at least every two years and then refill it with the correct mixture of water and anti-freeze.

PLATE 8.2 *The bonnet is raised to give easy access to the battery. The engine has replaceable element type fuel filters; the oil filler cap is near the radiator.*
(Ford New Holland)

Other items to be serviced on some tractors include the power steering oil reservoir, clutch and brake fluid reservoirs and the front wheel drive housing oil levels. The tractor instruction book gives full maintenance requirements and is essential reading for all tractor drivers.

Suggested Student Activity

With the aid of the instruction book carry out daily and weekly maintenance to a horticultural tractor.

Safety Check

To avoid unpleasant skin disorders through handling fuels and lubricants, it is a wise precaution to put barrier cream on your hands before starting work and wash with a good quality hand cleanser when the job is completed. Rubber gloves provide even better protection, especially when handling diesel fuel.

Chapter 9

PLOUGHS

The plough has been used in various forms for many centuries. It is the primary cultivation tool used to prepare the soil for seeds and plants. Good quality ploughing will reduce the after-cultivations required to break up the soil in order to achieve a good tilth.

The plough is an implement with one or more mouldboards which cut and turn the soil to bury weeds and crop residues and give a well broken surface for subsequent cultivations. Ploughs used in horticulture will have one or two furrows and occasionally three. Four and five furrow ploughs are popular with farmers and those with large fields and powerful tractors may use models with eight or more furrows.

For horticultural work, ploughs are attached to the hydraulic linkage of compact tractors or medium powered agricultural tractors. Single furrow ploughs are also available for some models of pedestrian operated garden tractor. There are two basic types of plough:

Conventional ploughs have mouldboards which turn the soil to the right. Time is required to mark out the field before work can start and there will be a number of ridges and furrows left after ploughing is completed.

Reversible ploughs have both left- and right-handed bodies allowing the tractor driver to work up and down in the same furrow. This type is heavier and more expensive than the right-handed plough, but it leaves a level surface, and the time needed to mark out the field is minimal.

The turnover mechanism used to bring either the left-hand or right-hand bodies into work may be manual, mechanical or hydraulic. Some ploughs are reversed by means of a handle which the tractor driver must use to remove one set of bodies from work and then to engage the opposite set before starting the next run. Others have a mechanical reversing mechanism which uses the weight of the plough to operate it. When the plough is lifted from work, its weight pushes forward towards the headstock. The weight is used, through a linkage, to rotate the plough when the driver pulls the turnover lever. After the

PLATE 9.1 *Two furrow reversible plough on a compact tractor.* (Wessex)

PLATE 9.2 *Two furrow reversible plough with hydraulic turnover ram.* (Krone)

plough has been reversed once, the mechanism must be reloaded before it can be used again. This is achieved by lowering the plough into work and pulling it forward to return the headstock in the loaded position.

Many reversible ploughs are turned hydraulically by a double acting ram attached to the headstock which rotates the plough to left or right. The ram is supplied with oil from the tractor hydraulic system and is controlled by a lever in the driver's cab.

The Parts of a Plough

The parts of the plough which work in the soil—the disc, the skimmer and the body—are attached to legs or stalks which are in turn bolted to the plough beam or frame.

The base of the plough body is called the *frog,* and the soil wearing parts are bolted to it.

The *share* cuts the bottom of the furrow slice. Shares are made of steel and when they become worn, it is important to fit new ones. Ploughs used to have chilled cast iron shares, which have a self-sharpening effect, but these have become very expensive and plough manufacturers now use steel shares.

The share is shaped so that it pulls itself into the soil. As it wears away, it becomes blunt and increases the draft of the plough, making it harder to pull through the soil. A plough body with a worn-out share will not have enough 'suck' (see Fig. 9.2) to enable it to penetrate the ground to its full working depth.

The *landside,* which is bolted on to the side of the frog, absorbs the side thrust of the plough against the furrow wall. A *heel iron,* usually bolted to the rear landside, helps carry the weight of the back of the plough. The landside and share are designed to give 'lead' towards the unploughed land, thus helping to maintain the correct width of furrow.

The *mouldboard* lifts and turns the furrow slice and buries the weeds, etc. There are

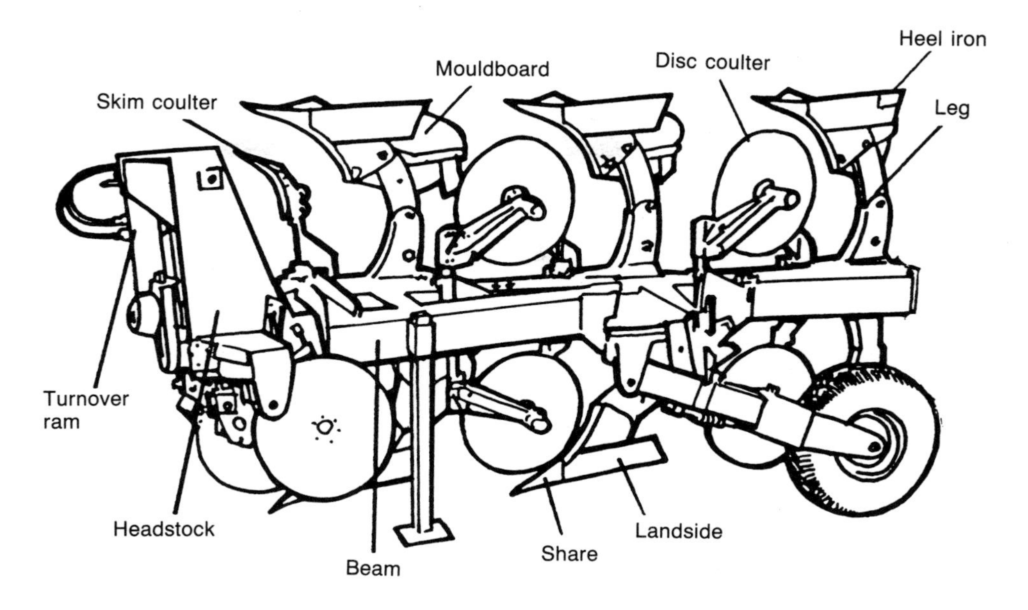

FIGURE 9.1 The parts of a reversible plough. *(Agrolux)*

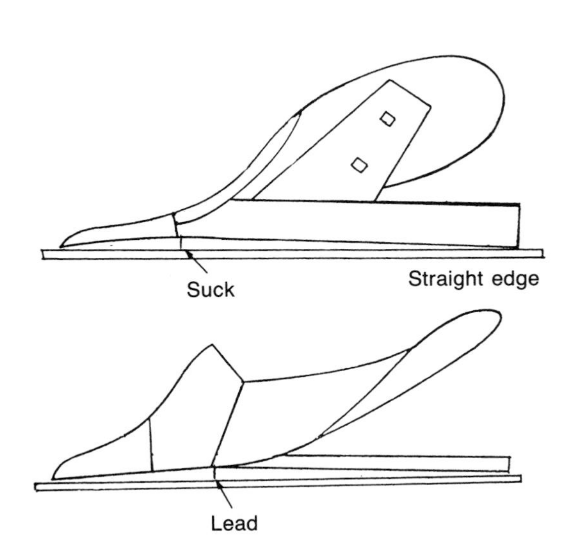

FIGURE 9.2 Suck is necessary for the plough body to maintain its working depth. Lead helps maintain correct furrow width. Both can be checked with a straight edge. A plough body without lead or suck needs a new landside and share.

numerous shapes of mouldboard, each producing its own type of surface finish. Some ploughs have *mouldboard extensions* or tail-pieces which help to press down the furrow slice. These are very useful when ploughing grassland.

The *disc coulter* cuts the side of the furrow about to be turned. When set correctly it will give a clean side to the furrow. The disc coulter bearing needs regular lubrication.

The *knife coulter* serves the same purpose as a disc. It is a large knife clamped to the frame which is pulled through the soil. It takes up very little space but it leaves a less defined side to the furrow.

The *skimmer* or skim coulter cuts a small piece of soil and vegetation from the corner of the furrow slice about to be turned, and throws it into the previous furrow bottom. This reduces the likelihood of rubbish growing up between the furrows.

Body Types

There are numerous designs of plough body with different shaped mouldboards. Each

PLATE 9.3 *Reversible plough with knife coulters and manual turn over mechanism.* (Wessex)

body has its own particular use. The main types are:

General purpose A low draft body with a gently curving mouldboard which turns a furrow slice three parts wide by two parts deep, e.g. 300 mm wide and 200 mm deep. It is useful for ploughing grassland and sets up the furrows for maximum exposure to winter frosts.

Semi-digger This has a shorter mouldboard with a sharper curve. It turns an almost square furrow and leaves a more broken finish. The semi-digger body is used for general ploughing work.

Digger This has a short, abruptly curved mouldboard which gives a very broken surface finish in most soils. It can turn a furrow which is deeper than its width and is particularly useful for the preparation of deep seedbeds for root crops.

It is not unusual for holdings to have ploughs with at least two different types of body. Many ploughs have a mouldboard design developed for European soils which produces good quality ploughing over a range of working depths. The ploughing speed will decrease as working depth increases. These continental design mouldboards have generally replaced the more traditional body types in recent years.

Slatted mouldboards are preferred by some growers. They consist of a number of curved slats bolted to a frog which are designed to improve the movement of soil over the mouldboard when working in sticky soils.

Bar point This is a body with a special design of point consisting of an extendable bar, often spring loaded, which can be moved forward as it wears. The bar point is ideal for ploughing soils where boulders are near the surface. The spring loaded point can recoil when it hits an obstruction beneath the surface. After passing the obstruction, the spring returns the point to its working position. Ordinary shares would break when working in such conditions so although the bar point leaves a very rough furrow bottom, it saves on the cost of share replacement.

Special ploughs
Although rarely used in horticulture, some ploughs have an automatic reset mechanism which allows a plough body to swing back if it hits an obstruction.

Variable furrow width ploughs are becoming popular, especially with farmers. The furrow width can be varied by the tractor driver by means of either a hydraulic ram or a turnbuckle system. Variation in furrow width is useful when ploughing in soils where the tractor has power to spare and can cope with wider furrows. It is also an advantage when changes in working depth are needed or where a higher ploughing speed is required to increase the soil crumbling effect of the mouldboards.

Controls and Adjustments

Hitching Connect the plough to the three point linkage in the correct order: the left-hand lower link, the right-hand lower link and finally the top link. Remember: 'Left—right—top'. When detaching the plough from the tractor, remove the linkage in the reverse order.

Set the top link in the mid-position when transporting the plough. When attaching a reversible plough, check that both lift rods are the same length. After attaching a conventional plough, set the right-hand lift rod to

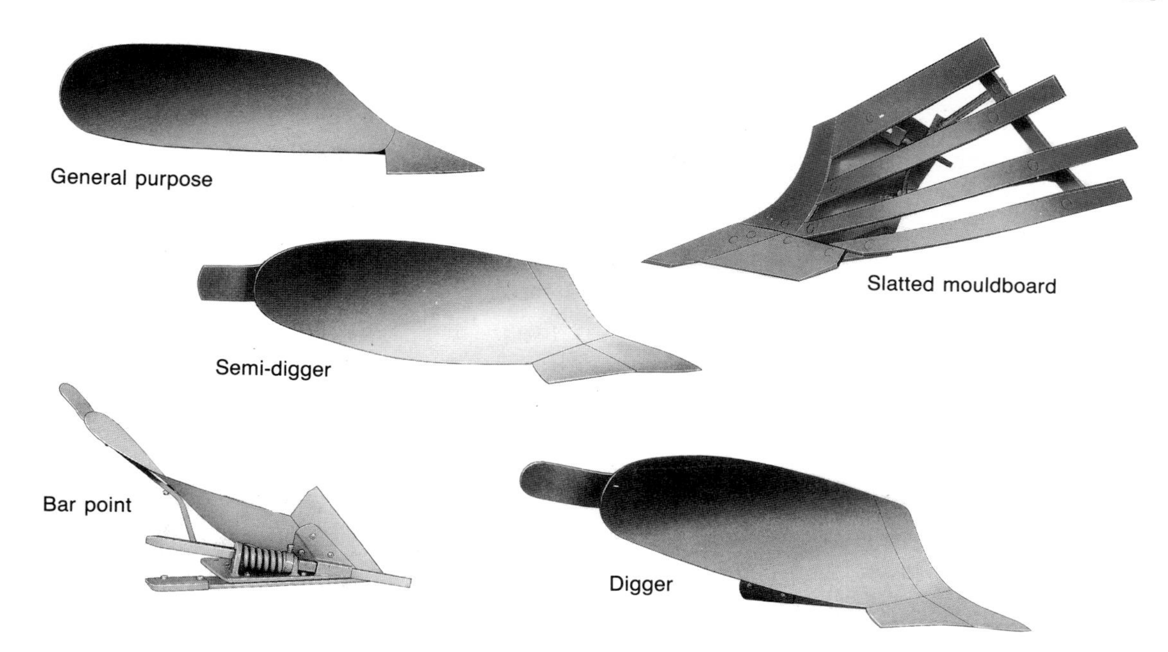

PLATE 9.4 *Types of plough body.* (Agrolux)

the mid-position; it will require readjustment when setting up the plough in the field.

Depth of ploughing is adjusted with the draft control lever and its stop. Some ploughs have an adjustable depth wheel.

Levelling Reversible ploughs have the lift rods set at the same length. Screw adjusters are provided on the plough headstock to ensure the plough bodies all work at the same depth to give a level finish. The levelling box on the right-hand lift rod is used to level the furrows when using a conventional right-handed plough.

Pitch This is adjustable on all ploughs by using the top link. Pitch adjustment relates to the angle of entry of the shares into the ground and is correctly set when the heel iron makes a slight mark on the furrow bottom. The top link needs lengthening if the heel iron rides clear of the furrow bottom, or shortening if the heel iron cuts a deep mark in the bottom of the furrow.

The plough will not enter the ground properly if there is too little pitch. Excessive pitch will cause the shares to dig in, resulting in a very rough finish. When ploughing in hard ground, share penetration can be improved by shortening the top link, and frequent share renewal will be necessary.

Front furrow width must be set to match the width of the remaining furrows. A wide or narrow front furrow will cause uneven work which can be seen on ploughed land by the presence of high or low furrows at regular intervals across the field. It is often assumed that the front furrow adjuster is incorrectly set, but a wide or narrow front furrow may be due to incorrect wheel track settings. Check the plough instruction book to ensure that the wheel track is correct for the required furrow width before using front furrow adjusters.

Reversible ploughs have front furrow adjuster bolts on the headstock, which, once set, rarely require adjustment. Incorrect front furrow width may also be due to faulty setting of the levelling adjustment on the headstock, and this should be checked before altering the front furrow width setting.

Front furrow width is altered on conventional ploughs by means of a lever or a crank handle connected to the cross-shaft. Slight rotation of the cross-shaft moves the plough to the left or right and in this way widens or narrows the front furrow.

Coulter setting

- The centre of the disc coulter hub should be set above the point of the share.
- The disc should be set 12 mm towards the unploughed land side of the share. A stepped furrow wall indicates that the disc is set too far towards the unploughed land. When the disc is set too far towards the ploughed land, the furrow wall will be very ragged.
- The height of the disc above the share will depend on the depth of ploughing. The bottom of the disc should be about 35 mm above the share. The disc will need to be raised when deep ploughing. It is important to make sure that the coulter is set high enough to prevent the disc bearing housing dragging along on the soil as this will cause undue wear.

A knife coulter should be set so that the tip of the knife is in the same position as the lowest point of a disc coulter.

Skimmer setting The skimmer or skim coulter should be set against the disc coulter so that it cuts a wedge of soil, about 50 mm deep, from the corner of the furrow slice about to be turned. The skimmer point should be below and behind the centre of the disc coulter hub. Excessive skimmer depth should be avoided, as it may cause problems in achieving the required ploughing depth, especially in hard land.

Plough Maintenance

- Grease all lubrication points every day during the ploughing season.
- Change all soil engaging parts when they are worn. Shares will need changing frequently, as blunt shares make the plough harder to pull, and this wastes power, fuel and money. Skimmer points, landsides and heel irons may last for a full season, depending on soil type and conditions.
- Keep nuts and bolts tight at all times.
- At the end of the season, replace any broken items, coat bright parts, especially the mouldboards, with a rust preventing solution and store the plough under cover.

Ploughing

No matter what sort it is, a plough which is not correctly adjusted will require extra power to pull it through the soil. This wastes fuel and time, which both cost money, so it pays to ensure that:

- The plough is correctly attached.
- The tractor wheels are at the correct track setting.
- The pitch is correctly adjusted.
- The mouldboards are parallel to each other. This can be done by checking that measurements from fixed points on adjacent mouldboards are equal. Misalignment can be rectified by adjusting the length or position of the mouldboard stays.
- The shares are in good condition.
- The discs and skimmers are set deep enough to do their work efficiently but not so deep as to add to the draught of the plough.
- The plough is suitable for the tractor and does not have too many furrows for the power available to pull it.

Reversible Ploughing

For a tidy finish, it is worth taking time to mark headland furrows before reversible ploughing a field. The distance from the headland furrows to the edge of the field will depend on the size of the plough. For example, the headland mark for a two furrow plough needs to be about 8 m from the hedge. Narrow headlands may take less time to plough but will make turning more difficult and increase wear and tear on the tractor.

It is best to plough the field longways to

reduce turning on the headland. Start at one side of the field and before making the first proper run, turn some shallow furrows away from the hedge to cut through and kill off any vegetation close to the hedge. Plough across the full width of the field and then finish off by ploughing the headlands. The headlands should be ploughed to and from the hedge in alternate years.

Conventional Ploughing

The aim is to get a level surface finish and keep idle headland running to a minimum. Conventional ploughs can be used to plough a field in 'lands' or round and round.

Systematic ploughing Using a right-handed plough, mark a headland furrow all round the field about 7 m from the hedge for a two furrow plough. Add 1 m for each extra furrow. Next, mark the field off in sections with a single, shallow furrow. Each section is called a land, which should be 10 m wide for each furrow on the plough, so a three furrow model will have a land of 30 m.

A ridge is made by ploughing up and down either side of the marking furrow at rather less than the required ploughing depth. Once all the ridges are made the plough is set for normal work and the field is ploughed by alternate 'casting' and 'gathering'.

Casting means working between two ridges with the tractor turning to the left on the headlands.

Gathering means ploughing round a ridge with the tractor always turning to the right on the headlands.

When the width of unploughed land between the ridges is reduced to three furrows (with a two furrow plough), the plough is reset to take shallower furrows to complete the work. With the main part of the field completed, the headlands are ploughed. Headlands should be ploughed to and away from the hedge in alternate years so that deep furrows do not occur. A well ploughed field will have neither high ridges nor deep finishing furrows.

PLATE 9.5 *Reversible ploughing.* (Kverneland)

Round and round ploughing can be from the hedge to the centre of the field, or from the centre outwards. Ploughing from the hedge best suits a field with reasonably square corners. First make a headland mark followed by diagonal furrows from corner to corner. The field is then ploughed round and round towards the centre, leaving an unploughed strip either side of the diagonal mark. This gives space for turning. The headlands and diagonal strips are ploughed last.

When ploughing from the middle to the hedge, the centre of the field must be found. To do this a helper walks round the outside of the field holding a cord tied to the tractor. The driver keeps the cord taut, making a shallow marking furrow as the tractor is driven round the edge of the field. The assistant next walks round the marker furrow with the tractor closer to the centre of the field. Eventually a small replica of the field shape will remain at the centre. This is ploughed first around a ridge and then the field is ploughed round and round to the hedge.

The centre of a regular shaped field can be found by measurement, leaving a small replica shape at the centre. The field can then be ploughed in the same way. If round and round ploughing is used, the work must be from the hedge and then the centre in alternate years to maintain a level field.

Suggested Student Activities

1. Read a specialist book on ploughing techniques to help improve your knowledge of this subject.
2. Learn the names of the main parts of a plough.
3. Find the adjustments on a reversible plough and identify their purpose.
4. Attach a plough to a tractor and make the basic settings for normal ploughing.
5. Make sure that when you use a plough, the mouldboards, shares and disc coulters are protected with a rust preventing solution at the end of each day's work.

Safety Check

Take great care when working close to posts carrying electricity or telephone cables, especially if they have stay wires. Driving too close to a stay wire can result in the implement catching the stay and probably pulling the post to the ground. Such accidents happen every year, and in the worst situation a live overhead electricity cable could fall on to the tractor.

Chapter 10

CULTIVATION MACHINERY

A wide choice of cultivation equipment in both type and size is available for use in nurseries, glasshouses and cropping on a field scale. There are implements suitable for all sizes of power unit, from pedestrian controlled cultivators to expensive power take-off driven machines designed for compact and larger tractors.

SUBSOILERS

Free movement of air in the soil combined with good drainage is necessary for healthy plant growth. Soil compaction and soil pans, mainly caused by tractor wheels, restrict soil air movement and impede drainage.

Subsoilers are used to break up compacted soil and pans, allowing free passage of air and water. They need plenty of power to pull them through the soil to a typical working depth of 400–500 mm. A subsoiler consists of a thin but strong leg with a share at the base. The share has a replaceable point, and on some models, the front vertical cutting edge (shin) can be renewed when worn. It is possible to fit wings to the share in order to give an extra shattering effect, but these

PLATE 10.1 *Rotary cultivating with a compact tractor.* (Dowdeswell)

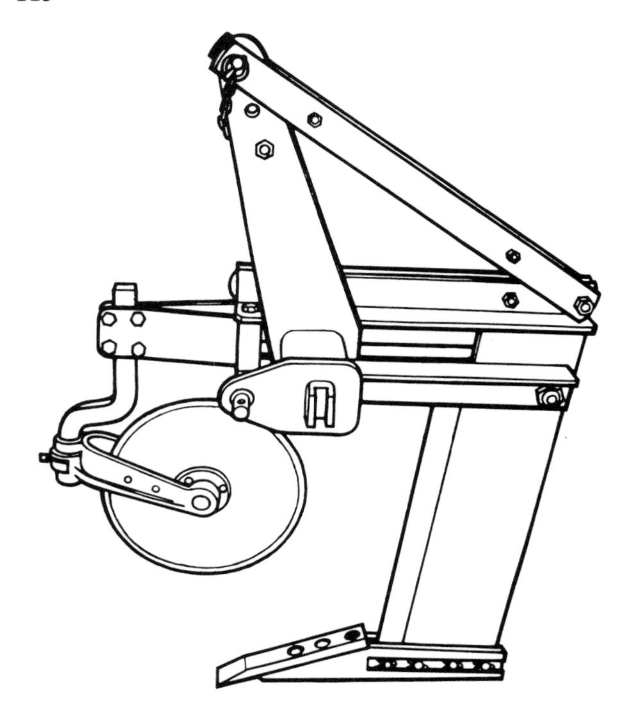

FIGURE 10.1 Single leg subsoiler with disc coulter and replaceable share.

require greater tractor power. Many subsoilers have a shear bolt or pin at either the top or base of the leg. If the share hits a heavy obstruction, the pin will break, allowing the leg to swing backwards and thus avoiding damage.

To achieve maximum shattering effect, subsoiling is best carried out when the ground is hard. It is usual to subsoil at right angles to the intended direction of ploughing or other heavy cultivation.

The most powerful models of compact tractor will be able to handle a single leg subsoiler mounted on the three point linkage in all but the toughest of soils. In more difficult conditions, up to 37 kW (50 hp) may be required to pull a single leg subsoiler. Few horticultural units other than large scale vegetable producers are likely to have such large tractors. Where soil conditions are tough, it is likely that a contractor using a subsoiler with three or more legs will need to be called in.

Turf Drainers

Turf drainers are similar in construction to subsoilers, but are designed for use on grassland where there are compacted areas indicated by poor growth or surface water lying in low places. Turf drainers do not work at the same depth as subsoilers, and they may have a number of legs, depending on the size of tractor used. The bullet-shaped point at the base of the blade leaves a small drainage channel at the bottom of the slit. A spring loaded rear roller consolidates the turf lifted by the drainer blade.

Maintenance

Replace the share point when it is worn, making sure that the bolts holding the point on to the leg are tight. Some subsoilers have a renewable shin on the leading edge of the leg, and this will also need replacing when it is worn out. Keep the share tight on the subsoiler leg at all times.

Chisel Ploughs

A chisel plough is a form of heavy cultivator with a number of legs, each having a share with a replaceable point. Many models of

PLATE 10.2 *Three leg chisel plough with reversible points. The model shown is suitable for the larger compact tractors; it has a maximum working depth of 250 mm.* (Wessex)

chisel plough have a safety shear bolt at the base of the leg. This allows the share to swing backwards if it hits an obstruction, thus avoiding a damaged or broken share. The chisel plough is considered by some to be an alternative implement to a mouldboard plough. It can be used to break up the soil, especially where there is little surface vegetation, before final cultivations are carried out in preparation for a new crop.

Chisel ploughs designed for large scale field work have a working width of 2–5 m and a power requirement of 45–120 kW (60–160 hp).

Maintenance

Replace the shares when they are worn, and check that the bolts which secure the shares are tight. Models with safety shear bolts must always be fitted with the correct type of shear bolt as supplied by the manufacturer.

TINED CULTIVATORS

A cultivator consists of a frame with a number of tines for breaking and mixing the soil. The tines have replaceable shares or points, and various designs are used for different types of work. Working depth is generally controlled by the tractor hydraulic system. Depth control wheels attached to the cultivator frame (see Plate 10.3) are required for some garden and compact tractors which have a very basic hydraulic system. Some tined cultivators have a crumbler roll attached at the rear which helps to break up clods, consolidate the soil and, at the same time, control the working depth.

Types of share include the reversible share used to prepare seedbeds after ploughing and for general work. Reversible shares can be turned when one end is worn. Duckfoot and broad shares are used mainly for weed control.

As with all cultivation equipment, working widths vary: small models used with low power tractors cover 1.2 m, whereas models for use with very powerful farm tractors can cover up to 8 m. The very smallest tined cultivators are used on pedestrian machines. A toolbar, attached to the power unit, can be fitted with a small number of cultivator tines for inter-row work.

Types of Tine

Rigid tines are used for heavier work, the tines staggered across the width of the toolbar to allow free passage of soil and reduce blockages where there is much surface vegetation. Alternatively, the tines may be grouped with spaces between each group for inter-row cultivations in potatoes and other row-crops. In hard conditions, share penetration is improved if weights are placed on the cultivator frame.

Spring loaded tines (see Plate 10.4) are rigid tines held in position by heavy tension springs. The tines are able to lift slightly when they hit large clods. The spring pressure pulls the tines down again and this action helps to shatter stubborn clods. This type of tine is also useful in soils where numerous stones may damage a rigid tined cultivator, as individual tines can lift and ride over large obstructions.

Spring tines are made from square or flat sectioned spring steel, sometimes with a coil at the top of the tine. The vibrating motion of the tines as they pass through the soil is

PLATE 10.3 *A 12 kW (16 hp) compact tractor with a spring loaded, rigid tine cultivator. It has depth wheels to control the working depth of the tines.* (Massey-Ferguson)

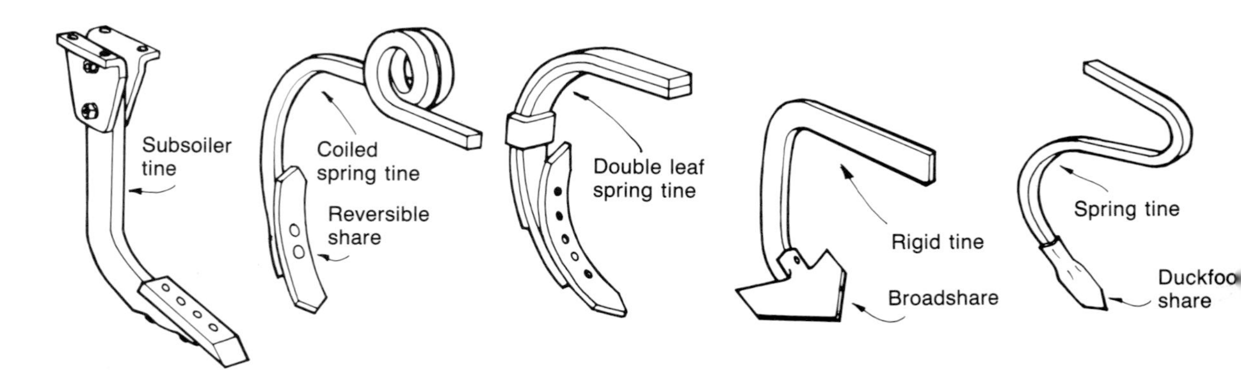

FIGURE 10.2 Types of tine and share.

PLATE 10.4 *Spring tine cultivator with depth control wheels.* (Massey-Ferguson)

very effective in breaking down clods. Spring tines will give fast preparation of seedbeds in most soil conditions.

Another type of spring tine cultivator, or harrow, has much lighter spring tines attached to a frame which runs along at ground level. The tines shatter the clods and the harrow frame has a levelling action on the soil.

Maintenance

As with all soil working implements, it is important to renew shares or points when they are worn. Never use a cultivator with a missing share, as the base of the tine will wear away and it will be difficult to keep the share tight when it is eventually replaced. Reversible points should be turned when the first end is worn out.

Depth wheels where fitted will require regular lubrication if they do not have sealed bearings.

Power Driven Cultivators

Various types of power driven cultivator are used in horticulture, including small domestic models with petrol engine or electric motor drive, pedestrian controlled machines for the professional user, and numerous models of tractor mounted rotary cultivator.

Rotary cultivators are used after ploughing for the rapid preparation of deep, loose seedbeds. They will also prepare a seedbed in most soil conditions without the need to plough at all. This is especially so with the pedestrian machines on nursery beds and in glasshouses. The soil is left in a very loose condition and, unless the ground is left for some time to settle, rolling will be necessary to consolidate the soil to ensure good seed germination.

Pedestrian Operated Cultivators

There are two designs of self-propelled pedestrian controlled rotary cultivator. One type, usually known as a power tiller, is propelled by the cultivator rotor. Transport wheels can be attached to facilitate pushing the machine from site to site. The second type has rubber tyred wheels which propel the cultivator forwards or backwards when turning at the end of a bout. There is a separate drive system to the cultivator blade rotor.

Power tillers

Power tillers are driven by air cooled petrol engines. They are made in various sizes, from the small machine illustrated in Figure 10.5 to the more powerful professional model shown in Figure 10.6. The small single speed domestic models have an engine of about 1.6 kW (2 hp). Heavy duty power tillers used by commercial growers have an engine with a power output of up to approximately 5.3 kW (7 hp), with at least two forward and one reverse speeds and working widths of up to 1 m.

The belt or chain driven rotor shaft has cultivating blades attached to each side of the chain case (see Plate 10.6). It is possible to fit one, two or three sets of blades on each side of the chain case to give working widths suitable for cultivating between different row spacings. Rotors with spike tines for preparing a fine tilth can also be used.

Power tillers require quite a lot of effort by the operator to prevent them from moving forward too quickly, as this will result in the

PLATE 10.5 *Domestic power tiller with a 1.6 kW air cooled petrol engine driving the single speed rotor. It has a working width of 590 mm with a maximum cultivating depth of 150 mm.* (Kubota)

PLATE 10.6 *Power tillers with twin blade rotors and transport wheels attached.* (Allen Power Equipment)

blades failing to work at the correct depth or to produce a satisfactory tilth.

A depth bar or leg at the back of the tiller limits the working depth. Forward movement of the tiller can be controlled by pushing the depth bar down into the harder subsoil below normal cultivation depth, thus helping to prevent the tiller from moving forward too quickly.

Many power tillers can be fitted with a small rowcrop toolbar. The blade rotors are removed and a pair of rubber tyred wheels are fitted to the rotor shaft to drive the machine forwards. The rotor shaft speed can be changed on most tillers by moving the vee-belt to an alternative pair of pulleys or with a speed change gearbox. A high speed of about 100 rpm is used for cultivating. The shaft speed must be reduced when the wheels are fitted for toolbar work; toolbar equipment, such as cultivator tines, hoe blades and a ridging body, can be used, and a small plough can be fitted to more powerful models.

PLATE 10.7 *A pedestrian operated rotary cultivator with a working width of 410 mm (16 in) and maximum cultivating depth of 200 mm. The handlebars are offset so the operator need not walk on cultivated ground.* (Dowdeswell)

Rotary cultivators

Wheel-propelled rotary cultivators are also made in various sizes. Plate 10.8 illustrates an example at the lower end of the power range. It has a 3.9 kW (5.2 hp) air cooled petrol engine which drives the wheels through a gearbox with two forward speeds and one reverse. The power is transmitted from the engine to the gearbox by a vee-belt. This system provides a simple clutch arrangement to disengage the drive to the gearbox when the operator wishes to stop or change gear. When the drive is engaged, belt tension is maintained by an over centre device in the jockey (tensioner) pulley linkage.

De-clutching is achieved by releasing tension on the vee-belt with a lever on the handlebars which moves the jockey pulley away from the vee-belt.

Drive to the blade rotor is from the gearbox input shaft through a chain drive to the rotor shaft. A typical 3.9 kW rotary cultivator has a rotor speed of 220 rpm when the engine runs at 3,000 rpm, and the forward gears give speeds of 0.9 and 2.9 km/h (0.6 and 1.8 mph).

Larger versions of wheel-propelled rotary cultivators have approximately 6.8 to 11.2 kW (9 to 15 hp) diesel engines which are mostly air cooled. Some models have air cooled petrol engines.

These machines are more robustly built, with a typical machine having a gearbox with three forward gears and one reverse, controlled by a plate type clutch. The final drive is a bull gear and pinion. The differential has a diff-lock which comes into operation when the rotor drive is engaged. The forward gears give speeds of 1.5, 2.3 and 4.9 km/h when the engine runs at 2,800 rpm. At this engine speed, the blade rotor, which is chain driven from the main gearbox, runs at about 180 rpm. It is fitted with 'L' shaped blades bolted in a scroll pattern on the rotor flanges.

The 'L' blades can be changed for narrow pick-tine blades with horizontal chisel points which can be used to break up hard ground and renovate grass. It is usual for a safety slip clutch to be included in the rotor drive to protect the machine from damage through overloading.

As with power tillers, it is possible on many wheel drive rotary cultivators to remove the

PLATE 10.8 *Rotary cultivator with toolbar and ridging body.* (Kubota)

rotor unit and fit other implements such as a plough, hoe blades or ridging body. A front counter balance weight may be needed on some models when using a plough.

Adjustments

Tilth can be varied on some models by changing the relationship of forward speed to rotor speed. Some models have a two speed rotor, and most, except very small machines, have two or more forward speeds. A low forward speed and a high rotor speed give a fine tilth.

The rotor shaft runs under a shield which has, in many cases, an adjustable hinged rear flap. The tilth will be fine if the flap rests on the ground and coarse if the flap is fully raised.

Rotor reverse is available on some rotary cultivators. A lever is used to select reverse rotor drive in order to improve performance when cultivating hard-packed ground or wet grassland. In forward rotor drive, the rotor turns faster than the wheels and in adverse conditions the tines are unable to dig in properly. This makes the cultivator difficult to handle as it tends to jump or skip forward.

In reverse rotor drive, the rotor turns in the opposite direction to the wheels, and the problem of skipping is overcome because the tines dig upwards to hold back the cultivator and give a balanced wheel-to-tine relationship.

Working depth is adjusted by means of a skid which limits the penetration of the rotor blades.

Handlebars on many models of rotary cultivator can be offset to either side of the machine. This is useful when working in rowcrops or where it is preferable that the operator does not walk on freshly cultivated soil. Handlebar height can be adjusted on many machines to give a comfortable working position.

Tractor Mounted Rotary Cultivators

Power take-off driven rotary cultivators are made for all sizes of tractor hydraulic linkage. Trailed machines are made mainly for large tractors. Rotary cultivators range from small models with a 760 mm (30 in) rotor to massive machines which cover 5 m in a single pass. Tractor mounted rotary cultivators need plenty of power, especially in tough conditions. This generally limits the working width of machines used in horticulture to about 1.3 m except for field scale production

PLATE 10.9 *Offset, tractor mounted rotary cultivator. It has a working width of 760 mm.* (Dowdeswell)

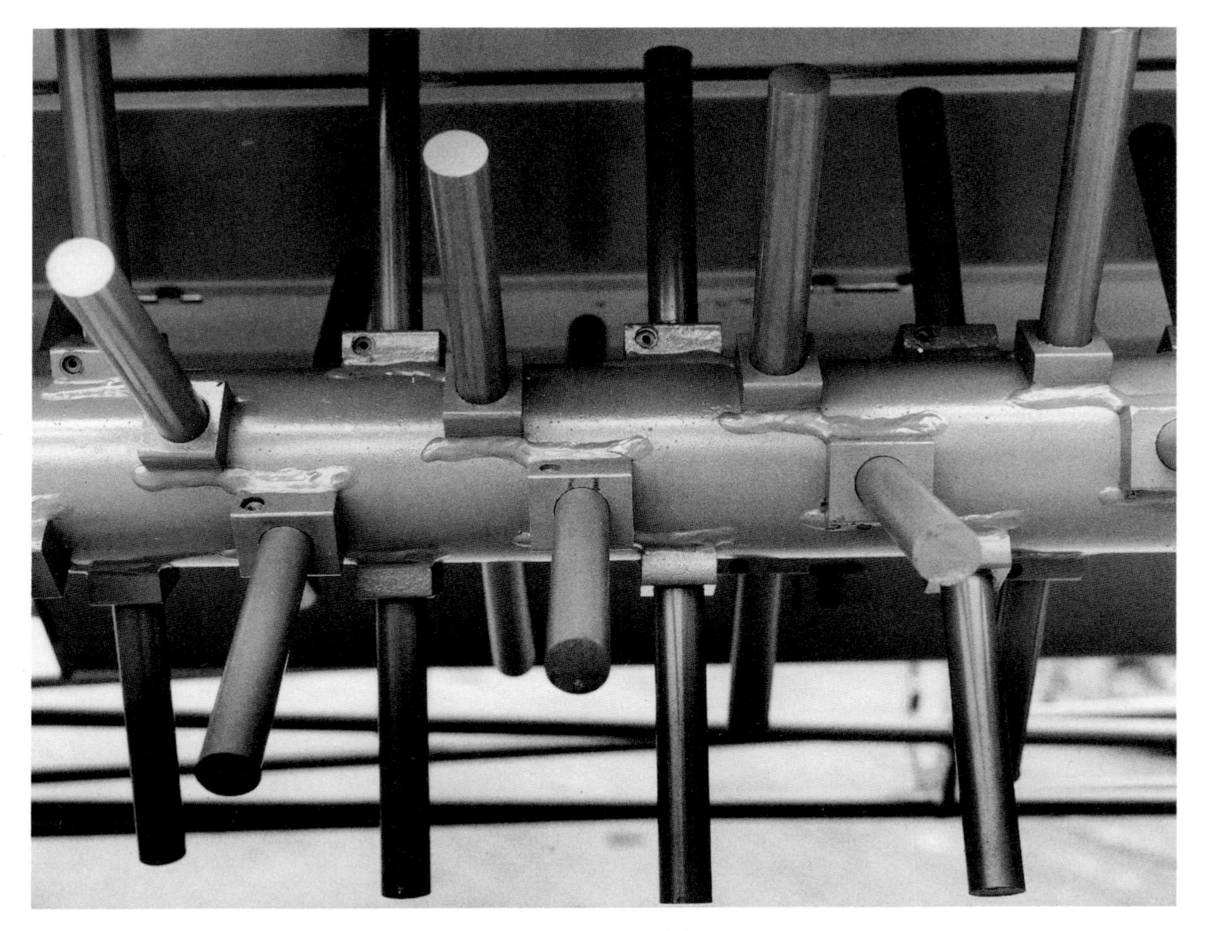

PLATE 10.10 *Spike rotor for tractor powered rotary cultivator.* (Dowdeswell)

where high powered tractors are available. A small cultivator with a 760 mm rotor needs from 9–26 kW (12–35 hp) at the power take-off shaft. Since the power requirement increases as soil conditions get tougher, the smallest compact tractors can handle a 760 mm machine in large glasshouses or nursery beds, but a more powerful tractor will be needed for the same cultivator when working in hard ground.

Rotary cultivators are mounted centrally on the hydraulic linkage. Offset models of the smaller machines are also made. These are ideal for tractors with a wheel track wider than the working width of the rotor, because it avoids the need for the tractor wheels to run on freshly cultivated ground.

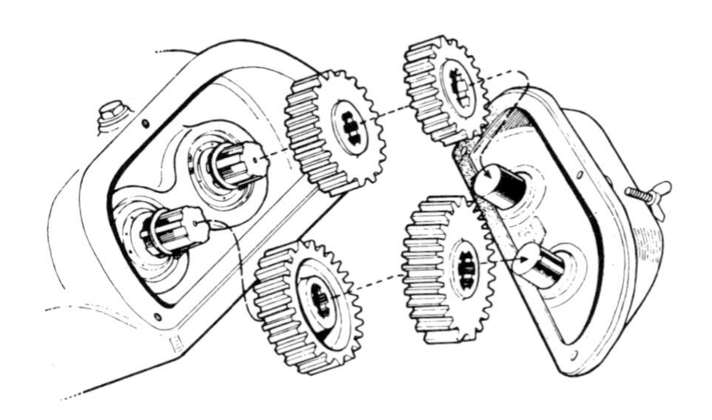

FIGURE 10.3 Rotary cultivator gearbox. The gears can be changed round to give a range of rotor speeds.

A typical drive arrangement is shown in Plate 10.9. The power take-off drive passes through a bevel gearbox on top of the rotor housing to the chaincase where a heavy duty roller chain transmits the power to the rotor. An overload slip clutch is built into the power drive shaft. Some machines have an extra gearbox which gives rotor speeds ranging from about 90–300 rpm when the power shaft runs at 540 rpm. The blades run at a speed of about 180–200 rpm on models which do not have a gearbox for changing rotor speeds.

Adjustments

Adjustments on tractor mounted and pedestrian controlled rotary cultivators are very similar.

Hitching follows the normal three point linkage procedure, but it is necessary to hold the machine rigid on the linkage with stabiliser bars or by tightening the external check chains.

Tilth depends on three factors:

- Tractor forward speed.
- Rotor speed, which can be changed on most models of rotary cultivator.
- Position of the rear flap. Plate 10.9 shows one in a raised position.

A coarse tilth will be obtained with a fast forward speed, low rotor speed and the hinged flap in the raised position. For the finest tilth, select a low forward speed, a fast rotor speed and set the flap in its lowest position.

Depth, which is usually between 150 and 250 mm, is controlled by a combination of depth wheels and skids. The depth limiting skid can be seen near the chaincase in Plate 10.9.

Maintenance

Blades must be maintained in good condition and correctly set on the rotor shaft according to the instruction manual. Keep the rotor blade bolts tight, and renew worn blades, as they will do poor work, especially in hard conditions, and waste power. Bent blades also absorb extra power and should be straightened or replaced if beyond repair. Take care to fit new blades in the correct sequence on the rotor.

Belt and chain drives should be checked for correct tension. The cultivator drive chaincase contains oil, the level of which should be checked regularly, as should all gearbox oil levels.

Weed cutters are fitted at each side of the rotor housing on some models to prevent weeds winding round the ends of the rotor shaft. The cutters should be set so that they just clear the end rotor flanges.

The power take-off shaft bearings and other grease points must be **lubricated** at the start of each working day.

Pedestrian controlled machines also require regular engine maintenance, especially the air cleaner, when working in dusty conditions. Small engine maintenance is described in Chapter 14. Tyre pressures should be checked weekly and the tyres inflated if necessary. When the cultivator is out of use, check the tyres at intervals as they tend to lose pressure when the machine is stored for long periods.

Power Harrows

There are two types of power harrow: one has vertical rotary tines and the other has reciprocating tine bars. Power harrows, like rotary cultivators are capable of producing, and in some cases forcing, a seedbed very quickly in most soil conditions.

Rotary Power Harrows

Rotary power harrows have vertical rotating tines fixed to a series of gear driven rotor heads across the full width of the machine. The drive is from the power take-off through a gearbox and system of gearing to the tine rotors. The gearbox gives a choice of rotor speeds, typically from 200–450 rpm;

PLATE 10.11 *Power harrow with rear crumbler roll.* (Lely)

250–350 rpm is used for most work. The working width ranges from 1.5–8 m.

The smaller power harrows used for horticultural work are of lighter design. Many have a rear crumbler roll which helps to break clods and speed up seedbed preparation. A typical 1.3 m rotary power harrow needs a tractor of about 22.5 kW (30 hp).

Adjustments

The tilth produced depends on the speed of the tine rotors and the forward speed of the tractor, a fine tilth being achieved with a low tractor gear and a high rotor speed.

Working depth is adjusted by raising or lowering the crumbler. Depth limiting skids serve the same purpose on machines without a crumbler roll.

Reciprocating Bar Power Harrows

This type of power harrow has gone out of fashion though many are still in use. They have two, or less often four, reciprocating tine bars driven by the power take-off through a gearbox and an eccentric unit which gives the reciprocating motion to the tine bars. Working widths range from 1.5–4.5. Depth is adjusted

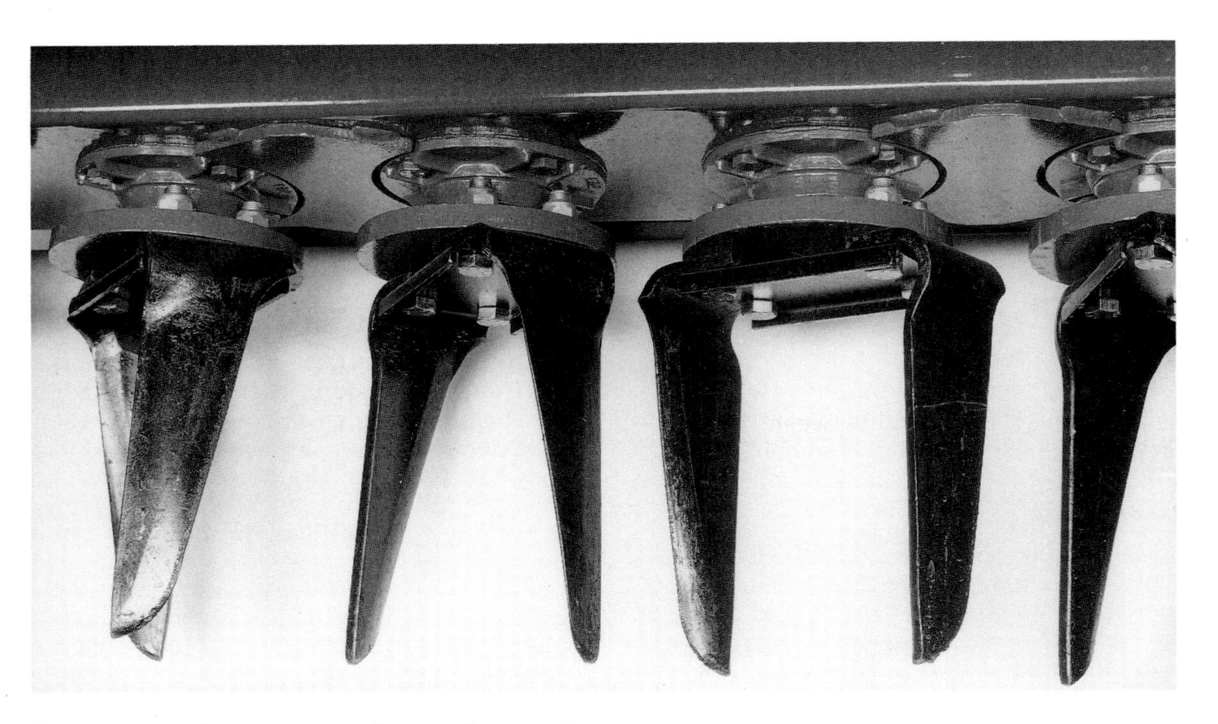

PLATE 10.12 *Arrangement of power harrow tines.* (Colchester Tillage)

with the crumbler roll if fitted, or by depth limiting skids.

Power Harrow Maintenance

Gearbox oil levels should be checked and regular lubrication is essential. All nuts and bolts must be tight, especially those retaining the tines. Worn tines will need to be replaced, tine life depending to a great extent on soil type and condition.

Disc Harrows

Disc harrows suitable for compact and low powered tractors have two and sometimes four sets of saucer-shaped discs carried on a rigid frame attached to the three point linkage. The discs are equally spaced on the shaft by metal spacing blocks. Heavy duty mounted and trailed disc harrows are made for larger tractors. A typical disc harrow suitable for horticultural work will have a working width of 1.1–2 m with 360 mm diameter discs spaced at 150 mm. Much larger models with disc diameters of up to 750 mm and working widths of up to 6 m or more are made for agricultural work.

Disc harrows cut and consolidate the soil. The gangs of discs run at an angle so that as the discs cut the soil it is moved sideways. The front and rear gangs throw the soil in opposite directions to each other in order to give a reasonably level surface. The running angle of the discs affects soil movement and surface finish.

Disc harrows are used to prepare seedbeds (especially after ploughing in grass, where a tined implement would pull turf to the surface) and also for cutting up surface trash and crop residues.

Adjustments

Disc angle is altered by either a hand lever or by manually changing the angle of the disc gangs after slackening the bolts which hold them on the harrow frame. Larger models may have a hydraulic ram for adjusting the cutting angle of the discs. When set at their widest angle to the direction of travel, there will be considerable soil movement, and a rather unlevel surface will be produced. There is less soil movement and some consolidation when the discs run straight.

Working depth is controlled by the tractor hydraulic system. Larger models have depth wheels, usually operated by an auxiliary hydraulic ram. Most disc harrows have a weight tray above each gang of discs, and it is often necessary to put weights in these trays to obtain adequate disc penetration.

Scrapers should be set close to the discs to keep them clean.

Maintenance

Regular lubrication of the bearings which support the disc shafts is important. Some disc harrows have sealed bearings which are maintenance-free. Nuts and bolts must be kept tight and the discs must not be allowed to work loose on their shafts.

Harrows

Harrows have rows of tines fitted to a frame which usually runs at ground level. The tines stir the soil and break up small clods after an initial pass with a cultivator, while the frame helps to level the surface of the soil. Some harrows are quite heavy and are suitable for breaking down ploughed land.

Spike Tooth Harrows

The spike tooth or zig-zag harrow consists of a frame with a form of trellis pattern with tines bolted to it where frame members cross. Spike tooth harrows are made in a variety of sizes and weights. The harrow sections can either be pulled by a trailed harrow pole or carried on a frame fitted to the three point linkage, making transport from field to field a simple matter. Heavier versions are used by some growers to break down ploughed land, while light harrows are useful for preparing seedbeds.

Harrow tines must be kept tight. If a harrow is used with loose tines, the holes will become enlarged and it will be impossible to keep the tines tight.

Levelling Harrow

A levelling harrow, also known as a Dutch harrow or float, is a useful tool for the preparation of nursery seedbeds and ground for turfing or reseeding. A levelling harrow has a hardwood scrubber board at the front (see Plate 14.13), which helps to level the surface and crush the clods. Three or four banks of tines stir the soil to give a finer tilth, and finally an adjustable crumbler roll compacts it. A clamp bolt secures the tines to the frame and can be slackened to allow tine depth adjustment when required or to restore the working depth as the tines wear. It is important to keep the tine clamp bolts tight.

A grader is another type of levelling harrow. It has three or four metal bars on a frame attached to the three point linkage. Some have tines on one or two of the bars, others have none. This implement is used for grading or levelling the soil in preparation for grass seeding, turfing, etc.

Levelling harrows for compact tractors in the power range of 9–26 kW (12–35 hp) are made in various widths from 1.2–1.8 m. Wider models are available for large scale field work.

PLATE 10.13 *Levelling harrow with a front levelling board.* (Wessex)

Chain Harrow

A chain harrow is similar in construction to a piece of chain-link fencing. Some versions have spikes attached to the links, others do not. The spikes may be on one side only, or there may be long spikes on one side and shorter spikes on the other. By using one side or the other, it is possible to vary the effect of this implement. Like spike tooth harrows, a chain harrow can be fitted to a three point linkage frame or towed behind a harrow pole. Working widths vary from 2–4 m for horticultural purposes with much wider models used for large scale field work. The wider mounted models can be folded for transport.

Although of limited horticultural use, a chain harrow may be used for turf renovation if more sophisticated equipment is not available. A smooth link chain harrow is a useful tool when a fine, level seedbed is required.

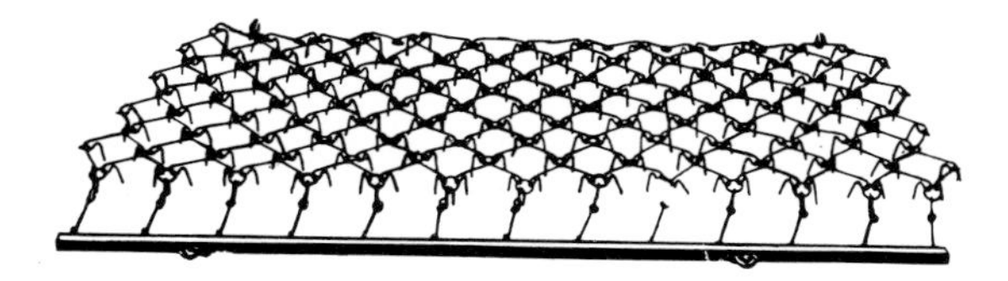

FIGURE 10.4 A tined chain harrow.

Rolls

There are two main types of roll: the flat roll and the Cambridge or rib roll. They are made in many sizes to suit tractors throughout the power range. Very large, multi-gang, hydraulically folded rolls are widely used for field scale crop production. Local authorities, nurserymen and landscapers are more likely to use single rolls, either three point linkage mounted or trailed, with a working width of 1.2–2.7 m (4–9 ft).

Flat rolls are used to level grassland and to push stones into the ground when preparing seedbeds. Most flat rolls consist of a hollow sheet steel cylinder with a diameter varying from 0.4–1.0 m or more. The wider models have two or more short cylinders to avoid

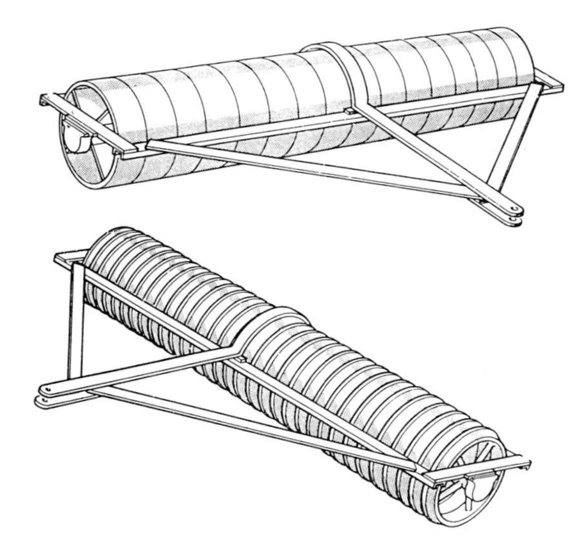

FIGURE 10.5 Top: Heavy flat roll. Bottom: Cambridge roll. *(Twose)*

scrubbing the turf when turning. An 11 kW compact tractor will easily handle a flat roll 400 mm in diameter and 1.2 m wide.

Another type of flat roll has a series of cast iron rings with flat outer surfaces which give far more consolidation than is possible with a hollow steel cylinder roll.

Flat ballast rolls can be filled with water, waste oil and sometimes sand, which adds extra weight to give maximum consolidation when rolling sports grounds and other grassed areas. A screw plug at the end of the rolling cylinder is used for both filling and draining the ballast. A typical ballast flat roll for a compact tractor with a 350 mm diameter and 1.2 m wide cylinder will weigh an extra 120 kg when filled with water. Scrapers to keep the face of the roll clean and stone-carrying trays for extra ballast are fitted to some rolls.

Cambridge rolls have a number of cast iron rings on the axle which leave a corrugated surface finish. They are useful for breaking clods and consolidating freshly ploughed land to conserve soil moisture. The corrugated surface is also ideal for broadcasting grass seed which is then covered with a light harrow.

The furrow press is another type of roll. It consists of cast iron rings carried on a frame pulled alongside a plough, and is used to consolidate the soil and conserve moisture. The press will be wide enough to consolidate the furrows turned by one pass of the plough. Tractors with a hydraulic front linkage system can have a front-mounted furrow press.

Maintenance

The axle bearings, either cast iron bushes or roller bearings, need regular lubrication when in use. Keep cast iron roll sections tight on their axle or else the sections may rub against each other and wear quite quickly. They can be kept tight by adding purpose-made washers to the axle shaft. Very loose cast iron rings may break, especially when the roll is towed along the road at speed.

TOOLBAR EQUIPMENT

Hoe blades, cultivator tines, ridging bodies, weeders and other tools can be used on tractor mounted toolbars or as attachments to pedestrian operated, self-propelled cultivators (see Plate 10.9). Beds for nursery stock and field grown vegetables can be prepared by using a tractor mounted toolbar with a combination of rigid deep tines, shallow spring tines and a crumbler roll at the rear.

Cultivator tines, hoe blades, ridging bodies, weeders, etc., are bolted to the toolbar so that the ground between the rows can be cultivated or hoed without damaging the crop. The wheel track setting must also match the row width. Plate 10.14 shows a toolbar with the tractor wheels running centrally between the ridging bodies. It is important to set up a toolbar so that the wheels and the toolbar equipment run between the rows.

Inter-row hoes may be fitted to front-, mid- or rear-mounted toolbars. Front-mounted hoes are very sensitive to slight movement of the steering wheel but allow good visibility of the

PLATE 10.14 *Ridging bodies on a tractor mounted toolbar.* (Kubota)

PLATE 10.15 *Inter-row hoe with plant shields.* (Stanhay Webb)

crop. Mid-mounted hoes are less sensitive to steering errors but crop visibility is not good. Rear-mounted hoes are rarely used as they need a second operator to steer the hoe. Some rear-mounted hoes are made for one person to use, but they require very accurate steering by the tractor driver.

'L' shaped or side hoe blades are used to cut the weeds close to the rows and 'A' hoes remove the weeds in the row centre. Some inter-row hoes have plant shields which prevent soil from being thrown over the seedlings by the hoe blades.

The number of rows hoed in one pass of the tractor should be the same as the number of rows drilled. Unless the joins between each drilling bout are perfect, some of the crop will be damaged or destroyed if, for example, a four row hoe is used in a crop planted with a five row drill.

Suggested Student Activities

1. Make a list of the cultivation equipment on a holding or nursery you know.
2. Look out for different types of toolbar attachment which can be used on pedestrian controlled cultivators.
3. Experiment with the adjustments on a rotary cultivator which affect the tilth produced and note the results.
4. Find out how hoe blades can be set on a toolbar to suit different row widths. Do not forget the importance of correct wheel track setting.

Safety Check

Always make sure that power driven cultivators and harrows are fully guarded before use. This applies to both tractor driven and pedestrian controlled machines.

Chapter 11

FERTILISER DISTRIBUTORS

Fertilisers can be applied in solid or liquid form. A number of designs and sizes of fertiliser distributor are available for horticultural purposes. Liquid fertilisers are usually applied with a crop sprayer for field scale crop production. Other types of specialist liquid feed equipment for glasshouses and nursery work are described in Chapter 16.

Hand Operated Spreaders

Hand pushed spreaders with a small hopper carried on wheels are used to apply fertilisers to lawns and other small areas of land. These machines can also be used to broadcast grass seed and to apply various treatments, such as powdered lawn dressings.

PLATE 11.1 *Mounted spinning disc fertiliser broadcaster on a compact tractor.* (Ford New Holland)

PLATE 11.2 *Hand operated fertiliser distributor on a bowling green.*
(Sisis Equipment Ltd.)

Roller Feed

This type of hand operated spreader has a wooden or plastic feed roller with grooves or slots in its surface. The feed roller, which rotates in the bottom of the hopper, is directly driven by a small land wheel at each end of the roller. The application rate can be varied on some models by fitting alternative rollers with larger or smaller slots so that more or less fertiliser is carried from the hopper. Where this option is not available, multiple dressings can be applied to achieve the desired treatment.

Spinning Disc

Spinning disc spreaders have a hopper, usually cylindrical in shape, with a small wheel driven spinning disc mechanism. Fertiliser is fed to the spinning disc through an adjustable sized hole in the bottom of the hopper. Application rate can be varied on some models by increasing or decreasing the size of this outlet. Alternatively more than one dressing can be applied to achieve the required application rate.

Belt Feed

A tapered rectangular hopper is carried on a pair of wheels which drive the feed belt. The belt carries the fertiliser from the hopper and spreads it on the ground. The outlet is at the front of the hopper and distributes the material across the full width of the machine. Models in common use have spreading widths usually from 600–900 mm. Many belt feed spreaders have small rollers at the back so that the hopper can be tilted on to the rollers and the drive wheels lifted off the ground at the end of each bout.

Tractor Driven Distributors

There are two types of tractor fertiliser distributor: broadcasters and full width machines, which are used to accurately spread fertilisers. Specialist top dressing machines mainly used on turf can apply precise amounts of fertiliser and other materials to promote even growth and quality of amenity grass areas and sports turf.

Broadcasters

Broadcasters have a hopper on a frame which is mounted on the tractor hydraulic linkage with either a spinning disc or oscillating spout spreading mechanism. Drive is usually from the power take-off shaft, but some spinning disc broadcasters have hydraulic motor drive. Small trailed models suitable for towing behind compact tractors, ATVs, etc., have land wheel drive to the spinning disc mechanism. Although unsuitable for most horticultural work, trailed broadcasters with a high capacity hopper and work rates to match are used by some growers for field scale vegetable crop production.

PLATE 11.3 *Twin spinning disc broadcaster.* (Lely)

Spinning disc broadcasters have one or two horizontal spinning discs which receive fertiliser from the hopper and spread it in an even pattern over the ground. The rate of flow of fertiliser from the hopper can be controlled to vary the application rate.

Twin disc machines have contra-rotating discs so that one spreads to the left of the machine and the other to the right. This achieves a more even spread pattern across the full width of the bout. Typical hopper capacity of a small trailed model suitable for use with a compact tractor is approximately 70 kg (150 lb). Mounted broadcasters come in many sizes with hoppers holding from 250 kg to 1 tonne. Hopper size must be related to tractor power and hydraulic lift capacity. Hopper capacities of trailed spinner broadcasters range from 1–2.5 tonnes or more.

The effective spreading width of broadcasters varies according to machine size, height of the spinning disc above ground level and the type of fertiliser. A typical spinning disc broadcaster with a 300 kg hopper capacity gives a spreading width of from 5 m for powder fertiliser to about 12 m for granular material. This compares with a typical 600 kg hopper machine designed for use with a medium powered tractor which gives a

PLATE 11.4 *Spinning disc unit; the gearbox is used to change the speed of the discs.* (Lely)

PLATE 11.5 *Spinning disc broadcaster towed by an ATV. The disc is wheel driven through a gearbox which can be seen beneath the fertiliser hopper.* (Lely)

spreading width from 10 m for light material to 16 m for granular fertiliser.

Pendulum or oscillating spout broadcasters are an alternative design to the spinning disc machine. The tractor mounted or trailed hopper has an oscillating (swinging) spout which distributes the fertiliser supplied from the hopper through an adjustable outlet.

The spout is driven by the power take-off through an eccentric gearbox which oscillates it. Hopper capacities vary from 250 kg for the smaller mounted models to about 2.5 tonnes for a large trailed model. Spreading width varies from 6–12 m depending on machine size and type of fertiliser.

PLATE 11.6 *Pendulum broadcaster.* (Vicon)

Hitching Fertiliser Spreaders

Once fitted to the three point linkage, the spreader must be held rigidly on the tractor by either fitting stabiliser bars or tightening the external check chains. It is important to set the spout or disc level with, and at the correct height above, the ground. This information is given in the instruction book for

PLATE 11.7 *Pendulum broadcaster mounted on a trailer unit being towed by an ATV. A small petrol engine drives the spreader through an extended power take-off shaft.* (Kawasaki)

the machine. The height of the disc or spout has a direct effect on the spreading width and pattern. Incorrect setting is likely to result in poor matching of the work, which often causes striping—light and dark strips—on the crop or grass.

Adjustments

Application rate is controlled by means of an adjustable slide in the bottom of the hopper. This varies the amount of fertiliser passing to the spinning disc or oscillating spout. When the slide is set wide open, maximum application will be achieved.

Forward speed also varies the application rate on all power take-off and hydraulic motor driven broadcasters. A fast forward speed, selected with the tractor gearbox, will give a low application rate compared with a slow forward speed. The power take-off shaft must run at the speed stated by the manufacturer; the throttle must not be used to vary forward speed as this will also change the power shaft speed.

It should be noted that changes in forward speed do not alter the application rate of wheel driven broadcasters.

Some spinning disc broadcasters have an adjustment which can be used to alter disc speed. Increased disc speed will give a wider spread, but it will also reduce the rate of application unless the forward speed is reduced or the adjustable slide is opened to increase the flow of fertiliser on to the disc.

The *shut-off lever* is used to cut off the flow of fertiliser to the spreading mechanism at the end of a bout.

Using Broadcasters

Equal matching of the work is very important. The instruction book will give the spreading width for different materials and application rates. Powders will not be thrown as far as granular fertilisers, and broadcasting on windy days will adversely affect the spreading pattern. The only certain way of good bout matching is to use some form of marking aid as a driving guide, such as stakes or empty fertiliser bags placed along the edge of the field at carefully measured intervals. It is also possible to fit the tractor with markers which leave a scratch mark on the seedbed.

Checking The Application Rate

Fertilisers are very expensive so they must be applied accurately to keep costs down and avoid waste or under-dosing.

The machine instruction book gives full settings for a wide range of application rates. Once set for a selected rate, the machine can be calibrated (checked) by either a static or a field test.

A simple field check can be made by putting the required amount of fertiliser in the hopper for a selected area. For example, for a rate of 100 kg per hectare, place 50 kg in the hopper and carefully measure out an area of 0.5 hectare in the field. Spread the fertiliser over the measured area. If the machine is accurate, the hopper should be empty only when the area is completed.

Broadcasters can also be calibrated when stationary. Some instruction books give the time taken to spread a stated area at a specified speed. The application rate can be checked by running the machine at the required setting with the power take-off at the correct speed for the stated time. If done on a concrete floor inside a building, the fertiliser can be collected and weighed so that actual output can be compared with that required.

There is a calculation to find the correct running time to cover a stated area if this information is not provided in the instruction book. Use this formula to find the time taken to cover one hectare:

$$\text{Time (mins)} = \frac{600}{\text{spreading width (m)} \times \text{speed (km/h)}}$$

For example, the time to cover one hectare at 6 km/h with a spreading width of 10 m will be:

$$\text{Time (mins)} = \frac{600}{10 \times 6} = 10 \text{ minutes}$$

The machine could be run for five minutes only to give a figure for application rate per 0.5 hectare. Then, of course, the weight of the fertiliser collected should be doubled to give the application rate per hectare. The spreading width for various fertilisers at different application rates should be given in the instruction book.

Full Width Distributors

Unlike broadcasters, full width distributor machines are almost as wide as their width of spread. A mechanical feed mechanism at the bottom of the hopper distributes fertiliser evenly over the land. They may be either tractor mounted or trailed; in either case the feed mechanism is land wheel driven, usually by a system of gears which provides a range of fertiliser application rates. The main benefit of full width machines is their ability to provide an even spread over the full width. However the wider models can present transport problems around the holding.

The main types of feed mechanism are:

Fluted roller feed consists of a roller the same width as the hopper with flutes or notches in its surface. The flutes expel fertiliser through a series of holes in the hopper bottom. Adjustable shutters over the fertiliser outlets can be opened or closed as required to vary the application rate. The speed of the feed roller can be altered by means of a gearbox in the drive train from the land wheels.

Another type of fluted roller feed consists of a series of small feed rollers spaced at intervals on a square shaft across the full width of the hopper. Drive from the land wheels is transmitted to the feed rollers by a system of chains and gears which provide a range of roller shaft speeds. An adjustable shutter controls the rate of flow of fertiliser from the hopper.

Agitator feed consists of a series of agitator discs on a shaft across the full width of the hopper. Rotation of the discs forces fertiliser through holes in the bottom of the hopper. Application rate is varied by the adjustable shutters which control flow of fertiliser from the outlets, and some models have a gearbox which is used to change the speed of the agitator shaft. Forward speed does not alter the application rate on land wheel driven machines, but excessive speeds on rough land can result in an uneven application rate due to bumping and bouncing of the machine.

Pneumatic Distributors

Although of limited horticultural use, except for field scale vegetable crop production, there are numerous models of pneumatic distributor on the market. These machines accurately meter fertiliser into a high speed flow of air, created by a power take-off driven fan, in a series of tubes. After the fertiliser has been introduced into the airflow, it is carried along the tubes to equally spaced outlets across the full width of the machine. Deflector plates below the outlet points distribute the material evenly over the ground.

Application rate is varied by changing the speed of the metering roll which feeds the fertiliser into the air stream. Variations in forward speed, achieved by using the gearbox, will also alter the spreading rate because the machine is power take-off driven.

Like the mechanical feed distributors, pneumatic spreaders can achieve a very even spread pattern because the fertiliser is discharged from outlet points across the full width of the machine.

Top dressers

The application of top dressings, including fertiliser, sand and other bulky material, to promote healthy growth is an important part of turf management. Top dressers may either be mounted on the three point linkage or trailed. They have a hopper with a hydraulically driven belt conveyor on its floor. An adjustable rear shutter is used to control the flow of material carried by the belt to the polypropylene brush which distributes the top dressing evenly over the turf. The brush, which is as wide as the hopper conveyor, is also driven by a hydraulic motor.

Smaller mounted top dressers have a roller at the bottom of the hopper which distributes the material on to the ground. Application rate can be varied by altering the setting of the adjustable shutter or by changing the forward speed with the tractor gearbox. Top dressers may either be mounted on the three point linkage or trailed. The different types of top dresser are illustrated in Chapter 15.

PLATE 11.8 *Pneumatic fertiliser spreader.* (Ferrag)

Suggested Student Activities

1. Look for the different types of fertiliser distributor described and locate the controls used to change application rate.
2. With the help of the instruction book find out how to set the application rate for a fertiliser broadcaster.
3. Trace the flow of fertiliser through a pneumatic fertiliser distributor.
4. With the help of an operator's manual, make a list of the range of materials which can be applied with a turf top dresser.

Safety Checks

Always check that power take-off guards are fitted and in good condition when using a fertiliser broadcaster or any other power driven equipment.

When lifting a sack of fertiliser remember to bend your knees and not your back. Many people injure their backs by lifting boxes and sacks in an incorrect way.

For your own protection, wear goggles, mask and gloves when filling a fertiliser spreader hopper.

Chapter 12

DRILLS AND PLANTERS

Nurserymen, growers and landscapers need to sow many types and sizes of seed, either by broadcasting or by sowing in rows. Vegetable growers, in particular, need to plant many thousands of bare rooted plants and seedlings grown in soil blocks. Various types of seed drill and mechanical planter are used in horticulture to speed up the planting process and reduce labour costs.

PRECISION SEEDER UNITS

A seeder unit used for vegetable and other crops has a feed mechanism which places seed in the soil at various row widths and spacings. A number of seeder units, usually between three and six, are attached to a tractor mounted toolbar. For large scale vegetable production, twelve or more units may be fitted to a very wide toolbar. At the other extreme, a single seeder unit can be fitted with a pair of handles and pushed along, or a pedestrian controlled cultivator may be fitted with one or two seeder units and used to sow a few rows of seed in a nursery or small market garden.

Each seeder unit consists of a frame carried on two wheels with a coulter, seed covering device, hopper and feed mechanism. The rear wheel firms the soil after the seed has been planted, while the front wheel on single units and some toolbar mounted units is used to drive the feed mechanism. Where a number of seeder units are attached to a tractor toolbar, a master wheel drive system is normally used; the toolbar land wheels drive the seeder unit feed mechanisms by means of a system of chains or belts and a gearbox.

Seeder Unit Feed Mechanisms

Precision seeder units have a feed mechanism which gives an even spacing of seed in the soil. Seed spacing can be altered to suit all types of seed and production methods.

Belt feed
A small, endless rubber belt with holes punched in it carries the seed from the hopper to the outlet point. A repeller wheel turning in the opposite direction to the seed belt helps to ensure that only one seed is carried in each hole. The seed belt is stretched slightly at the outlet point and the seeds fall through the holes into a shallow

PLATE 12.1 *Seeder units with master wheel drive on a small tractor.* (Stanhay Webb)

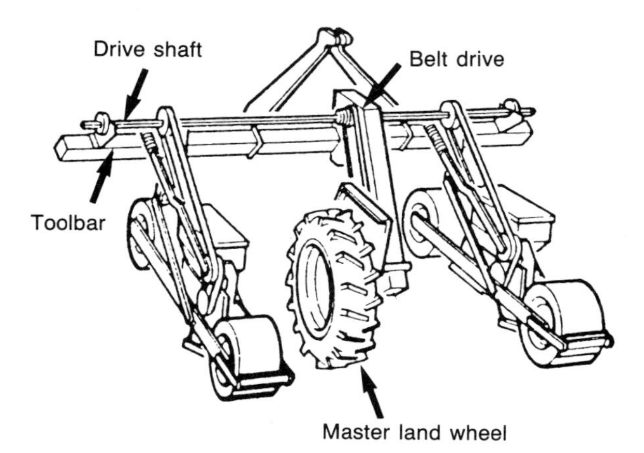

FIGURE 12.1 Two wheel master drive seeder unit. *(Stanhay Webb)*

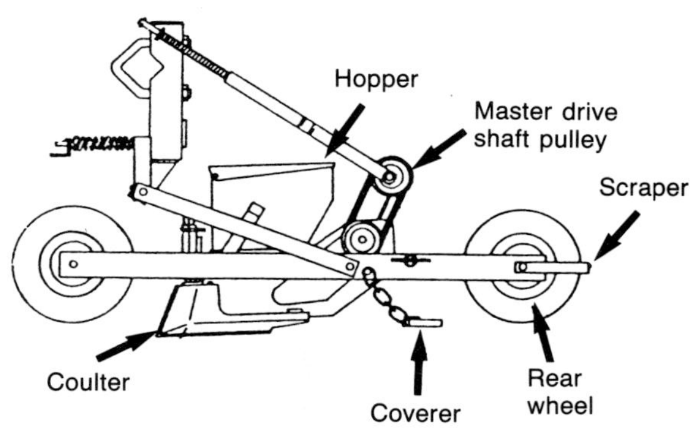

FIGURE 12.2 Belt feed seeder unit. *(Stanhay Webb)*

PLATE 12.2 *Hand pushed belt feed seeder unit.* (Stanhay Webb)

furrow made by the coulter. A choke in the hopper outlet restricts the flow of seed from the hopper to the seed belt, and various sizes of choke are available for sowing different sized seeds. Large seeds need a small choke so that they can pass through the hopper outlet, while small seeds will need a larger choke.

Seed rate and spacing are determined by:

- Changing the seed belts. A range of belts is available with holes punched at different intervals. Seed belts with different sized holes are made to suit a variety of seed types and sizes. Some seed belts have two or three rows of holes (see Figure 12.1) for sowing carrots and other crops. A different sized choke may also be needed when changing seed type.
- Seed belt speed can be altered on master wheel drive drills. This is achieved by using the gearbox lever to change the speed of the output shaft and vee-belt pulleys which drive the feed mechanism seed belts.

With master wheel drive it is usually possible to obtain a full range of seed rates and spacings for a particular crop without the need to change the seed belts.

Cell wheel feed
Holes in the rim of the cell wheel collect single seeds from the hopper. A repeller wheel, turning in the opposite direction to the cell wheel, helps to ensure that only one seed enters each hole. The seed is carried by the cell wheel to the outlet point where an ejector plate prises the seed from the holes and they fall into a shallow furrow made by the coulter.

Seed rate and spacings can be altered in a similar way to belt feed by varying the speed of the cell wheels or by changing them. Cell wheels with different sized holes are used for different types of seed.

Vacuum feed
This recently introduced feed mechanism has a power take-off driven fan on the toolbar

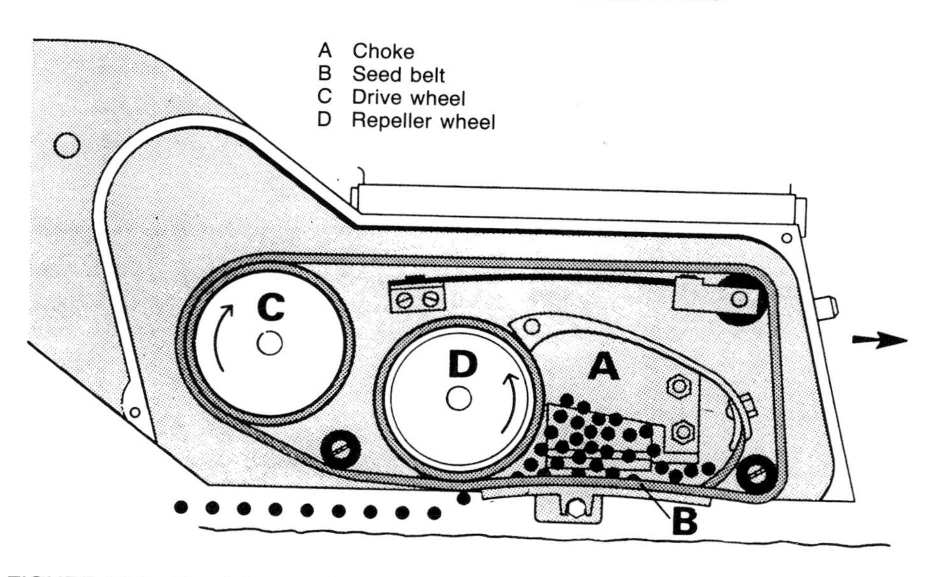

FIGURE 12.3 Precision seeder belt feed mechanism. *(Stanhay Webb)*

PLATE 12.3 *Two belt feed seeder units on a pedestrian controlled tractor. The rear press wheels have the cage wheel pattern which is recommended for light soils.* (Stanhay Webb)

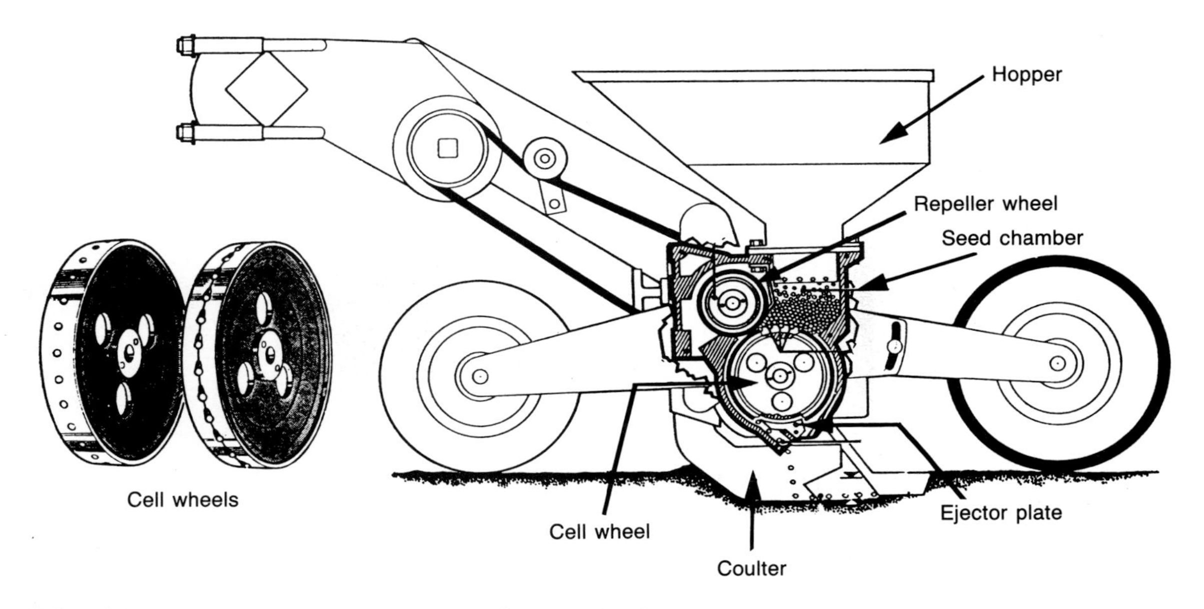

FIGURE 12.4 Precision seeder cell wheel feed mechanism. *(Stanhay Webb)*

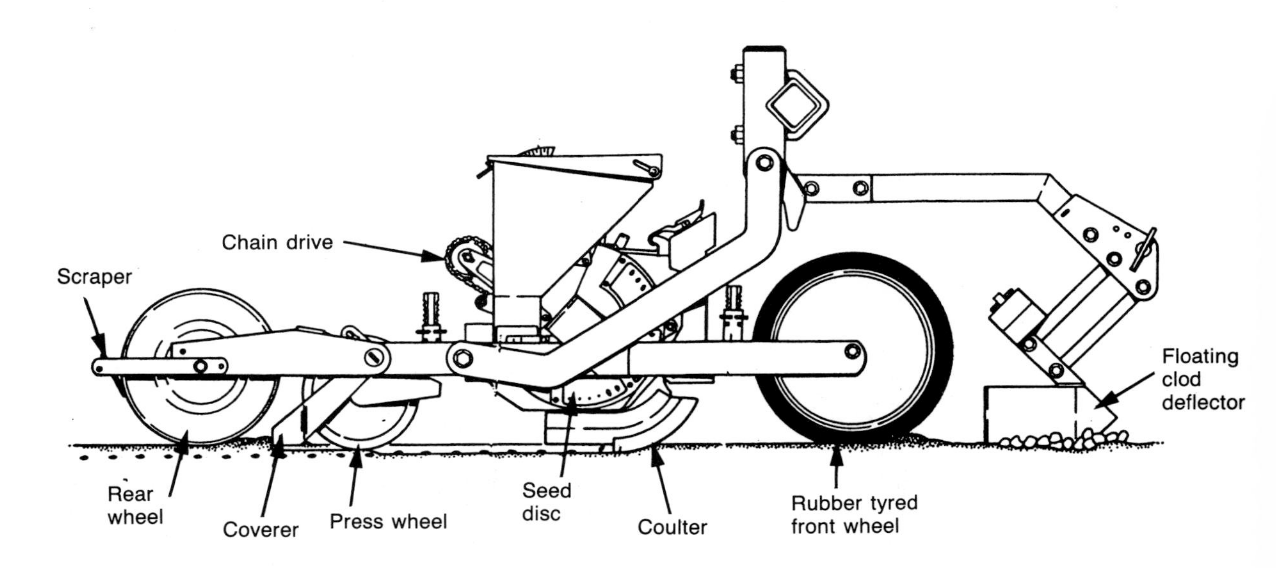

FIGURE 12.5 Vacuum feed seeder unit. *(Stanhay Webb)*

which creates a vacuum in each seeder unit. The feed mechanism consists of a flexible seed disc with regularly spaced holes, which turns in a housing under the seed hopper. One side of the seed disc is under vacuum, while the other side receives seed from the hopper. The vacuum pulls individual seeds against the holes in the disc which carries them to a discharge slot above the coulter. At this point a blanking plate cuts off the

PLATE 12.4 *Tractor mounted seeder unit with vacuum feed system. The fan is centrally mounted on the toolbar. The outfit has three granule applicators with four feed spouts from each granule hopper.* (Stanhay Webb)

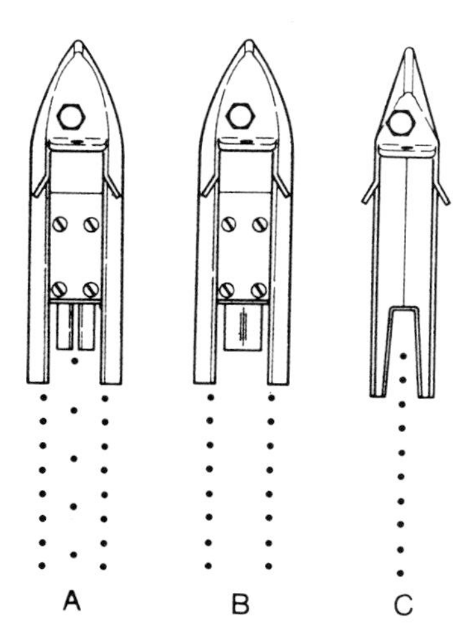

FIGURE 12.6 Seeder unit coulters for three row (A), double row (B) and single row seeding (C). *(Stanhay Webb)*

vacuum and the single seeds fall to the ground. A flow of air from the fan ventilates the feed mechanism to keep dust out of the seed disc housing and to clean the seed holes in the flexible disc.

Flexible seed discs with one, two or three lines of holes can be used to drill seed in single, closely spaced double or triple rows for such crops as carrots. Vacuum feed is suitable for many types of seed including celery, parsnip, parsley, red beet and the entire brassica family.

Seed spacing in the row is altered by changing the speed of the feed discs by means of the master land wheel drive gearbox.

Warning Systems

Toolbars fitted with a number of seeder units usually have warning lights to tell the driver if any of the seed belts or cell wheels are not working. Another indicator light warns the driver that more seed is needed in the hoppers.

Using Seeder Units

When two or more seeder units are used on a toolbar, they can be spaced to give the required row width to suit any crop. The tractor wheel track setting must suit the row width and the unit placed on the toolbar so that if possible, the seeder units do not run in the tractor wheelings.

Markers are used to help the driver join his work accurately. This is very important when drilling rowcrops which will be mechanically hoed and harvested. Wide or narrow joins may result in damage to the young plants or harvested produce.

Markers, one each side of the drill, have a pointed tine on a bar which pivots on the drill frame. It makes a mark to show the driver where to drive on the next run across the field. The marker tines can be adjusted to provide an accurate guide for joining the work with any row width and front wheel track setting.

To set drill markers, follow this method:

1. Measure the distance from the centre of a tractor front wheel to the outer seeder unit coulter (measurement A). This may be either within or outside the wheel track depending on the number of units on the toolbar.
2. Measure the row width (measurement B).
3. If the drill is wider than the front wheel

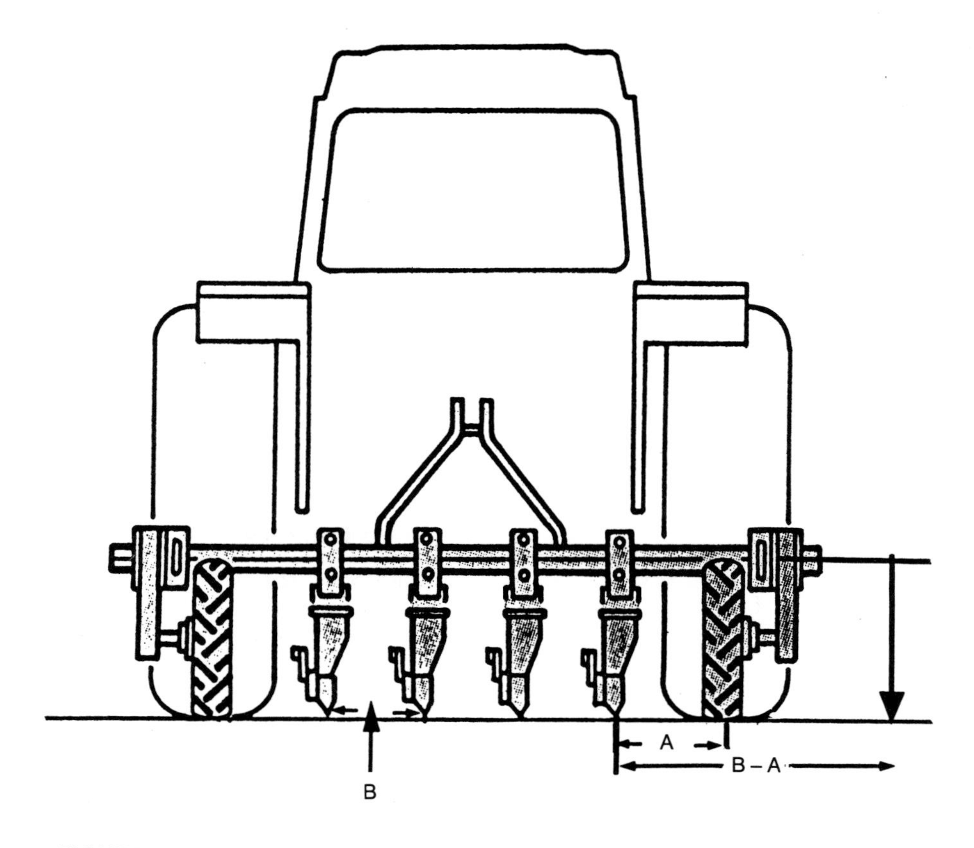

FIGURE 12.7 Setting markers. *(Stanhay Webb)*

track, add measurements A and B together and set the marker this distance from the outer seeder unit coulter. If the drill is narrower than the wheel track, subtract measurement A from B to find the distance the marker point should be from the outside coulter.

Drills for Grass

Grass seed may either be applied to bare soil to establish a new turf or drilled into existing turf to renovate worn or tired areas. A number of machines are used to sow grass seed, which is very expensive and needs to be accurately placed on the soil and covered to protect it from birds. Some machines are designed to overseed existing turf by placing the seed directly into the ground, then covering and consolidating it to promote rapid germination.

Broadcasters

Both pendulum and spinning disc type broadcasters can be used to sow grass seed on well prepared soil. After it has been broadcast, the seed must be covered and consolidated to retain moisture and aid germination.

The quantity of seed needed to establish turf is very small, so before work commences, the broadcaster should be set as directed in the instruction book. This usually means having the hopper outlet almost closed and selecting a relatively high forward gear, taking care not to exceed the correct power take-off speed.

Full Width Grass Drills

The most common type of grass drill consists of a hopper with a wheel at each side, either attached to the three point linkage or towed from the tractor drawbar. A shaft driven by the land wheels across the full width of the hopper, carries a series of closely spaced rotating brushes. Small quantities of grass seed are brushed through adjustable sized openings at the bottom of the hopper from which it falls to the ground.

Grain Drills

Most types of drill designed for grain crops can also be used to sow grass seed. Seed placed in the hopper is carried by a fluted or studded feed roller from the hopper to flexible tubes which convey it to coulters spaced at intervals of 90, 120 or 180 mm. The coulters, either disc or Suffolk type, make a shallow furrow in the soil for the seed, which is then covered. The advantage of sowing seed with a grain drill is that it is placed in the soil, but a grain drill is not really suitable for establishing top quality turf as the plants are rather widely spaced in the rows.

Overseeders

An overseeder is used mainly to restore worn or thin turf. Grass seed is broadcast over the ground and falls into shallow slots or grooves cut into the ground. Some overseeders have flexible tubes which direct the seed down to the slots. A rear press roll then consolidates the surface to ensure the seed is in close contact with the soil. (See Chapter 15.)

Planters

There are two types of planter used for horticultural crops. One is mainly used for planting potatoes, but it is suitable for some other crops such as bulbs and tuberous crops. Transplanters are used to set brassica, celery and many other crops, either bare rooted or in blocks, and also young trees, nursery stock plants, etc., for growing on.

Planting Potatoes

A deep seedbed, free from large clods and stones, is required for a successful potato crop. For growers with a small area of potatoes, a power harrow or rotary cultivator will produce a suitable seedbed for mechanical planting. Large scale potato growers usually have specialist machines for seedbed preparation, not only to aid the planting operation but also to avoid problems when the crop is mechanically harvested.

The bed system is preferred by many large

scale producers. The land is prepared with a bed former which has two large ridging bodies used to make a ridged bed of soil 1.5–2 m wide. The ridged bed is then passed over a stone and clod separator where a rod link conveyor removes any large stones and clods from the soil before it is returned to the ground to form a deep seedbed ready for the mechanical planter. The stones and clods are carried sideways on a conveyor which deposits them in the furrow formed by the ridging body when the bed was originally made. This process ensures that when the crop is mechanically harvested, the stones will not pass through the machine.

Potato Planters

Most potato planters are fully automatic machines mounted on the three point linkage with a land wheel driven feed mechanism. A few models of planter are hand fed, with one operator required for each row, but tuber spacing in the row is still automatic. Some high output planters which set four rows at a time are trailed with a hopper capacity of about 2 tonnes of unchitted seed. Most mounted machines plant two rows at a time, the coulter making a furrow for each row. The hopper has a capacity of 500 kg to a tonne or more depending on model. An automatic feed mechanism picks single tubers from the hopper and places them in the furrow at regular intervals. After the potatoes are placed in the furrow, they are covered by concave discs or ridging bodies.

PLATE 12.5 *A two row bed former.* (Grimme)

Many planters have a fertiliser placement unit, complete with coulters, which places metered amounts of fertiliser in the soil close to the tubers. The fertiliser should not be allowed to come into direct contact with the tubers as this may scorch the shoots.

Planting depth is controlled either by the tractor hydraulic system or by depth wheels. The depth of the coverers can also be altered by lifting or lowering them on the planter frame. This will also change the shape of the completed ridge.

Row width can be altered by moving the furrow openers and coverers on the toolbar. Row widths of 610–910 mm are in common use.

The feed mechanism of the automatic planter shown in Figure 12.8 has a land wheel driven endless chain with a series of cups. Each cup collects a single potato from the hopper and carries it to an outlet point just above the furrow opener. The potato falls to the bottom of the furrow and is then covered by a pair of angled discs at the rear of the planter.

Spacing of potatoes in the row can be altered by using different sized gears or chain sprockets to change the speed of the feed cups. A typical machine can plant at spacings of 150–450 mm.

Some machines are hand fed but have automatic spacing of the tubers in the row. An operator is required for each row. Potatoes are placed in cups on an endless conveyor belt which carries them to an outlet point above the furrow openers. From there they drop into the soil and are covered by discs or ridging bodies. Spacing of potatoes in the rows can be altered by changing the speed of the feed cups. This type of machine is useful for planting chitted seed as it is less likely to damage the tender shoots.

Growers with a small area may use a simple hand fed machine which also needs one operator for each row planted. Potatoes

PLATE 12.6 *A two row potato planter with fertiliser placement unit and covering bodies.* (Kverneland)

are dropped by hand down a spout behind the furrow opener into the ground. The potatoes are then covered by discs or ridging bodies. Spacing between tubers is controlled by a bell which rings at set intervals to tell the operator to drop a potato down the spout. The frequency of the bell ringing can be altered to achieve different tuber spacings. This type of planter is also suitable for chitted seed.

Polythene Laying Machines

Potatoes and some other crops which are sensitive to frost can be planted and then covered with polythene to protect them from late frosts. This is an expensive process, but early maturing of a protected potato crop, for example, should bring high cash returns.

Polythene layers are tractor mounted with provision to carry a roll of the material which is laid over, for example, two potato ridges. The sides of the sheet are pushed into the ground by a pair of discs and firmed into the soil. Removal of the sheeting is a task which must be done by hand.

Transplanters

There are two basic types of transplanter: one for bare rooted material and the other for plants grown in various types of soil blocks or modules. Each transplanter unit plants one row. One to seven transplanter units may be fitted to a three point linkage toolbar, with tractor power the limiting factor in the number used. Some multi-row transplanters are trailed machines.

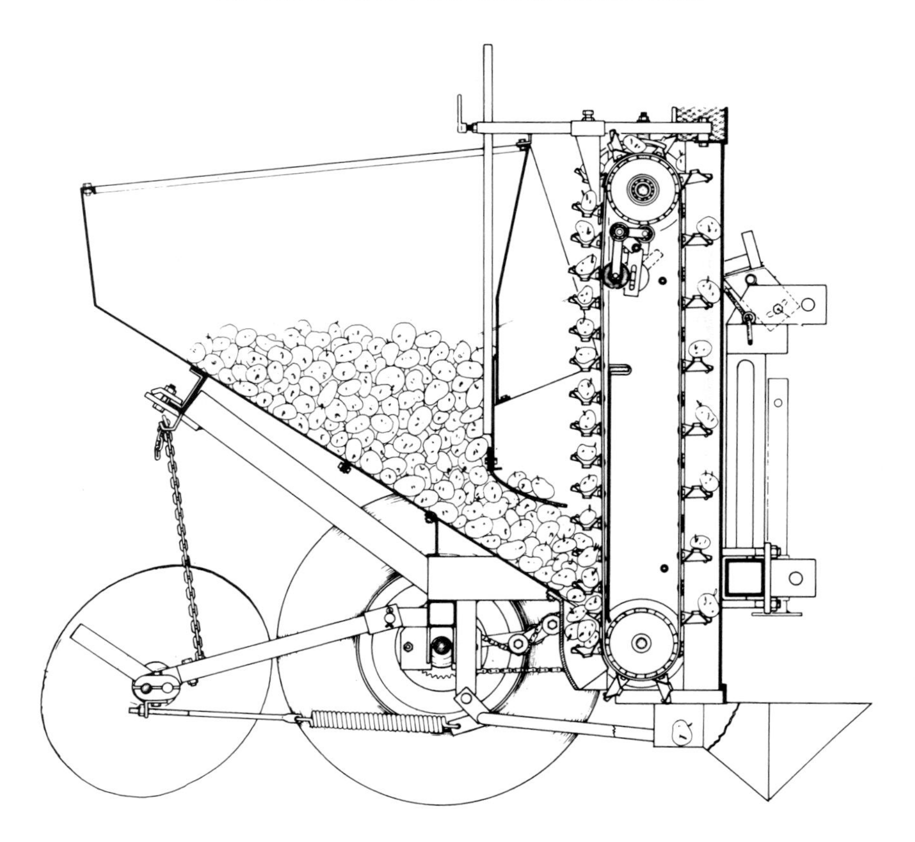

FIGURE 12.8 Sectional view of a cup feed potato planter. *(Standen)*

Bare Root Transplanters

The main parts of a bare root planter are a furrow opener, plant conveyor and two rear packer wheels. The operator places single plants in the gripper pockets equally spaced on a chain which conveys them to a narrow slot in the soil made by the furrow opener. The plant is released from the conveyor pocket as it is placed in the furrow, and the rear packer wheels, which run at an angle to each other, firm the soil around the plant. The conveyor is chain driven by one of the rear packer wheels. The operators, one for each transplanter unit, sit with their backs to the tractor, filling the conveyor pockets with plants which are carried in a box on the machine.

Depth of planting is adjusted by raising or lowering the furrow opener. The spacing between the plants in the row is changed by using different chain sprockets to alter the speed of the conveyor which carries the plants down to the furrow. Spacing between the rows is varied by moving the transplanter unit on the toolbar. Where narrow row spacings are required, it is necessary to have a tandem arrangement with two bars so that the units can be staggered (see Plate 12.10). A typical bare root planter can be adjusted to set plants at spacings of 300–500 mm (12–20 in) at row widths from 460 mm (18 in). Narrower row spacings can be achieved with a tandem toolbar.

A second type of bare root planter has two flexible discs which run at an angle. The discs are arranged so that the rims are apart at a point close to the operator who is required to place a plant between the rims at marked intervals. Further rotation of the disc rims

PLATE 12.7 *Transplanter at work.* (Ford New Holland)

PLATE 12.8 *Bare root transplanter unit. The plant pockets and drive chain can be seen in the open position ready for a plant. The pockets are closed as they enter the pocket guide so that the plants are held gently as they are carried down to the slot in the soil prepared by the furrow opener.*
(Colchester Tillage)

brings them close together to gently grip the plant and carry it round to a slot in the soil made by a furrow opener. Having released the plant, the discs firm the soil around the roots.

Module Transplanters

This type of transplanter is similar in construction to the bare root model, with the exception of the mechanism which conveys the plants to the soil.

The operator fills the four conical shaped revolving cups with plants grown in blocks or modules. When the plant is positioned above the drop tube, the bottom section of the cup opens and the plant falls into the tapered inner section of the furrow opener shoe where it is held upright. To complete the operation, the kicker arm moves the plant into the furrow and the packer wheels firm the soil around the plant.

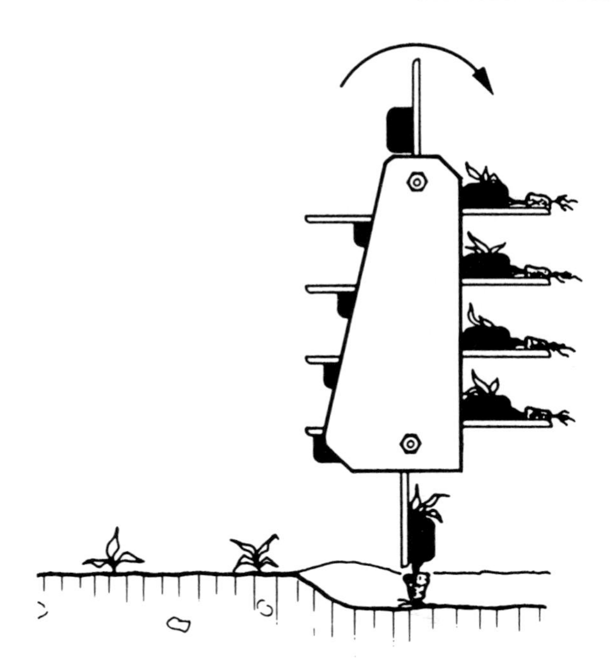

FIGURE 12.9 Transplanter mechanism for bare root and module grown plants. The plants are carried down to the soil after being gently gripped by cam operated carriers. *(Stanhay Webb)*

PLATE 12.9 *Module transplanter unit. The cone shaped cups which rotate above the drop tube can be seen close to the operator's seat. This planter unit also has a carousel unit which will hold four boxes of plants.* (Colchester Tillage)

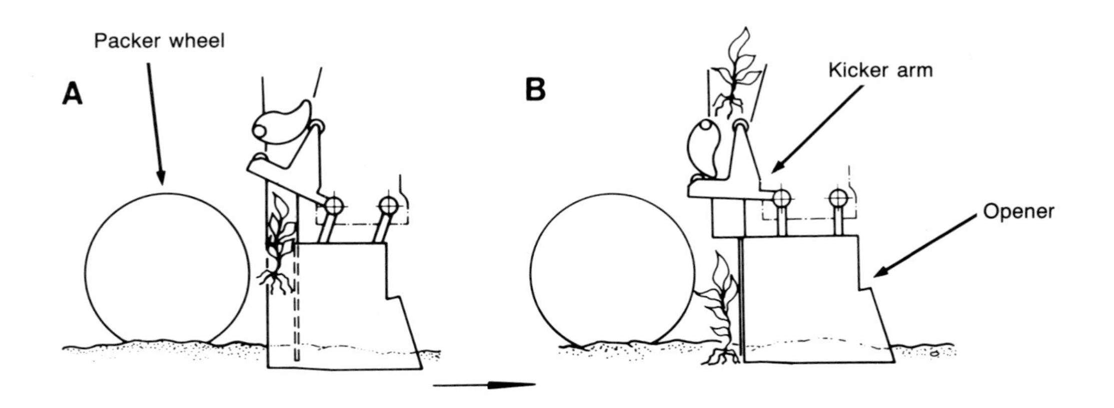

FIGURE 12.10 Module transplanter kicker mechanism. The plant is held in the opener (A) until the kicker is engaged. At this point, the kicker moves the opener forward (B) and the plant is released into the soil. At the same time, the next plant falls down into the opener ready for planting. *(Stanhay Webb)*

Rear-mounted rotary mower on compact tractor. (Kubota)

Four wheel drive front mower. (Kubota)

Ride-on triple cylinder mower. (John Deere)

Turf roller which can be filled with water or waste oil for ballast. (Twose)

Four wheel steer ride-on mower. (Kubota)

Seven unit trailed gang mower with hydraulic drive to cutting cylinders. (Ransomes, Sims & Jefferies)

Hollow tine aerator on compact tractor. (Kubota)

Tractor mounted turf slitter. (Charterhouse)

Pedestrian controlled leaf sweeper. (Bomford Turner)

Lawn tractor with mid-mounted rotary mower and trailer. (John Deere)

Compact tractor with mounted sprayer. (Massey-Ferguson)

Barrow sprayer with spraybar attachment. (Allman)

Brush cutter in use. (Sachs-Dolmar)

Flail hedge cutter on a compact tractor. (Twose)

Trench digging in progress with a compact digger. (Kubota)

Turf slotter. The drainage material is carried in the hopper. (Charterhouse)

Rotary power harrow and crumbler roll. (Lely)

PLATE 12.10 *A five row tandem toolbar arrangement with three transplanter units at the front and two on the second bar.* (Colchester Tillage)

The drive arrangement for the planter mechanism may be either through a master wheel drive system, similar to that used for precision seeders, or by individual drive from the packer wheels to the feeder unit and kicker arm. In both cases, the plant spacing in the rows is varied by changing the speed of the mechanism by using different chain and sprocket drive combinations. Row width depends upon the position of the transplanter units on the toolbar. A typical machine provides plant spacings in the row from 130–590 mm (5–23 in) and rows widths from 350 mm (14 in) upwards.

Various attachments are available for use with transplanters. Carousel tray holders give maximum plant storage on some machines, while others have additional plant trays. It is also possible to have a watering attachment complete with tank and pipes which deliver water to the plants as they are set. A fertiliser attachment, a granule applicator and bout markers are other items of optional equipment fitted to transplanters.

Using transplanters

Tractors used with transplanters must have a suitable wheel track setting to match row width, and the transplanter will need to be held rigidly on the hydraulic linkage with either stabiliser bars or tightened check chains. Front end weights will usually be required to counterbalance the weight of the machine and its operators.

Planting speed is limited by the pace at which the operators can fill the plant conveying mechanism and, for most work, a very low forward gear will be required.

Planting depth will depend upon the type of plant being set; it is important to ensure that the furrow is deep enough to take the full length of the plant root. As such a wide variety of plants and soil blocks can be handled, it is important to carefully reset the

PLATE 12.11 *A two row transplanter with watering attachment with a carousel holder fully loaded with plants.* (Colchester Tillage)

depth of the furrow openers when changing from one species to another.

Suggested Student Activities

1. Look for the different types of precision seeder described in this chapter and examine their feed mechanisms.
2. Find out how to change the precision seeder belts and cell wheels.
3. Follow the line of drive of a transplanter unit and find out how to vary the spacing of the plants in the row.
4. Study an automatic potato planter to find out how the feed mechanism operates and how tuber spacing is changed.

Safety Check

Avoid wearing loose clothing such as long flapping overcoats, loose belts and long scarves when working on or near machines such as transplanters which have many moving parts. Any item of clothing caught in a shaft, chain or other mechanism can have very unpleasant, if not fatal, consequences, especially if it is power take-off driven. Be safe: wear a boiler suit and safety footwear with steel toecaps and non-slip soles.

Chapter 13

SPRAYING MACHINERY

Sprayers are used to apply chemicals, usually diluted with water, to the soil or growing plants to control weeds, pests and fungal diseases. Liquid fertiliser and plant growth regulator treatments can be applied with most types of sprayer.

A good spraying machine will apply a wide range of chemicals evenly and accurately to the soil or plants. Failure to achieve this objective will result in either uneven, and therefore ineffective, treatment or waste of chemicals with possible crop damage through overdosing.

Sprayers used in horticulture range from the simple hand held applicator to sophisticated machines with computerised, electronic controls. They include:

- Hand held applicators
- Mist blowers and dusters
- Knapsack sprayers
- Pedestrian operated, engine driven sprayers
- Tractor mounted and trailed sprayers
- Orchard sprayers

The Parts of a Sprayer

Except for the very basic hand held applicators, which are little more than a small plastic container and hand operated pump, all have similar components which vary only in size and design. They include:

The *tank* is made of non-corrosive material such as polythene or reinforced fibre glass. It has a large filler opening with a well fitting lid, filter basket and provision for draining when necessary.

The *pump* may be hand, engine or power take-off driven, depending on the type of sprayer. All modern pumps are made from corrosion resisting materials and have a relief valve which controls the pump output pressure.

Filters are a vital part of any sprayer. They collect small pieces of dirt and other matter which will quickly block the small holes in the jets or nozzles. Most sprayers have filters in the tank filler opening, pump circuit and nozzles.

The *control unit* varies from a simple tap to a complicated electronic unit which allows the driver to operate all controls from the tractor seat.

The *nozzles* atomise the diluted chemical before it is applied to the soil or plants. There are several designs of nozzle and atomiser which will apply the chemical at different rates and in different spray patterns. The

PLATE 13.1 *Small crop sprayer with a petrol engine driven pump towed by a garden tractor. The hose reel for a hand lance attachment is fitted at the front of the sprayer tank.* (Cooper Pegler)

PLATE 13.2 *Hand held rotary atomiser for applying herbicides.* (Micron)

number of nozzles varies from just one on a knapsack sprayer to, for example, twenty on a tractor mounted crop sprayer with a 10 m spraybar.

Hand Held Applicators

The most basic applicator for horticultural chemicals consists of a small plastic container with a hand lever operated pump built into the cap. It is only suitable for occasional spot treatment with pesticides and fungicides.

Hand held rotary atomisers have a torch battery powered motor which drives a small, high speed spinning disc. Diluted chemical from a plastic container on the applicator handle is supplied to the spinning disc, where centrifugal force created by the disc atomises it into tiny droplets. The operator handles the sprayer in a similar way to a hand lance nozzle so that the chemical from the atomiser disc reaches its target. The chemical is usually gravity fed to the spinning disc through a restrictor valve which controls the rate of flow.

Rotary atomisers have a low water requirement and may be used to apply herbicides, fungicides or insecticides.

Mist Blowers and Dusters

Mist blowers and dusters may be hand held or knapsack, and they are used to apply chemicals as a mist or a fine powder. Some machines can be adjusted to apply chemicals in both mist and powdered forms, whereas others are single purpose machines only capable of applying chemicals as a dust or as a mist.

The knapsack version has a small two stroke engine to drive a fan which, in calm weather conditions, can blow the dust or mist a distance of about 15 m vertically and 17 m horizontally. Mist blowers/dusters can be used in applying insecticides and fungicides to top and soft fruit and to crops under glass or in polythene structures. They can also be used to apply disinfectants in buildings.

Hand held mist blowers and dusters are used mainly in glasshouses. They have a small fan creating an air blast which distributes the chemical atomised by a small rotary atomiser over the foliage or other surfaces. The fan may be driven by a two stroke engine, or by an electric motor powered by either the mains or a battery.

Knapsack Sprayers

The knapsack sprayer is a very common item of horticultural equipment, used to apply herbicides to paths, small areas of weeds, etc., and for spraying insecticides and fungicides on to trees, shrubs and other plant material.

There are three main types of knapsack sprayer: diaphragm and piston pump sprayers are operated with a hand lever; the third is pressure operated. The main components of a knapsack sprayer are a tank, generally

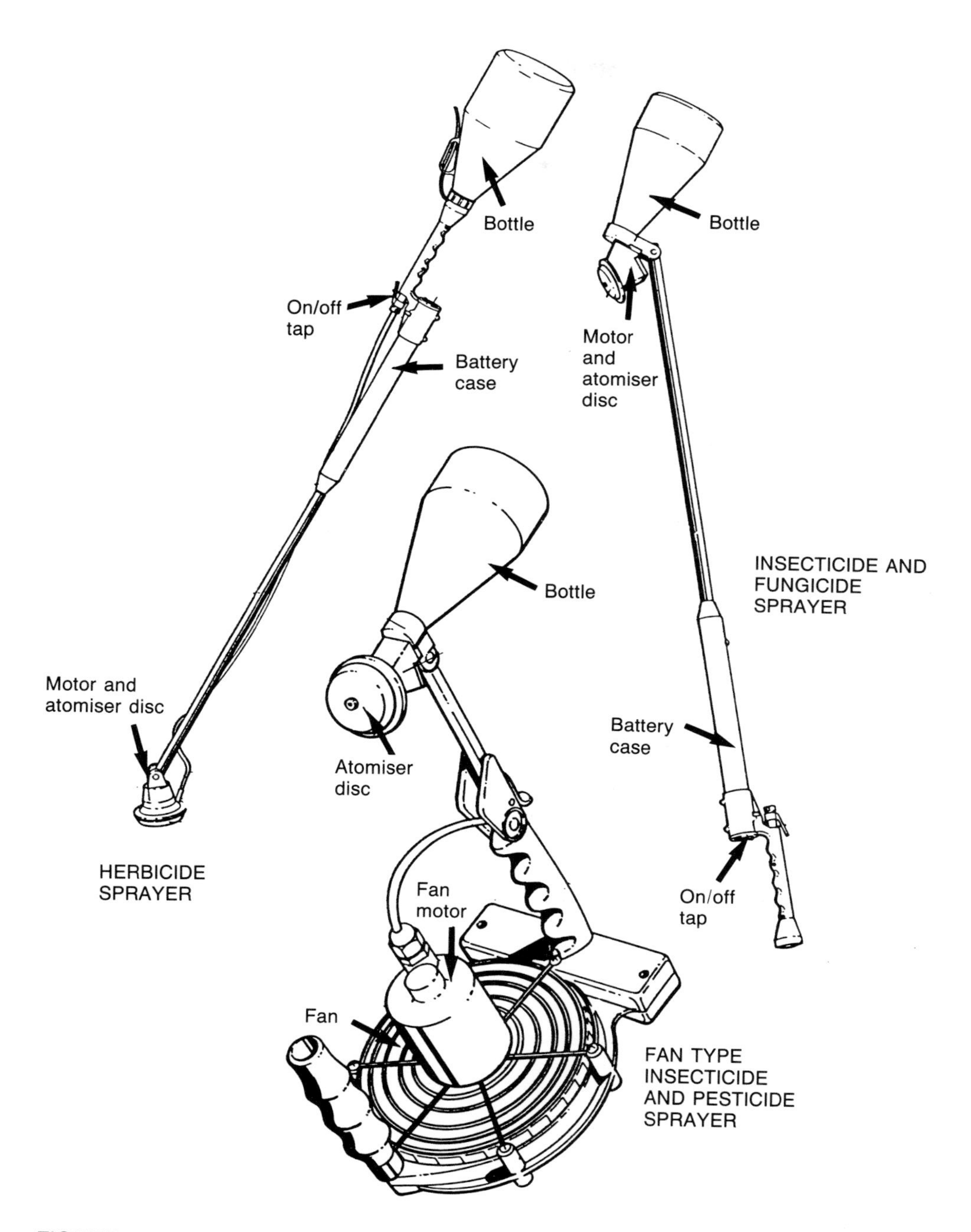

FIGURE 13.1 Types of hand held rotary atomiser. *(Micron)*

PLATE 13.3 *Knapsack sprayer with a small petrol engine driven pump and fan suitable for the application of liquids, dusts and granules. It has a horizontal spraying range of about 11 metres and a vertical range of about 9 metres.* (Cooper Pegler)

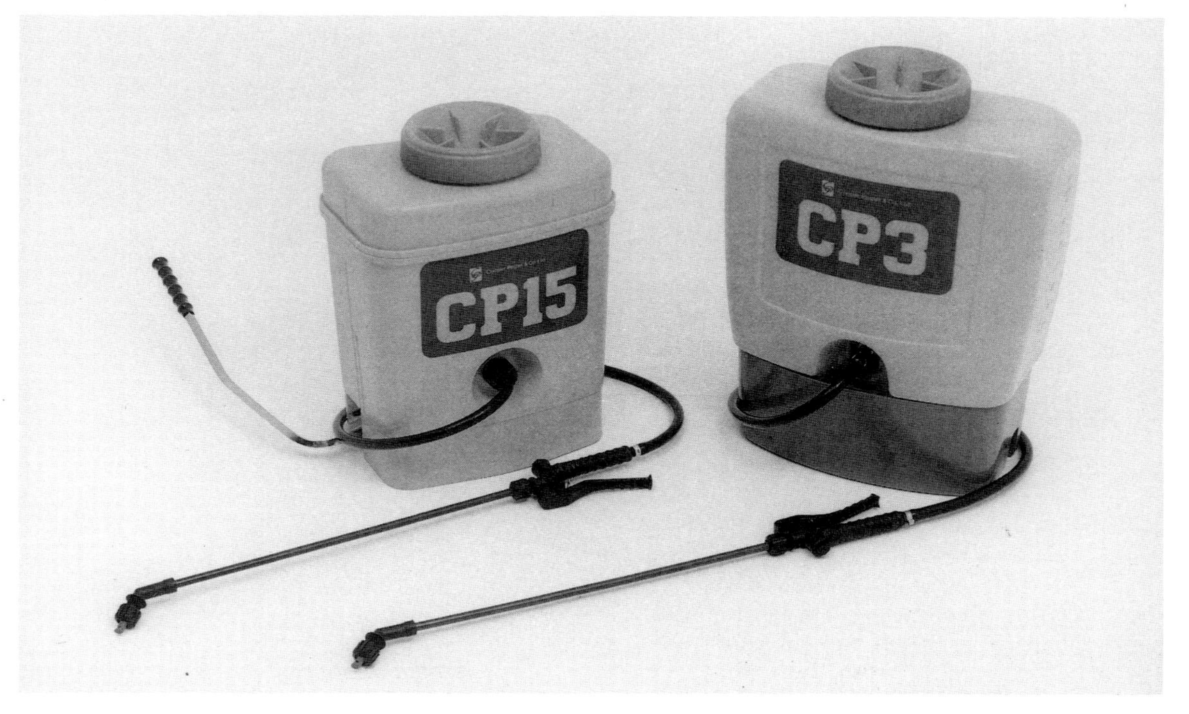

PLATE 13.4 *Knapsack sprayers; both have a lever operated diaphragm pump.* (Cooper Pegler)

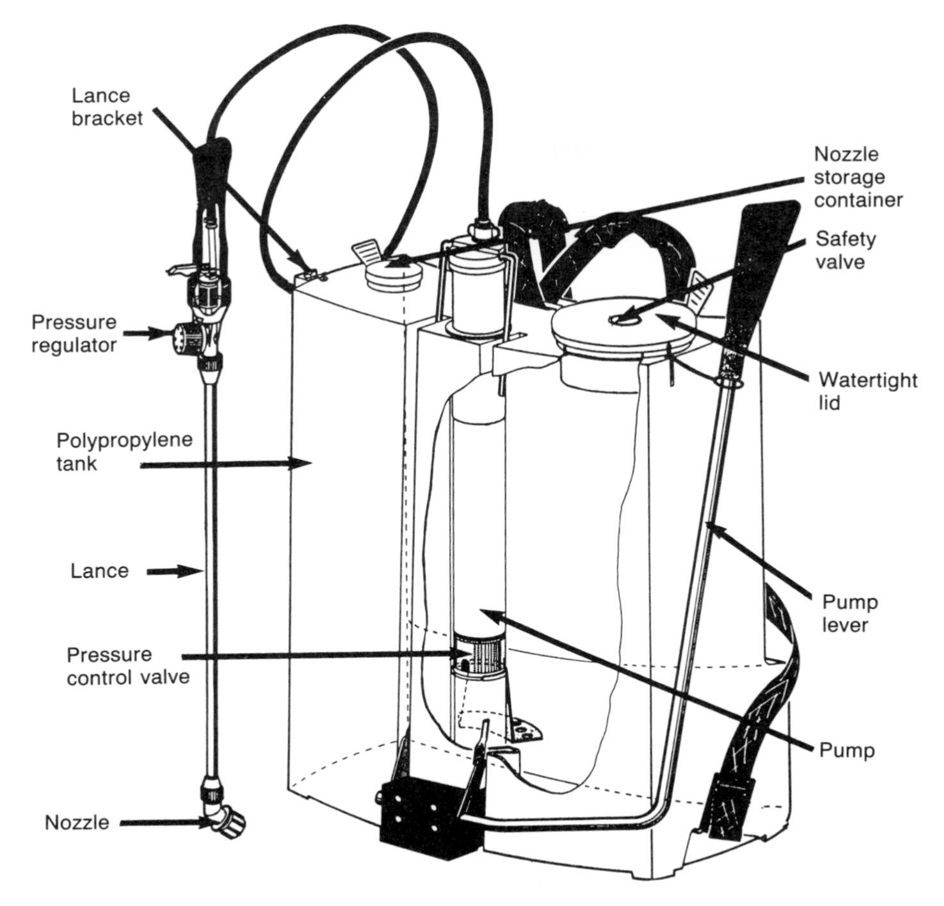

FIGURE 13.2 Piston pump knapsack sprayer. *(Tecnoma)*

made from a plastic material (though older sprayers with galvanised steel tanks are still used), a hand lance and hose, pump, filters, pressure gauge and carrying straps. The filters, usually located in the tank filler, lance handle and nozzle must be cleaned every time the sprayer is used.

Lever operated, piston pump sprayers have a piston and cylinder inside the tank. When the piston is moved upwards by a linkage connected to the hand lever, liquid is drawn from the tank through an inlet valve into the cylinder. On the return stroke of the hand lever, the piston moves downwards and forces the liquid from the cylinder into a pressure chamber. The liquid reduces the air space in the pressure chamber and therefore increases the air pressure. When the hand lance trigger is used to open the release valve, the pressure causes the atomised chemical to be discharged from the nozzle.

Knapsack sprayer piston pumps work at pressures of up to about 5 bar (75 psi) at which very small droplets are formed which are liable to drift. The main problem with piston pumps is a high rate of wear if abrasive chemicals or dirty water is used. For this reason, it is important to keep the filter in the tank filler opening clean at all times.

Lever operated, diaphragm pump sprayers are, with the exception of the pump unit, similar to piston pump knapsack sprayers. The

PLATE 13.5 *A selection of pressure operated knapsack sprayers.* (Lurmark)

pump consists of a cylinder with a diaphragm made of rubber or a similar material at its base. There are two valves arranged so that the outlet valve is pushed open when cylinder pressure increases, and the inlet valve is sucked open when cylinder pressure decreases.

With the hand lever on the upward stroke, the diaphragm is drawn downwards increasing the volume of the pump cylinder. This creates suction which opens the inlet valve allowing liquid to flow from the tank into the cylinder.

The downward stroke of the hand lever moves the diaphragm upwards. This pressurises the liquid in the cylinder, causing the outlet valve to open and release the liquid into the pressure chamber. At the same time the pressure increase in the cylinder closes the inlet valve. The pressure chamber is connected to the hand lance and nozzle. A few strokes of the lever build up pressure in the chamber and when the lance trigger is operated, the atomised chemical is discharged from the nozzle. A relief valve controls operating pressure which is usually maintained with a pumping speed of about thirty strokes per minute.

Compression sprayers have a tank which is partly filled with diluted chemical but has about one third of the volume left as air space. A hand pump built into the tank lid is used to pressurise the contents so that when the lance trigger is operated, atomised chemical is discharged from the nozzle. As spraying proceeds, the pressure will drop and the pump must be used to restore it to the required level. Many compression sprayers have a pressure gauge and

PLATE 13.6 *Small barrow operated sprayer for turf.* (Allen Power Equipment)

PLATE 13.7 *Close up of barrow sprayer showing pressure gauge and low pressure anvil nozzles. The ground wheel driven pump is situated below the tank. The nozzles are protected with a bumper bar.* (Allen Power Equipment)

a pressure regulating valve which maintains a constant output pressure. Some knapsack sprayer tanks have a decompression valve. This should be used to release tank pressure before the lid is removed for cleaning or refilling.

Care must be taken when removing the lid if it does not have a decompression valve. It should be removed very slowly and a face shield must be worn to protect against chemical splashes.

BARROW SRAYERS

Much larger areas can be covered with pedestrian operated barrow sprayers. They range from simple two wheeled machines with a land wheel driven gear pump, up to large four wheeled trolley sprayers with a petrol engine driven diaphragm pump, hand lance and sometimes a short spray-bar. A typical four wheeled barrow sprayer has a 320 litre (70 gal) tank and a 2.25 kW (3 hp) pump. Some barrow sprayers have an electric motor to drive the pump. The single nozzle hand lance has nine metres of delivery hose. Some barrow sprayers have a spraybar, usually about three metres wide, which may be used to treat small areas of rowcrops, turf, etc. Another option is to have a drawbar rather than a handle on the barrow so that the sprayer can be towed by an ATV or small garden tractor.

PLATE 13.8 *Barrow sprayer with a petrol engine driven pump. The hose reel and lance can be replaced with a short spraybar if required.* (Allman)

Nozzles Knapsack and barrow sprayers may have *fan* nozzles, which give a flat fan shaped spray pattern, or *cone* nozzles, which produce a hollow cone spray pattern (see Figure 13.4). A third design is the *anvil* or deflector nozzle which produces a wide angled, flat fan pattern with large sized droplets and little drift.

Hand lances are usually from 600 mm to 1 metre long with a hand controlled trigger and a single nozzle. When large areas have to be sprayed, this can be replaced by a short spraybar with three or four nozzles.

Filters are located in the filler opening, lance handle and often in the nozzle. Some knapsack sprayers may also have a filter in the suction line side of the pump. Filtration is arranged so that the filler opening has a coarse mesh, the lance filter is medium mesh and the nozzle filter is very fine. This ensures that minute particles of dirt do not block the very small hole in the nozzle. Filters must be kept clean to ensure long service from the pump—the filler opening has a filter basket which must always be in position when the tank is filled; the lance filter can be removed for cleaning by unscrewing the lance handgrip; and the nozzle filter is retained by the nozzle cap.

Calibration

It is important to calibrate a knapsack sprayer so that the correct dose of chemical is applied. The factors which affect the application rate are:

- spraying pressure
- size of nozzle
- walking speed of the operator
- height of the nozzle above the target

There are various ways of calibrating a knapsack sprayer, but the following method is most common. First, the sprayer must have the correct size of nozzle and spraying pressure for the required application rate. Information on pressures and nozzles for different application rates will be found in the instruction book.

Before a knapsack or barrow sprayer can be calibrated it is necessary to find out the walking speed of the operator, the width of the spray pattern and nozzle output. Make these checks with clean water in the sprayer tank.

Walking speed (km/h) is calculated by dividing 360 by the time it takes (seconds) to spray over a distance of 100 m. For example if the time required is 90 seconds, walking speed is $360 \div 90 = 4$ km/h.

Spray pattern width is best checked on dry concrete with the nozzle held at the height required when spraying. For this example, assume the spray pattern is 1.6 m wide.

Nozzle output is calculated by spraying water into a measuring jug for one minute. For this example, assume the output is 2 litres/min.

When a barrow sprayer or hand lance has more than one nozzle, the output from all nozzles must be measured and the total output used.

Calculate sprayer output using the following formula:

$$\text{Volume (litres/hectare)} = \frac{600 \times \text{nozzle output (litres/min)}}{\text{spray width (m)} \times \text{speed (km/h)}}$$

If nozzle output = 2 litres/min
spray width = 1.6 m
walking speed = 4 km/h

$$\text{Volume} = \frac{600 \times 2 \text{ litres/min}}{1.6 \text{ m} \times 4 \text{ km/h}}$$

$$= 187 \text{ litres per hectare}$$

One-tenth of this quantity will be applied to an area of 1,000 sq.m.

Tractor Mounted Crop Sprayers

Sprayers mounted on the three point linkage or trailed are used mainly for field scale crop production and turf maintenance. Very large self-propelled sprayers and equally large trailed models are produced, but are not used in horticulture except by large scale vegetable growers.

Mounted sprayers are made in many sizes ranging from small machines with a tank capacity of up to 600 litres, which are ideal for horticultural crops and grass, to very sophisticated machines with a tank capacity of 1,300 litres or more.

A small sprayer with a 200 litre tank and a 6 m spraybar is made for tractors in the 12–15 kW (16–20 hp) range. It has been common practice to classify sprayers according to the quantity of liquid they can apply per hectare (or acre). Most modern chemicals are applied at fairly low rates so most sprayers can handle a wide selection of chemicals.

Ultra low volume application means that as little as 25 litres of diluted chemical is applied per hectare (2 gallons per acre). This offers the advantage of a reduction in the amount of water to be carried, covering more hectares per tankful and cutting the cost of chemicals. These very low application rates are achieved with rotary atomisers of a type

PLATE 13.9 *Lightweight crop sprayer for field crops.* (Lely)

similar to those used with hand held applicators.

Low volume application means up to 250 litres per hectare (20 gallons per acre), which is suitable for most horticultural chemicals. Low volume sprayers are by far the most common tractor operated type used in horticulture.

Medium volume sprayers will apply from 250–500 litres per hectare (20–40 gallons per acre) and *high volume* models up to about 1,200 litres per hectare (100 gallons per acre).

The Parts of a Low Volume Sprayer

The *tank* on all modern sprayers is made from a non-corrosive plastic material, though some sprayers have a galvanised steel tank. For field operations, some growers have a sprayer with an additional tank at the front of the tractor to give extra capacity. A pump is used to transfer the chemical to the main tank.

The *pump* is driven by the power take-off shaft and may be a roller vane, piston or diaphragm type. The pump capacity will vary with the size of sprayer. Typical examples are a roller vane pump on a 200 litre tank sprayer, which will deliver about 50 litres per minute, and a larger 1,000 litre tank model with a diaphragm pump, which has an output of 150 litres per minute.

Sprayer pumps should run at the standard power take-off speed of 540 rpm. It is important to run the pump at a constant speed, so changes in tractor forward speed must be achieved by using the gearbox. Pump operating pressures vary from 1–6 bar or more, but high pressures produce very small droplets from the nozzles which will result in drift. This must be avoided, especially when working in light winds. Spraying should not proceed when there is more than a very light breeze, as neighbouring crops may suffer from scorch.

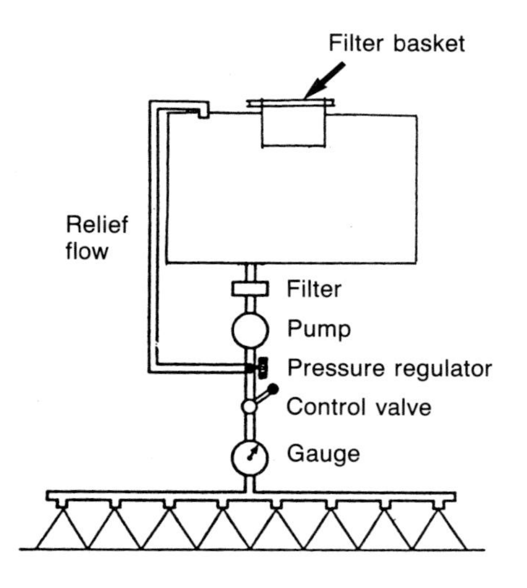

FIGURE 13.3 Basic crop sprayer circuit.

The *relief valve* provides a means of adjusting the pump operating pressure, which is one of the factors determining application rate. The relief valve also allows excess chemical not required at the spraybar to return to the tank. The return flow agitates the tank contents, keeping them well mixed.

Filters are a vital part of any crop sprayer. The small holes in the nozzles will soon block if dirt is allowed into the system. The number and location of filters varies with different sprayers. The first filter is in the tank filler, usually in the form of a basket under the tank lid. Tank contents are next filtered through the suction filter before they reach the pump and many sprayers have another filter on the pressure side of the pump. Many sprayers have a further stage of filtration at the spraybar where each nozzle has a small, fine mesh filter.

Some sprayers have a self-fill hose which uses the sprayer pump to fill the tank from a clean water supply. The self-fill hose also has a filter.

The *spraybar* is made of non-corrosive material, usually plastic, which is supported by a lightweight metal frame. The larger sprayers have three or five sections which are folded for transport either hydraulically or manually. Isolator taps allow the driver to turn off one or more sections of the spraybar when required. This is useful when, for example, treating a strip which is narrower than the full spraybar width.

Spraybar widths of 3–7 m are commonly used in horticulture, but much wider spraybars of up to 24 m or more are used by contractors and large scale vegetable growers. The method of mounting allows the spraybar to be raised or lowered to suit various crop heights.

A spring loaded break-back device is built into the spraybar hinge points, allowing the side sections to swing backwards and thus preventing damage if the end of the spraybar hits an obstruction.

A major problem, especially when spraying on rough ground, is swaying or yawing and bouncing of the spraybar. This alters the position of the spraybar and nozzles in relation to the spray target, causing overdosing or missed strips. The spraybar is attached to some machines by a special linkage which absorbs much of the movement of the tractor on uneven surfaces, keeping the spraybar parallel with the ground.

The *nozzles* or jets are equally spaced across the spraybar, usually at intervals of 500 mm. The nozzle body is made of plastic or metal, whereas the nozzle itself may be brass, plastic or ceramic. The two common types of nozzle give a flat fan or a hollow cone spray pattern. To ensure even application of the chemical, the flat fan nozzle must be fitted to the spraybar so that the spray pattern is exactly parallel with the spraybar.

Many nozzles have an anti-drip device which prevents chemical dripping on to the crop or soil when the machine is turned off. One type consists of a small, spring loaded valve which seals the nozzle outlet when the sprayer is turned off. Another type has a small diaphragm in the nozzle which automatically closes the outlet when the operating pressure falls below a certain level.

More than one set of nozzles is required to give a full range of application rates. Some sprayers have multi-head nozzles which can be rotated when a change of nozzle size is required.

The *control system* is operated either by a hand lever or electrically from the tractor cab.

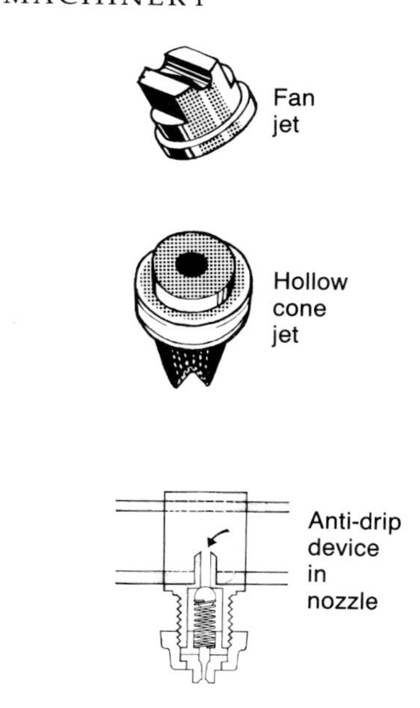

FIGURE 13.4 Types of crop sprayer nozzle.

The control unit has three positions: spray on, spray off and mix. When switched off, the pump draws liquid back from the spraybar to prevent drips. If the control is left in the spray off setting, there is a risk with some sprayers of drawing air into the tank, causing the contents to foam. This is undesirable so, after a few seconds, the control should be

PLATE 13.10 *Five-way nozzle giving instant jet change by rotating the unit.* (Tecnoma)

set in the mix position to allow the pump to circulate and mix the tank contents.

Sprayer Adjustments

Application rate
Three factors affect the application rate of a crop sprayer:

- *Tractor forward speed* A low forward speed will give a high application rate. By halving the forward speed, using the gearbox, the amount of chemical applied will be doubled.
- *Spraying pressure* Higher application rates are achieved by adjusting the relief valve to increase the operating pressure. However, this will reduce droplet size and increase the tendency for the chemical to drift. This must be avoided, as drift can damage nearby crops and it also threatens the environment by harming trees, hedgerows, wild flowers, etc.
- *Nozzle size* A nozzle with a large hole will give a high application rate. Operating pressure and forward speed will give small changes in application rates but for major changes, different nozzles are required.

  To ensure that the application rate remains at a constant level, some sprayers have electronic control systems which compensate for slight changes in forward speed or spraying pressure.

Spraybar height
The spraybar must be parallel with the ground when in work, with the spray pattern meeting just above the target. When spraying weeds which are taller than the crop, the weeds must be considered the target. The height of the nozzles above the target is determined by their spray pattern. Figure 13.5 illustrates how incorrect spraybar height results in missed strips or overdosing. The height of the spraybar is adjusted by moving it up and down on the frame. On small sprayers, this is usually a manual task, but the larger models often have mechanical or hydraulic adjustment.

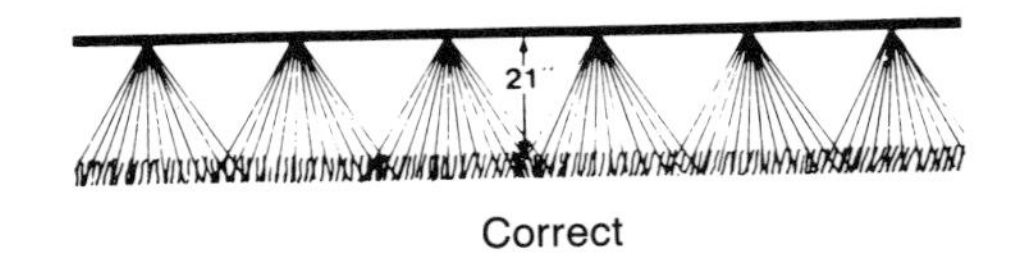

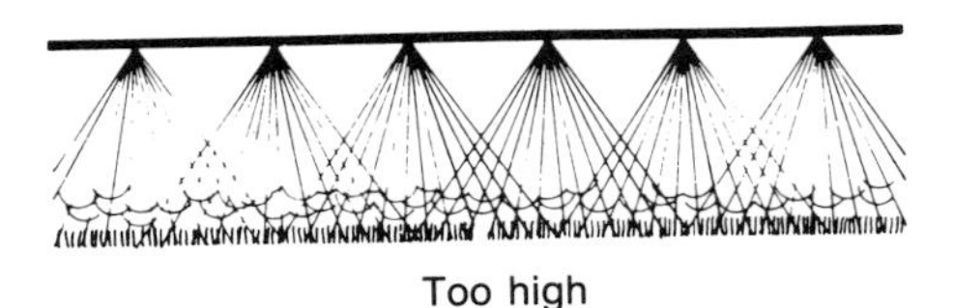

FIGURE 13.5 The effects of incorrect spraybar height.

Chemical Mixing Equipment

The process of mixing chemicals with water is one of the more hazardous parts of the spraying process. The mixing devices fitted to many sprayers make it no longer necessary for the operator to come into close contact with undiluted chemicals.

Some sprayers have a suction probe, operated by the sprayer pump, which will draw chemical from its container into a small calibrated mixing tank on the side of the machine. The chemical is then transferred to the main tank, where it is thoroughly mixed with water by the sprayer pump with the control unit at the mix setting.

Powdered chemicals can be pre-mixed with a small amount of water in a mixing bowl which is fitted to some models of sprayer. When the powder and water are well mixed, the concentrated chemical solution is transferred to the main tank where the sprayer pump completes the mixing cycle.

For large scale spraying, a separate mixing tank may be used. This consists of a large capacity tank on wheels with its own drawbar. It has a mixing tank unit similar to

PLATE 13.11 *Chemical induction device and container washer on a field crop sprayer.* (Vicon)

that found on some sprayers and a power take-off driven pump. The tank is filled with water and the chemical is introduced and mixed in readiness for pumping into the sprayer tank with a flexible hose. The advantage of the system is that the sprayer does not have to return to the yard every time a refill is required, and chemical handling by the tractor driver is minimal.

USING CROP SPRAYERS

Preparation for Use

The sprayer should be checked over before it is required so that any necessary spares can be obtained and repairs completed in good time. When fitting the sprayer at the start of the season:

- Turn the sprayer pump by hand to ensure it has not seized.
- Attach the machine to the hydraulic linkage and connect the pump to the power take-off shaft, making sure the guards are in place.
- Wearing the proper protective clothing, wash the tank, both inside and out, remove the nozzles and flush the machine by pumping a tankful of clean water through it.
- Wash and replace the filters, clean the nozzles and flush the sprayer with another tank of clean water. Check the spray pattern from each nozzle and replace any found faulty. They wear with use and will in time need to be replaced. All nozzles should have the same output. To check this, run the sprayer and collect the water from each nozzle for half a minute in a measuring jug. Replace any nozzles with high or low outputs.
- Finally, check the sprayer for leaks from hose connections.

Spraying

The sprayer should be set to give the required application rate. The instruction book will give the required nozzle size, spraying pressure and tractor forward speed. Sprayer application rate, or calibration, should be checked at the start of the spraying season by either a field test or a static check.

Static calibration is carried out by running the sprayer for a calculated period which is equivalent to the time taken to spray one hectare. The following formula is used to calculate this:

$$\text{Time (mins)} = \frac{600}{\text{spraying width (m)} \times \text{speed (km/h)}}$$

For example, the time taken to spray one hectare at 6 km/h with a spraying width of 10 m will be:

$$\text{Time (mins)} = \frac{600}{10 \times 6} = 10 \text{ minutes}$$

Calibration is carried out by setting the sprayer to the required pressure and fitting the correct nozzles to give the chosen application rate. The tank should be completely filled with clean water. Then with the tractor engine set to give 540 power take-off rpm, run the sprayer for, in this example, ten minutes. Refill the tank to its original level, measuring the amount of water needed. This quantity will be the application rate per hectare at these settings.

Field calibration With this method, the tractor is driven for a calculated distance equivalent to spraying 0.5 hectare using the forward speed, spraying pressure and nozzle size for a chosen application rate.

The following formula can be used to find the distance to travel to cover 0.5 hectare:

$$\text{Distance (m)} = \frac{5000}{\text{spraying width (m)}}$$

The spraying width is calculated by multiplying the number of nozzles by the nozzle spacing. For example, a sprayer with 20 nozzles spaced at 0.5 m has a spraying width of 20 × 0.5, which equals 10 m.

Distance to travel (m) to cover 0.5 hectare =

$$\frac{5000}{10} = 500 \text{ m}$$

With a full tank of water, spray for a distance of 500 m which could, for example, be five trips over a marked distance of 100 m. Refill the tank, carefully measuring the quantity of water required. Double this amount to find the application rate per hectare.

It is important that tractor forward speed is correct in order to achieve an accurate application rate. Some proof meters indicate forward speed in various gears and at various engine speeds, while others have a chart as a guide to forward speed at certain engine speeds. By measuring the time taken for a tractor to travel 100 m in any gear or throttle setting, its speed can be checked with this simple formula:

$$\text{Speed (km/h)} = \frac{360}{\text{time (seconds) to travel 100 m}}$$

To check forward speed:

- Accurately measure a distance of 100 m and mark it with stakes.
- Select the gear which gives the required speed with the engine throttle set at a power take-off speed of 540 rpm.
- Carefully measure the time taken to drive 100 metres. Start some way before the first marker stake, so that the tractor has reached its operating speed when it passes the first stake.
- Finally calculate tractor speed. In this example assume the tractor took 60 seconds to travel 100 m.

$$\text{Speed (km/h)} = \frac{360}{60} = 6 \text{ km/h}$$

Now that the sprayer has been checked and set for the required application rate, the tank should be filled with the correct mix of chemical and water. Instructions on the container must be followed and protective clothing worn as directed on the label. In many situations, spraying may only be carried out by a person who holds a spray operator's certificate.

If the sprayer does not have a mixing device, partly fill the tank with water before adding the chemical. Then add the remaining quantity of water to complete the mix. After filling, agitate the contents by pumping them round the system to achieve thorough mixing.

Once in the field, spray two bouts around the headland and then start on the main part of the field, working from one side to the other. Direction of spraying will depend on the crop, wind direction and shape of the field. Do not spray when there is more than a very light breeze, if there is the slightest risk of the chemical drifting to other crops or neighbouring property.

To help with matching the work, a long length of stout twine can be tied to each end of the spraybar. When turning on the headland the end of the twine will stay in position until the spraybar passes over it on the return bout. Foam marker attachments which deposit blobs of foam from a nozzle at the ends of the spraybar are another method of leaving a guide for the driver on the next bout. When spraying rowcrops the rows themselves provide the guide.

Sprayer Maintenance

Cleaning the sprayer is an important task which must be done when changing chemicals or preparing it for storage. Small residues of chemical left in the sprayer may damage other crops at a later date.

For thorough cleaning, first put on your protective clothing and then:

- Remove the tank drain plug and empty the tank, being careful to dispose of the drained material where it can do no harm to livestock or crops and cannot contaminate streams or other water supplies.
- Scrub the sprayer with a stiff brush, especially the underside of the top of the tank, and rinse the machine with a hosepipe.
- Replace the drain plug, partly fill the sprayer with clean water and add some washing soda. If the last chemical used was oil based, use a liquid detergent (to check this, read the instructions on the chemical container). Pump the cleaning water around the sprayer circuit, spray it out and repeat the process.
- Next, rinse the machine by spraying out two tankfuls of clean water.
- Remove and wash the filters and nozzles in soapy water to ensure they are really clean and then replace them in readiness for the next spraying task.

At the end of the day empty the tank of chemical, and then half fill it with clean water. Clean the sprayer inside and out, and spray out the washing water onto waste land, making sure it cannot damage other crops, harm livestock or contaminate water supplies. Finally remove, clean and replace the filters and nozzles. Remember to wear proper protective clothing while carrying out this work.

At the end of the season remove all traces of chemical from the sprayer. The nozzles should be stored separately in a container after cleaning, and the pump drained to avoid frost damage. Release the relief valve spring pressure and store the sprayer under cover. Check for damage, order spare parts and carry out necessary repairs so that the machine will be ready for the next season.

Band Sprayers

Similar in construction to a basic crop sprayer, a band sprayer may be used either in conjunction with precision seeders or as a separate machine. Although band sprayers are not in common use, they have the advantage of only spraying a narrow strip or band of soil and, in so doing, give considerable savings in chemical costs.

When used with precision seeders, the band sprayer only differs from a crop sprayer by having a single nozzle attached to the rear of each seeder unit instead of a spraybar. These spray a pre-emergence herbicide onto a strip of soil where the young seedlings will appear. Without competition from weeds, the young plants will grow faster and it is possible to see the rows clearly when tractor hoeing the crop for the first time. Carefully calibrated and matched nozzles are needed for band spraying to ensure accurate application of the chemical.

Band sprayers can also be used to apply herbicides, pesticides, etc., to growing rowcrops, but this is not a common practice.

Granule Applicators

Some crop protection chemicals, including herbicides and pesticides, can also be applied in granule form, which has the great advantage of eliminating drift.

The main types of granule applicator include hand held applicators used to treat beds, borders and other small areas of land; small, barrow mounted, hand pushed machines for treating larger areas; and tractor mounted models for field crops.

Hand held granule applicators are used in glasshouses. They have a crank handle which drives a fan to create an air stream. Granules from a small hopper are metered into the air stream which carries them over a distance of up to 3 metres to the ground.

PLATE 13.12 *Two row granule applicator. The feed mechanism is driven from the land wheel.* (Stanhay Webb)

Tractor mounted granule applicators apply chemicals in narrow bands on or in the soil in a position where they will achieve maximum effect. The granules are carried in small hoppers on a toolbar. The hoppers contain a metering mechanism which delivers carefully controlled quantities of chemical through flexible tubes to nozzles, spaced at appropriate intervals across the toolbar, which direct the granules to the ground. It is also possible for the tubes to direct the granules to coulters which place them in narrow bands in the soil. The application rate is altered by varying the rate of flow of granules from the hopper.

ORCHARD SPRAYERS

Orchard sprayers are specialist machines for applying chemicals to top fruit and some soft fruits. The basic principle is to introduce finely atomised chemical into an air blast which carries it into the crop.

The chemical is applied in mist form to trees or bushes by the air blast, which must be strong enough to ensure good penetration through dense foliage to give effective contact of the chemical on both surfaces of the leaves. Because the fruit crop is normally grown in rows 2–4 metres apart, orchard spraying machinery must be compact; at the same time it must be effective when spraying small bushes or trees up to a height of about 6 metres.

Orchard sprayers may be tractor mounted, semi-mounted or trailed, and some specialist machines for large scale production are self-propelled. The larger semi-mounted and trailed sprayers are widely used in top fruit where medium or high volume spraying is carried out.

Tractor mounted sprayers normally have a tank capacity from 200–600 litres with a power take-off driven diaphragm pump mounted under the tank. Pump output is about 50–80 litres per minute with operating pressures up to 30 bar. An axial fan unit, 600–1,000 mm in diameter, at the rear of the machine is also power take-off driven through a gearbox. Mounted sprayers have a tractor power requirement of 15–25 kW (20–33 hp).

Semi-mounted sprayers have the fan and pump unit attached to the three point linkage whilst the tank, with a capacity in the range of 700–1,400 litres, is trailed.

Fully trailed orchard sprayers have a similar tank capacity and are normally power take-off driven, although some have a separate engine. Self-propelled orchard sprayers have a robust chassis with a power unit, tank, fan unit and cab for the operator. The machine is very compact with smooth external surfaces so that it can pass between narrow rows of fruit without damaging the crop.

Air blast sprayers have axial or centrifugal fans which create large volumes of air. An axial fan runs in a cylindrical casing, and the air enters, flows through and leaves the housing in a path which is parallel with the fan shaft. Centrifugal fans have the air inlet at right angles to the fan shaft; this means that the fan blades need to draw in the air and turn it through 90 degrees before it leaves the fan housing.

Orchard sprayers require large volumes of air, an output of up to 1,000 cubic metres per minute being common. The air carries the droplets of spray chemical, produced from a ring of nozzles, into the foliage. The nozzles are arranged in a circular pattern in the fan unit for about 100 degrees to either side of the twelve o'clock position. Most machines

PLATE 13.13 *Air blast orchard sprayer. The power take-off driven pump supplies chemical to jets in the fan housing. The air blast from the fan blows the atomised chemical on to the foliage.* (Tecnoma)

can be adjusted to spray to the left or right or to give complete coverage to both sides.

Ultra low volume orchard sprayers are also made; they have outputs of 50–150 litres per hectare which means that much less water is required. A high pressure stream of air passes through a venturi, at which point a jet introduces the chemical. The air pressure splits the liquid flow into extremely fine droplets. The air carries the finely atomised liquid through ducting to a number of outlets from which it is directed on to the bushes or trees. Some models of ultra low volume sprayer can be converted to dusters which apply finely powdered chemicals.

Many of the chemicals applied to fruit crops require constant agitation while in the spray tank to ensure they remain in solution—totally mixed with water. The chemicals may be mixed mechanically with some form of paddle blade or hydraulically by continuously circulating the contents of the tank.

Safety When Spraying

There are important regulations concerning the safe use of pesticides. Under these regulations, a pesticide means all types of agricultural chemical including herbicides, pesticides, fungicides, plant growth regulators, steriliants and seed dressings.

Hand held equipment, crop sprayers and orchard sprayers are all covered by the Application of Pesticide Regulations. The regulations also specify that all people who handle or apply pesticides must have proof that they have received adequate instruction in their use. A Certificate of Competence is required for anyone using pesticides on land not in their occupancy or in the occupancy of their employer. This, of course, includes anyone working for a spraying contractor, for example at a nursery, sports ground or public park. All persons born after 31 December 1964 require a Certificate of Competence to

PLATE 13.14 *Mist blower for top and soft fruit. This machine applies a fine mist spray at very low volume. The high level outlets or cannons are used to apply finely atomised chemical to the upper part of trees and bushes. The machine also has two bottom outlets or hands, which apply chemical to the underside of the leaves. The mist blower illustrated can have a dust hopper fitted making it possible to apply powdered chemicals.* (Tecnoma)

apply pesticides on their own land or their employer's holding, as well as for carrying out contract spraying for other people.

This information is no more than a basic summary of the requirements of the Application of Pesticide Regulations, which you should study to find out how they affect you when handling or applying the wide range of horticultural pesticides.

Some important points to remember when handling horticultural chemicals include:

- Always read the instructions on the container before opening it, and wear protective clothing as directed on the label.
- Pay special attention to any instructions on the label concerning the action to be taken should you be splashed with the chemical.
- Dispose of the containers safely, and wash them out first to thoroughly dilute remaining traces of chemical.
- Do not transfer chemicals into another container, as other people may not know what you have done.
- Chemicals must be kept in a locked store and recorded in a stock book with their full details and eventual use.
- Never eat, smoke or drink while spraying. Always wash thoroughly before doing any of these things after spraying.
- Never use your mouth to blow at a blocked nozzle in an attempt to clear it. Keep spare, clean nozzles with you as replacements for any which become damaged or blocked. Do not try to clear a blocked nozzle with a pin as this may enlarge the hole, making it useless.

- Clean all protective clothing after use. It is a good idea to wear a boiler suit made of a lightweight fabric with sealed seams designed for applying spray chemicals.
- Remember that spray chemicals can enter tractor cabs. A respirator is needed for some of the more toxic chemicals even when the driver is seated in a modern safety cab.

Suggested Student Activities

1. Look for the different types of chemical application equipment on a horticultural holding. Find out what each type of sprayer is used for.
2. Obtain and study the latest information concerning the safe use of horticultural spray chemicals.
3. Calibrate a knapsack sprayer with clean water only, using the procedure described in this chapter or in the operator's manual.
4. Practice using a knapsack sprayer with clean water only in a decontaminated machine. Aim for constant pumping and walking speeds and keep the nozzle at a constant height.

Safety Check

Make sure you comply with the Application of Pesticide Regulations which require that all users of spray chemicals have proper training and, in many situations, a Certificate of Competence. This applies to all types of sprayer and applicator including knapsack sprayers and hand held equipment. Find out what materials come within the scope of the regulations, and what protective clothing must be worn when handling horticultural chemicals before and after they have been diluted with water.

Chapter 14

GRASS CUTTING MACHINERY

There are many types of grass mower including pedestrian controlled, self-propelled, ride-on and tractor mounted mowers with cylinder, rotary, reciprocating knife or flail cutting mechanisms. Hand held grass trimmers driven by a two stroke engine or an electric motor are used to cut grass in confined spaces, on banks, around the base of fences and trees, etc.

Types of Mower

Hand mowers may have either a cylinder or a rotary cutting mechanism. These small domestic mowers have a cutting width of 250 or 300 mm (10 or 12 in). With the exception of some wheel or roller drive cylinder mowers, they are hand pushed and have the cutting

PLATE 14.1 *Ride-on mower with grass collector.* (John Deere)

PLATE 14.2 *Pedestrian controlled cylinder mower.* (Ransomes, Sims and Jefferies)

mechanism driven by a small mains operated electric motor. As with all mains operated garden equipment it is very important to ensure that the operator is protected by a residual current operated circuit breaker in the mains power supply.

Hand pushed models are designed for domestic lawns, narrow garden paths and similar grass areas where a larger, engine driven model would either be too cumbersome or too costly.

Motor mowers may be petrol, diesel or electric motor driven; with the exception of the narrower cut rotary models, they are self-propelled. Domestic models of cylinder mower have a cutting width of between 300 and 610 mm (12–24 in). The more expen-

PLATE 14.3 *Cylinder mower for cutting long grass.* (Ransomes, Sims and Jefferies)

sive professional models with wider cutting widths are used in large private gardens, parks, playing fields and sports grounds where a much higher workrate is required. A typical example of a large professional cylinder mower with a cutting width of 910 mm (36 in) has a 4 kW (5.5 hp) air cooled petrol engine with electric starting and a trailing seat mounted on a roller.

Some domestic cylinder mowers are driven by an electric motor powered by a 12 volt rechargeable battery. Battery electric mowers are usually limited to the narrower cutting widths. A mains operated battery charger is supplied with the mower.

There are many types and sizes of rotary mower ranging from small, hand pushed models with a two or four stroke engine or a mains electric motor, to the large self-propelled machines designed to cut large areas of grass and be capable of working on rough surfaces.

Domestic rotary mowers may be hand pushed or self-propelled with cutting widths from 410–520 mm (16–21 in). The smaller machines have a petrol engine of about 2.2 kW, and many have a grass collecting box or bag.

Rotary hover mowers are popular with both the home gardener and the professional user. The smaller domestic models with cutting widths from 250–490 mm (10–19 in)

PLATE 14.4 *Self-propelled rotary mower with rear roller drive and a grass collector.* (Hayter)

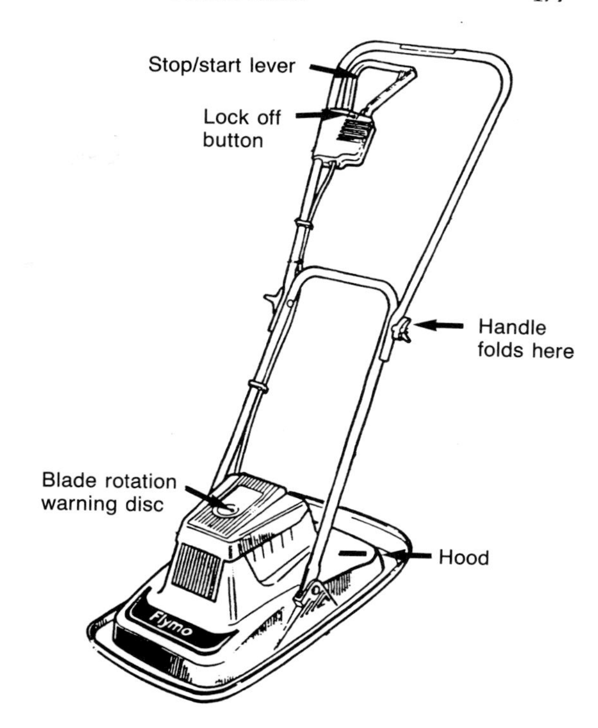

FIGURE 14.1 Electric hover mower. *(Flymo)*

may have an electric motor or a two stroke petrol engine. Professional models with a cutting width of up to 750 mm (30 in) have two or four stroke petrol engines. There are two types of hover mower: one has a grass collection system; the other leaves the grass on the lawn.

Hover mowers are useful for cutting steep banks and other sloping areas of grass. Great care must be taken when using a hover mower, especially when cutting grass on banks and slopes. Never walk backwards with the mower in work, avoid standing on steep slopes and always wear safety footwear with non-slip soles.

Pedestrian controlled flail mowers and reciprocating knife mowers are less common than cylinder and rotary models, but are well suited to cutting areas of rough grass.

Ride-on mowers may have rotary, cylinder or flail cutting mechanisms. Cutting widths and drive arrangements vary from a basic model for the large garden to expensive professional machines with hydrostatic transmission, four

PLATE 14.5 *Lawn tractor with mid-mounted rotary mower. It has four wheel steering allowing it to turn around a 300 mm circle.* (Massey-Ferguson)

wheel steering and hydraulic motor drive to the cutting mechanism, which may be front-, mid- or rear-mounted on the power unit. The ultimate ride-on mower is remote controlled by radio for complete safety when cutting steep banks and slopes. The machine has many built-in safety features to protect it from damage and the operator is completely safe from injury should the mower overturn.

Gang mowers may be self-propelled, towed by a tractor or ATV or mounted on the three point linkage. Self-propelled ride-on models are made in *two* versions: one type is used for cutting large areas of turf on sports grounds, etc.; greens mowers are designed to cut fine turf, especially on golf greens, and they have a grass collecting box for each cutting cylinder.

PLATE 14.6 *Ride-on greens mower with three front- and two mid-mounted hydraulically driven cutting cylinders.* (Toro)

Greens gang mowers are used almost daily to ensure the turf provides a top-class playing surface.

Cylinder Mowers

Pedestrian Controlled Machines

The cylinder cutting mechanism is designed to give a good quality finish to lawns and fine turf. Except for some gang mowers and professional machines, a cylinder mower is only suitable for cutting short grass.

The main parts of a cylinder mower are the cutting cylinder or reel, bottom blade, front roller and a rear roller. The front roller supports the front of the mower and provides a means of adjusting the height of cut. The rear roller, which propels the mower, is usually in two sections which can run at different speeds when turning a corner to provide a form of differential action. This helps to prevent damage to the turf when turning on damp grass.

The cutting cylinder has a number of cutting blades in a spiral formation on circular discs or flanges, fixed to the main shaft. The spiral arrangement gives a cutting action similar to that of a pair of shears, as the cutting blades pass across the bottom blade at high speed.

The number of blades on the cylinder varies according to the work required. Cylinder mowers designed for cutting rather rough grass may have three or four blades. A typical example has a 200 mm diameter, four blade cylinder which gives 46 cuts per metre of travel.

Mowers for domestic lawns and amenity areas have five or six blades on a cylinder usually between 330 and 610 mm (14–24 in) wide. These cylinders make more cuts per metre, typically 70–85, to give a good quality finish to a lawn. For very fine turf, especially on golf greens, a ten or twelve blade cylinder giving 140–160 cuts per metre is required.

Cutting cylinders are chain, gear or vee-belt driven from either the wheels or the rear roller. Motor mowers have a chain drive from the engine to the rear roller, which in turn drives the cutting cylinder. Clutch levers on the mower handles control the drive to roller and cutting cylinder.

The bottom blade is about 5 mm thick and extends across the full width of the cylinder to provide the stationary part of the cutting mechanism. The front edge of the bottom blade is sharpened with a slightly undercut angle similar to the cutting edge of a pair of scissors. The blade is thinner on greens mowers which are designed for golf greens and cricket pitches where a very close cut is required.

Controls

Starting is usually with a cord operated recoil starter after setting the throttle lever between a quarter and half open, making sure the fuel tap is on. The choke must be used when starting an engine from cold. Some mowers have electric starting. The engine is stopped with an ignition cut-out switch on the handlebars or by moving the throttle control lever to the stop position.

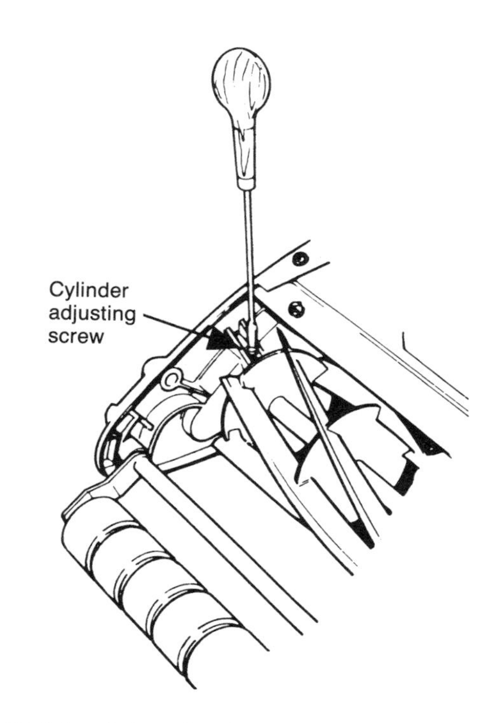

FIGURE 14.2 Adjusting cylinder to bottom clearance.

Cutting cylinder drive is engaged with a clutch control lever on the handlebars. There are various systems: some mowers have a small lever connected by a cable to the clutch unit; others have a dead man's handle which is moved towards the handlebars and must be held there until the operator wants to stop the machine.

Rear roller drive is engaged with a clutch control lever or with the dead man's handle on the handlebars. It is usual for the same control to perform both functions so that the cutting cylinder is running at full speed before the mower moves off. Some models have a throttle operated, centrifugal clutch which engages drive to the rear roller when the engine reaches a pre-set speed, and disengages it when the engine returns to tick-over.

Some cylinder mowers have a centrifugal clutch which engages the drive to the cylinder and rear roller when the engine reaches a certain speed. Machines at the top end of the range may have hydrostatic drive to the roller. This gives a stepless range of forward and reverse speeds controlled by a single lever.

FIGURE 14.3 Testing the clearance between cylinder and bottom blade. The paper should be cut cleanly across the full width of the cutting cylinder.

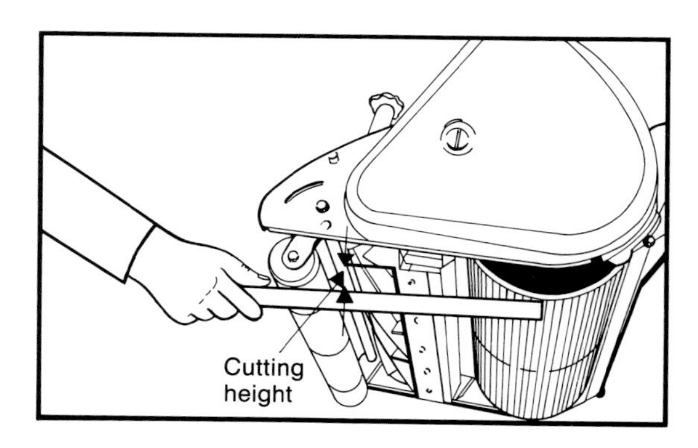

FIGURE 14.4 Use of a straight edge to measure cutting height of a cylinder mower.

Adjustments

The correct setting of the bottom blade in relation to the cutting cylinder is very important. The bottom blade must be set very close to and parallel with the cylinder blades. On some models, a screw adjuster at each end of the cylinder moves it closer to or further from the bottom blade. On others, the cylinder is fixed and the adjusters move the bottom blade. When correctly set, the cylinder blades will cut a piece of paper held against the bottom blade. Blade setting should be checked by using the paper cutting test at intervals along the full length of the bottom blade.

When the bottom blade is set too close to the cylinder, it will be difficult to drive and will cause excessive wear. If the setting is too wide, the mower will not cut properly. Stray stones on the grass will damage the blades. Minor damage can usually be remedied with a carborundum stone or a fine file, but more serious damage may necessitate regrinding of the cylinder and bottom blade, a job for a lawn mower mechanic.

Cutting height is altered by adjusting the position of the front roller, with either a screw handle or a lever. Height of cut can be checked by placing a straight edge under the front roller across to the rear roller, and measuring the distance from the straight edge to the cutting edge of the bottom blade. This check should be made at both sides of the bottom blade to ensure that the blade is not damaged or distorted.

Maintenance

At the start of the season, drain the fuel tank and refill with fresh petrol. Last season's petrol will be stale and may cause problems when starting the engine for the first time.

At the end of the season, change the engine oil, give the mower a thorough clean and, if necessary, have the cylinder and bottom blade resharpened in readiness for the following spring.

Pedestrian controlled mowers should be cleaned after use and stored in a dry shed. During the season the following maintenance is required:

- Check engine oil level each time the mower is used.
- Remove grass clippings and other rubbish from the engine cooling fins. It will be necessary to remove the engine cover occasionally so that the fins can be thoroughly cleaned.
- Remove the air cleaner filter and clean it as directed in the instruction book.
- Lubricate the cutting cylinder and rear roller bearings and also the clutch control linkage.
- Change the engine oil at intervals of 50 hours or as directed in the instruction book.
- At intervals of one month or 200 engine running hours, remove, clean and reset the sparking plug and clean the fuel filter bowl.
- Check the clearance between the cylinder and bottom blade from time to time and adjust if necessary.
- Check drive chain tension and clutch operation, and readjust when necessary.

Gang Mowers

Gang mowers have the same cutting mechanism as pedestrian controlled and ride-on mowers, but the cutting cylinder has a larger diameter and both it and the bottom blade are more robust. Tractor gang mowers may have three, five, seven or nine cylinder units.

Trailed gang mowers may have each unit separately driven by its wheels. Others are power take-off driven, either through a gearbox and drive shaft to each unit, or by an independent hydraulic pump supplying a hydraulic motor on each cutting cylinder shaft.

A typical example of a seven unit, trailed, wheel driven gang mower has a cutting width of 4.9 m, and will cut approximately 2.8 hectares per hour at a working speed of 8 km/h (7 acres at 5 mph). It has six blade cutting cylinders, 760 mm wide, which make 36 cuts per metre of travel. Each cutting unit has a clutch mechanism built into the driving wheel, so that drive can be disengaged for transport. The wheels may be either pneumatic tyred or steel.

Tractor mounted gang mowers with hydraulic motor drive usually have five or seven cutting units. A typical five gang hydraulic drive mower can be set to give between 37 and 88 cuts per metre, while one with a four blade cylinder has a range of 19–43 cuts per metre. Smaller three gang units are used on compact and other small tractors. The hydraulic drive system consists of a power take-off driven hydraulic pump supplied with oil from a reservoir on the mower frame. A hydraulic motor at one end of each cutting cylinder enables the driver to select the best cylinder speed for the conditions at the time. The control unit allows the driver to turn off the outer gangs, or those at one side if required, and also to vary the cutting cylinder speed. The cylinders can also be reversed to clear blockages. By using a combination of cylinder speeds and tractor forward speeds, hydraulic gang mowers can give a range of cuts from a rapid, light trim to a heavy cut.

The gangs can be lifted and lowered hydraulically, and the outer units can be raised to an almost vertical position for transport.

Ride-on gang mowers are built around a four wheeled power unit, and the cutting cylinders are hydraulically lifted and driven. A typical five gang mower has three cutting units mounted at the front and one at either side of the 28 kW power unit. This example has either six or eight blade cylinders with a total cutting width of 3.5 m. Smaller versions suitable for compact tractors have three cut-

PLATE 14.7 *Trailed gang mower with individual land wheel drive to each cutting cylinder. The tractor has grassland tyres with diamond pattern tread.* (Ransomes, Sims and Jefferies)

PLATE 14.8 *Hydraulic motor on gang mower cylinder.* (Ransomes, Sims and Jefferies)

ting units, one on each side and the third either at the front or the rear of the tractor.

Greens gang mowers are very specialist, self-propelled machines with hydrostatic transmission and hydraulic drive to the cutting cylinders. They are used for fast, fine cutting of golf greens and have grass boxes fitted to each cutting unit to collect the clippings. A typical machine has three cylinders with nine cutting blades and is mounted at the front of a 9 kW (12 hp), water cooled, diesel engined power unit on wide grassland tyres. The machine can cut at speeds of up to 8 km/h (5 mph) to give a wide choice of cuts per minute. At 4 km/h the cylinders make 240 cuts per metre, and at 8 km/h the cutting ratio is reduced to 120 cuts per metre.

PLATE 14.9 *Mounted gang mower with hydraulic motor drive.* (Ransomes, Sims and Jefferies)

PLATE 14.10 *Greens triple mower for fine turf.* (John Deere)

Gang Mower Controls

Trailed, wheel driven units have a clutch, operated with a lever, which engages and disengages the drive from the wheel to the cutting cylinder.

Before moving a set of gang mowers to another site, the rear rollers must be lifted clear of the ground and locked in position to prevent damage occurring to the cutting cylinders. The units must be hitched one behind the other in transport position before they can be moved.

Tractor mounted hydraulic driven gang mowers have isolator valves which allow the operator to disengage the drive to individual cylinders, and a main valve to control the complete machine. The units are also lifted hydraulically to a vertical position for transport.

Self-propelled models have the usual driving controls found on other types of power unit. Hydraulic valves are used to control oil flow to the cutting cylinder hydraulic motors. The gangs are also lifted and folded hydraulically. Less expensive models have a manual lever operated lift linkage for the cutting cylinders which are vee-belt driven.

Gang Mower Maintenance and Adjustments

Self-propelled models require similar engine maintenance to that described in Chapter 8. The most important maintenance tasks for trailed and self-propelled gang mowers are described below. For further details, see the instruction book.

- Hydrostatic and hydraulic drive system oil levels should be checked and topped up when necessary with the correct grade of oil. Filters must be changed as directed in the instruction book.
- Cutting cylinder bearings, rear roller bearings, lifting pivots and cutting height adjusters need regular lubrication.
- Clearance between the bottom blade and the cylinder must be checked and reset when necessary.
- Periodic back-grinding or back-lapping is necessary to keep the cutting edges sharp. This is achieved by putting special grinding paste on the blades and then rotating the cylinder backwards to hone the blades and bottom plate. A special tool is available to rotate the cylinder backwards when it requires back-grinding.
- Trailed gang mower units must be hitched so that the cutting unit frames are level, and alternative drawbar hitching positions are provided for this purpose. Cutting height is altered by adjusters on the rear roller of each unit.
- Cutting height adjustment with greens gang mowers needs special care. The gangs are set to a cutting height of 3–19 mm and inaccurate adjustment when cutting very close can ruin the turf and damage the cutters.
- Cutting cylinder speed is directly related to forward speed on wheel driven machines, but with hydraulic motor drive, infinitely variable cylinder speeds are possible on many models at the turn of a handle.

Rotary Mowers

'Walk-behind' rotary mowers, whether hand pushed or self-propelled, are driven by either a petrol engine or an electric motor. Ride-on rotary mowers vary from relatively inexpensive domestic models with a 7.5 kW petrol engine, to large professional machines with a cutting width of at least 5 metres and hydrostatic transmission, driven by a large, 60 kW or more, diesel engine.

The grass is cut by horizontally rotating blades, running at speeds in excess of 3,000 rpm. Some machines have two, three or four equally spaced blades, either swinging or fixed around the circumference of the rotating disc. Others have a hardened metal strip with an integral cutting blade at each end. Although manufacturers advise regular blade sharpening, any keen edge is soon ruined by the stray stone, which, of course, should not be on a lawn. Both types of blade can be sharpened on a grindstone but will eventually need to be replaced.

The cut grass is left on the ground by some rotary mowers, while others have some form of grass collecting system. This may be a simple grass box or bag behind the rotor under the handlebars, or a more sophisticated vacuum collection system on ride-on mowers, with a large plastic bag for the trimmings.

PLATE 14.11 *Pedestrian operated self-propelled rotary mower with electric starting.*

Pedestrian Controlled Mowers

Domestic models of rotary mower have a single cutting unit, either a disc or a bar blade, with a cutting width of 410–540 mm (16–21 in) driven by a petrol engine of approximately 2.25–3.75 kW. There are even smaller mowers, only suitable for small lawns, driven by a 1 kW electric motor. Cutting height can usually be set within a range of 20–75 mm, and is adjusted with a lever which alters the position of the mower wheels in relation to the rotor housing.

Many rotary mowers have a grass collecting bag or box attached under the handle. The operator has the choice of using the collector or putting the clippings on the ground, either at the rear or to one side of the mower. Choice usually depends on the type of grass: clippings will probably be left on a rough area, while it may be preferable to collect them when cutting a lawn.

Self-propelled walk-behind rotary mowers are wheel or roller driven. The roller, similar to that on a cylinder mower, provides a means of propulsion and gives a striped finish to the lawn.

Some pedestrian controlled rotary mowers pulverise the trimmings before putting them back on the ground. These machines, known as mulcher mowers, have a specially shaped rotor hood which retains the cut grass for a longer period so that the blades can pulverise it before it is discharged.

Rotary hover mowers have no wheels. They are supported by a downdraft of air created by an engine driven fan which draws air from outside the rotor blade housing. This creates air pressure on the underside which lifts the mower so that it floats on a cushion

PLATE 14.12 *The underside of a pedestrian controlled rotary mower deck.*
(Ransomes, Sims and Jefferies)

of air just above the surface of the lawn. The hover action allows the operator to move the mower in any direction, useful when cutting grass around irregularly shaped beds, beneath overhanging shrubs and on sloping banks, which would be difficult to cut with a wheeled mower.

Some models of hover mower have a grass collecting box. The grass clippings are lifted by the flow of air under the rotor hood which circulates into the grass box carrying the clippings with it.

Hover mowers have cutting widths ranging from 250 mm for small gardens and narrow paths up to contractor's machines with a 750 mm cut. Some of the larger hover mowers have a pair of wheels which make them easier and safer to use. The smaller models may be electric motor or petrol engine driven, though it is usual for mowers with a wide cutting width to have a petrol engine. Unlike electric models, petrol engined mowers are not limited in their use by the availability of an electric power point.

Controls and Adjustments

Starting is usually with a cord operated recoil starter. Great care must be taken to ensure that your feet are well clear of the rotor housing before attempting to start the engine. Use

PLATE 14.13 *Rotary hover mower with a snorkel air cleaner on the handles to prevent grass trimmings clogging the filter.*
(Allen Power Equipment)

PLATE 14.14 *An electric hover mower suitable for small lawns.* (Flymo)

the procedure described for cylinder mower petrol engines.

The cutting rotor is situated at the lower end of the engine crankshaft. On most machines the rotor will turn whenever the engine is running. Some mowers have a clutch and brake mechanism built into the drive from the crankshaft. This gives added safety since it is possible for the operator to leave the engine running with the blades stationary.

Electric rotary mowers are started and stopped by a single lever on the handle, which has a lock-off button to prevent accidental starting. The mower should be tilted slightly before it is started, and not lowered on to the lawn until the rotor is running at full speed.

Cutting height on most petrol engined rotary mowers is altered by adjusting the wheel setting. Hover mowers have washers between the rotor blade and the engine crankshaft. By increasing or reducing the number of washers above the blade, the cutting height can be changed. This must not be done until the machine has been unplugged or the lead removed from the sparking plug.

Maintenance

Petrol engined rotary mowers will benefit from some simple regular maintenance:

- Check the engine oil level every time the mower is used and add more oil when needed.
- Keep the cooling fins clean. Do not allow grass clippings to clog them or the engine will overheat and this shortens its life.
- Clean the air filter element at frequent intervals; this is very important in dry, dusty conditions.
- Remove the sparking plug at regular intervals, clean off the carbon and reset the gap. This will make the engine easier to start.
- Clean the mower after use. Excessive build-up of grass deposits under the rotor housing may result in clippings being left on the ground rather than being blown into the collecting bag or box.
- Keep the blades in good condition and replace them when worn. Never work on the rotor until you have made sure the engine cannot start.
- The blades can be removed and sharpened on a grindstone, but avoid overheating the cutting edges. Uneven blade sharpening can cause rotor vibration, especially on mowers with small, bolt-on swinging blades. This type of blade must always be replaced in complete sets to maintain rotor balance. It is possible to check the balance of one-piece bar blades by pivoting them on a short length of pipe or rod through the centre bolt hole. If one end is heavy, it will drop to the lowest point.
- At the start of the season, drain any stale petrol from the tank and refill with fresh fuel. Drain and renew the engine oil, and fit a new sparking plug for easy starting.

Electric rotary mowers require less maintenance. However for safety's sake it must not be neglected:

- Check the condition of the cable and plug. Replace damaged cables immediately.
- Check that the residual current operated circuit breaker (RCCD) is working every time the mower is used.
- Ensure that the blades are in good condition. Work on them only after unplugging the machine from the mains.

Ride-on Rotary Mowers

There are various designs of ride-on rotary mower with a front-, mid- or rear-mounted mower deck (cutting unit), with either two or three rotors driven by a power take-off shaft or a vee-belt from the mower engine.

Ride-on mowers for domestic use have a rear-mounted single cylinder petrol engine with a recoil starter. The engine drives the rear wheels and the mid-mounted mower deck (see Plate 1.5). Some ride-on mowers can be used to tow a small trailer.

A typical ride-on mower for a medium sized garden has a 4.5 kW (6 hp) petrol engine, and a gearbox with seven forward and one reverse gears. The single rotor blade mower deck has a cutting width of 660 mm

(26 in). Another example has a 6.7 kW (9 hp) single cylinder petrol engine with electric starting and a single rotor blade with a cutting width of 760 mm (30 in).

The mower deck has a side discharge which can be fitted with a grass collecting system. The cutting rotor creates a vacuum under the hood, which lifts the trimmings up through a delivery tube into a plastic or fine mesh carrying bag suspended at the back of the mower. When full, the bag can be emptied after driving the mower to the compost area or rubbish heap.

Lawn Tractors

Lawn tractors are more powerful and expensive than ride-on mowers. They are suitable for maintaining large gardens and public amenity areas. Lawn tractors have a conventional tractor layout, with a single or two cylinder 7.5–13.5 kW (10–18 hp) petrol or diesel engine with electric starting. Most models have hydrostatic transmission, controlled by a single lever or pedal, to give an infinite range of forward and reverse speeds within a typical forward speed range of 0–9 km/h (5.6 mph) and 0–4 km/h in reverse. Top of the range lawn tractors have four wheel steering which gives a very small turning circle, making it easier to cut grass around trees and small borders.

Lawn tractors have a mid-mounted two blade mower deck with belt or shaft drive, depending on the model. A three blade deck is used on mowers with an extra wide cut. Cutting width varies within a range of 965 mm to 1,170 mm (38–46 in). The deck is supported by wheels or rollers which also provide adjustment of cutting height, usually within a range of 25–100 mm (1–4 in). The mower deck is raised and lowered with a hand lever, except for some professional lawn tractors which have a hydraulic ram.

PLATE 14.15 *Lawn tractor with mid-mounted mower and grass collector.* (Hayter)

A grass collecting system similar to that used for domestic ride-on mowers can be used with lawn tractors. It operates in the same way, by using the air flow from the cutting blades to lift the clippings and blow them into a plastic or mesh carrier, suspended from a frame attached to the back of the tractor.

Compact Tractors

Compact tractors often have a mid-mounted rotary mower attachment, so that, in effect, they become high output ride-on mowers. Some manufacturers sell their low power compact tractors as garden tractors with a mid-mounted mower fitted as standard equipment.

A typical compact tractor has a three blade mower deck with a cutting width of 1.8 m (6 ft). The mower blades are driven by a central power take-off shaft through a transfer gearbox and twin vee-belts. The mower is attached by driving the tractor's front wheels over the deck and securing it to the lift linkage. The driving shaft is then attached to the power take-off. The mower is raised and lowered by the tractor hydraulic lift, and height of cut is adjusted with the support wheels on the mower deck.

A grass collecting unit can be attached to the back of the tractor if required. The tractor can be used for other work, using the drawbar or three point linkage without removing the mower.

Front mowers with a rotary cutting deck are mainly designed for professional use, but there are smaller and less expensive domestic ride-on front mowers suitable for large gardens. Front mowers have considerable advantage in their manoeuvrability and visibility of

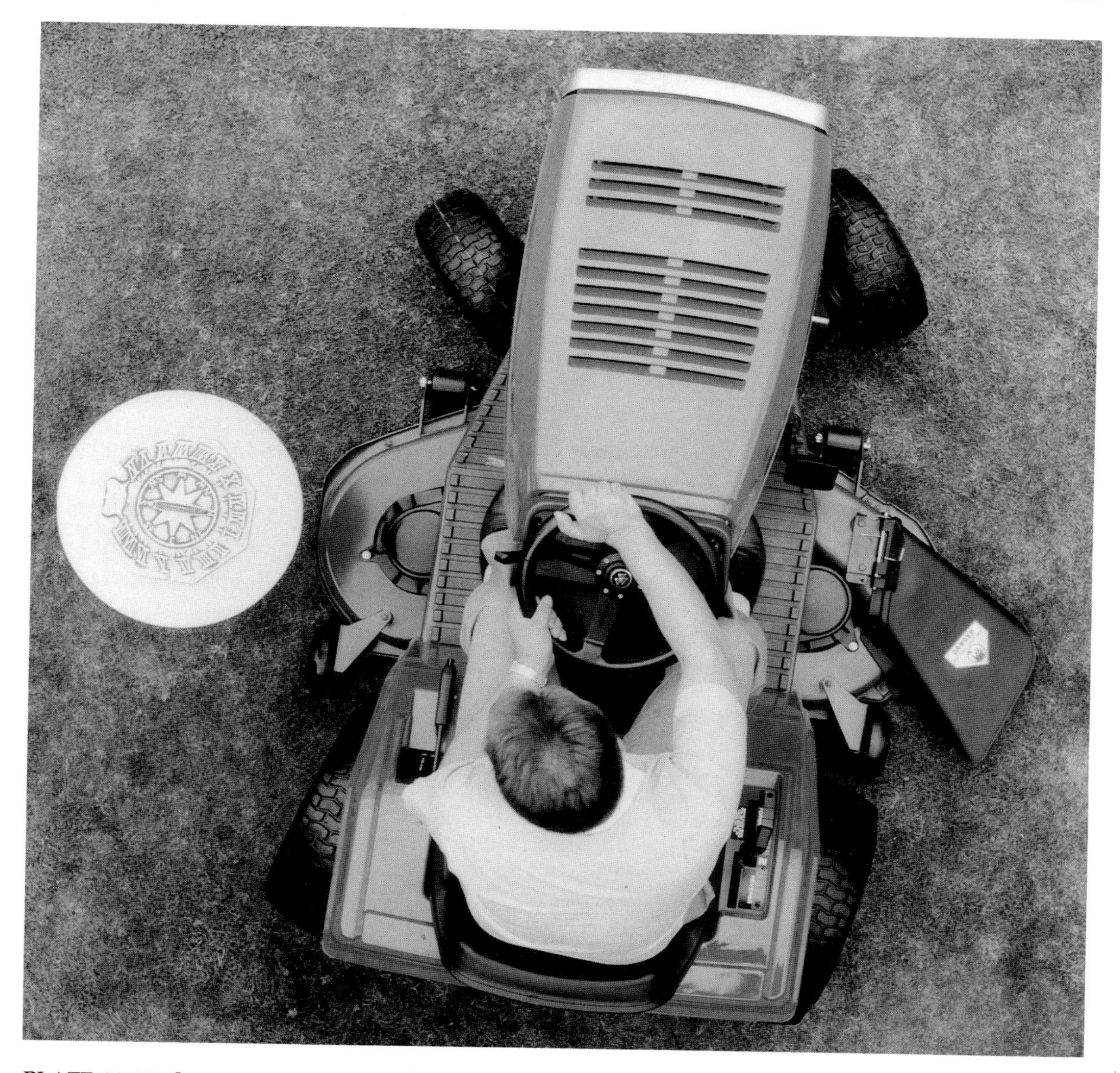

PLATE 14.16 *Lawn tractor with four wheel steering showing its small turning circle.* (Massey-Ferguson)

the grass about to be cut, making them ideal for trimming housing estate verges, parks and amenity areas, golf courses, etc.

Most professional front mowers have a diesel engine in the range of 12–18 kW (16–24 hp), and three rotors giving a width of cut from 1.2–1.8 m (48–72 in) depending on the model. Even wider front mowers are available; a typical machine has five rotors giving a cutting width of 2.25 m (88 in), and a 34 kW (45 hp) water cooled engine.

Almost without exception, front mowers have hydrostatic transmission. Some have four wheel drive. They are rear wheel steered

PLATE 14.17 *Ride-on front rotary mower.*
(John Deere)

with either a power assisted or hydrostatic steering system. Some front mowers have articulated, four wheel steering which gives even greater manoeuvrability. Professional front mowers have a hydraulic lift system to raise and lower the mower deck, which is designed to float so that the blades can follow the ground contours when in work. Cutting height is adjusted with the mower deck wheels which also prevent the blades from scalping the grass.

Front mowers are no less versatile than mid-mounted models. Many have a rear drawbar for light trailer work. Various implements, including a rotary brush, aerator, dozer blade, lawn sweeper and disc edger, can be attached to the front hydraulic linkage. Flail or cylinder cutting units can also be attached to certain models so that one power unit can operate a full range of grass cutting equipment.

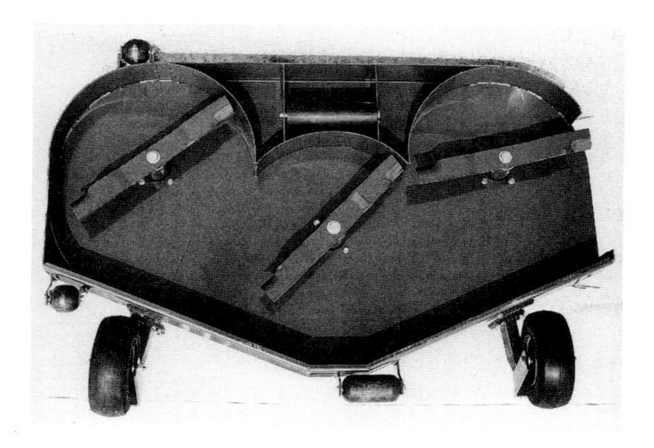

PLATE 14.18 *The underside of a multi-blade rotary mower deck.*
(Ransomes, Sims and Jefferies)

Domestic front mowers are less common. One model has a 10.4 kW petrol engine and two rotors which give a cutting width of 970 mm. They have three wheels, the front two being driven through a hydrostatic transmission system, and rear wheel steering to give the mower a very small turning circle.

Rear-mounted rotary mowers are made in various sizes for compact and other four wheeled tractors. They are mounted on the three point linkage, and the deck is supported by castor wheels or skids. A height limiting chain from the back of the tractor to the mower frame allows it to float over unlevel ground, and at the same time prevents blade damage or scalping. Cutting height is varied by adjusting the setting of the castor wheels. Most tractor mowers have three rotors driven by vee-belts from the power take-off shaft through a gearbox.

Another type of rear-mounted rotary mower has full width rollers at the front and rear to enable the machine to follow the ground contours. The front roller also pushes stones into the turf before they can be struck by the blades. The mower is power take-off driven through a gearbox and twin vee-belts, and it has limiting chains which allow it to float over the ground without damaging either the blades or the turf. Two sizes are made; one has three blades with swinging cutters at each end, giving a cutting width of 1.8 m, and needing an 18–30 kW tractor; the wider model with a 2.3 m cutting width has four blades, and is suitable for a 30–40 kW tractor.

Cutting height is altered by adjusting the position of the front and rear rollers. One particular advantage of this mower is that the rollers allow the mower to cut in both forward and reverse directions which can be an advantage in confined spaces.

ATVs and some small tractors do not have a power take-off shaft or hydraulic lift, but they can pull a trailed, engine driven rotary mower. One example has three blades giving

PLATE 14.19 *Power take-off driven rotary mower on a compact tractor. Height of cut is controlled by the rear castor wheels and the height limiting chains from the mower frame to the tractor.* (Wessex)

PLATE 14.20 *Tractor mounted rotary mower with rollers at the front and back of the mower deck.* (Dowdeswell)

a cutting width of 1.2 m and is driven by a 9 kW petrol engine mounted on the mower frame. It is towed from the vehicle drawbar, and height of cut is varied by raising or lowering the mower wheels.

Maintenance

For trouble-free running, ride-on rotary mowers need regular maintenance.

- Engine servicing must not be neglected, especially the air cleaner, engine and transmission oil levels, tyre pressures and cooling system.
- It is most important to keep the cooling fins on air cooled engines free from grass trimmings so that an unrestricted flow of air will ensure the engine does not overheat. Water cooled engine radiators must also be kept free from grass and other debris. A blocked radiator can cause overheating and will result in engine damage.
- Regular lubrication is required for the power unit and the mower; the tasks set out in the maintenance section in the operator's instruction book are very important and should be completed at the stated intervals.
- Drive belts must be correctly tensioned. When replacing the belts on twin vee-belt drives, both must be renewed to ensure that maximum power is transmitted to the blades. If only one is replaced, the remaining belt will have stretched and will be of little use.
- Blades must be kept in good condition and sharpened at regular intervals. (Use the procedure described in the instruction book for pedestrian type rotary mowers.) When replacing swinging blades, they must always be renewed in sets to keep the rotor in balance. Never attempt to work on the rotor or blades without first ensuring the engine cannot be started accidentally.
- Always clean the machine after use, making sure that all grass has been removed from the collecting bag if used and that the mower deck is clean.

Using Ride-on Rotary Mowers

- The mower engine should only be started when the operator is in the driving seat. Make sure that no other person is near the mower deck when engaging drive.
- The mower deck should be level so that it will give an even surface finish. After checking tyre pressures, set up the mower deck on a hard, level surface by adjusting the wheels until the deck is parallel with the ground.
- Cut grass only when it is dry; wet grass may block the mower and leave a ragged finish. It is better to mow the grass frequently, because short clippings will decay quickly. When cutting tall grass, it is better to cut it in stages, setting the blades high for the first cut and then lowering for the second and even third passes. Alternatively, set the mower to give the finished height of grass but use only half the cutting width.
- When using a grass collector, mow the grass when it is not too long. Drive the mower at a slow forward speed with maximum engine speed and the highest possible cutting height so that there is plenty of air to lift the grass to the collecting bag. Empty the bag frequently, as the mower will not pick up the grass properly if the bag or box is almost full. It is safer to stop the engine before emptying the grass bag.
- Take great care when mowing sloping surfaces—work up and down the slope, never across it. Maintain a slow and steady speed and avoid driving uphill with the grass collector bag almost full as this will make the mower unstable. Front wheel weights will improve stability on slopes, especially when using a grass collector.

Flail Mowers

Flail mowers are useful for cutting coarse or long grass, as they break up the material more thoroughly than is possible with a rotary mower. Flail mowers have free-swinging flail blades around a horizontal rotor which turns

PLATE 14.21 *An engine driven flail mower behind an ATV.* (Wessex)

at about 2,500 rpm. The flails are almost completely enclosed by the rotor hood. The back of the mower is supported by a pair of castor wheels or by a steel roller across the full width of the machine. Height of cut is adjusted with the roller or the castor wheels.

On tractor mounted models, the power take-off shaft drives the flail rotor through a gearbox and vee-belt. Flail mowers have various cutting widths ranging from a 1 m machine with a minimum power requirement of 10.5 kW (14 hp), while a heavy duty mower with 104 flails on a 1.8 m rotor needs at least 34 kW (45 hp). Flail mowers with even wider cutting widths are used by farmers and contractors.

An example of a trailed flail mower for use with an ATV has a single cylinder 8.5 kW (11 hp) petrol engine driving a 1.1 m flail rotor with 60 free-swinging flails.

A grass collecting version of a rear-mounted flail mower is also available with a grass hopper placed above a modified mower hood. The flail rotor has special fan blades at intervals along the shaft to create the extra draught

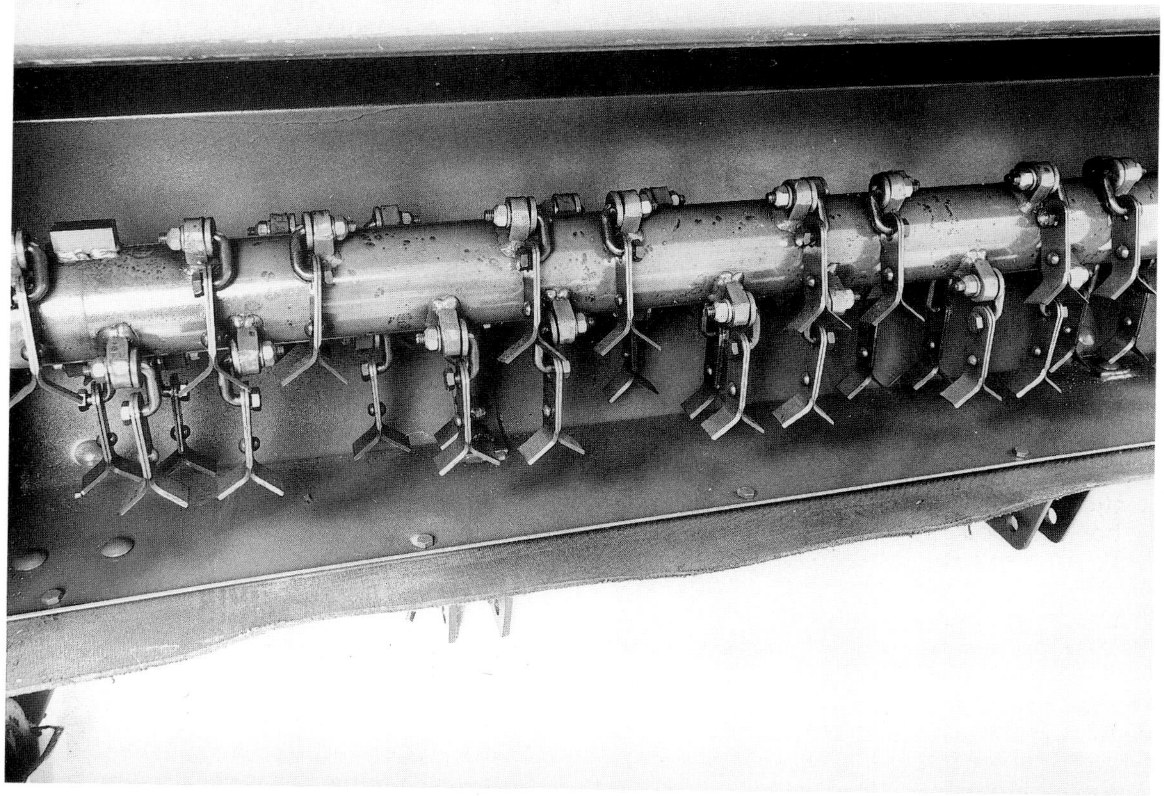

PLATE 14.22 *A flail mower rotor.* (Ransomes, Sims and Jefferies)

FIGURE 14.5 Types of flail for a tractor mounted flail mower. *(Bomford Turner)*

PLATE 14.23 *Rear-mounted flail mower.* (Bomford Turner)

PLATE 14.24 *Flail mower with a large capacity grass collector. The front castor wheels and rear roller control height of cut. The rear door is opened to empty the grass container.* (Wessex)

necessary to blow the grass trimmings into the hopper. When the hopper is full, the grass is discharged through the rear door which is opened hydraulically.

Reciprocating Knife Mowers

A reciprocating knife mower works very well in tall material. It is ideal for cutting vegetation where it is preferable to remove the material after it has been cut rather than converting it to a mulch which is left on the ground. But with the exception of one or two pedestrian controlled machines with a cutting width of about one metre, this type of mounted mower is rarely used.

The cutting mechanism consists of a series of fingers bolted to a cutter bar, with four fingers to each 300 mm of its length. Each finger has a ledger plate which provides a sharp cutting edge. A knife consisting of a number of triangular blades riveted to a narrow metal bar reciprocates through the fingers, and the knife stroke is 75 mm (3 in). The knife is held closely against the ledger plates by knife clips spaced at intervals along the cutter bar.

As the knife moves backwards and forwards across the ledger plates, there is a scissor-cutting action, and the grass or other vegetation is cut quite cleanly, provided the knife is sharp. A typical pedestrian operated knife mower has a small petrol engine which drives a pivoted sway bar which gives approximately one thousand knife strokes per minute. The engine also provides the power to propel the mower forwards, and the operator steers the mower with a pair of handlebars.

To ensure efficient cutting, the knife requires frequent sharpening with either a fine-cut file or a small angle grinder. The cutting mechanism should not be lubricated as the oil will mix with grit, causing rapid wear. The cutter bar must be protected with a safety guard when not in use.

Suggested Student Activities

1. Look for the different types of cylinder mower, taking note of the number of blades and the method of adjusting the

clearance between the cylinder and bottom blade.
2. With the aid of the instruction book, get to know the driving controls on a ride-on mower making sure you know how to stop the engine.
3. Follow the line of drive from the power unit to the mower deck on a rotary mower and find out how the drive belts are tensioned.
4. Using the instruction book, locate the lubrication points and components which require regular maintenance. Find out how often the oil should be renewed in the engine and gearboxes.

SAFETY CHECKS

Take great care when using grass cutting machinery on slopes. With pedestrian machines, avoid standing on steep banks and other sloping surfaces; keep your feet on level ground.

When driving a ride-on mower on slopes, remember to drive up and down or, if necessary, at a slight angle to the slope. If the tyres begin to slip, the slope is too steep or too wet for safe operation, so alter the angle of climb until the tyres regain their traction. Do not drive across a slope as the mower will be unstable.

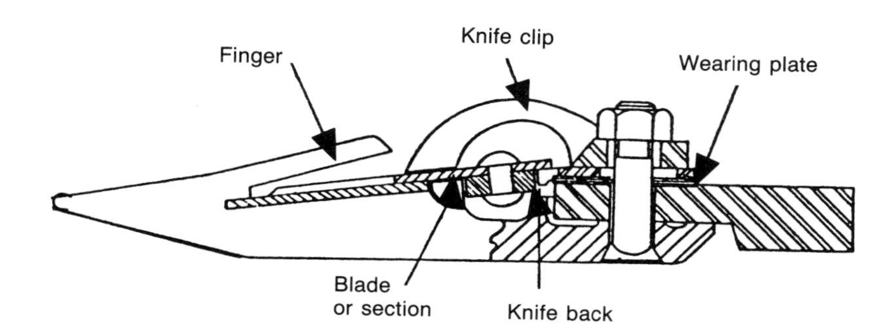

FIGURE 14.6 Cross section of a reciprocating knife cutter bar.

Chapter 15

TURF CARE EQUIPMENT

Intensively used areas of turf require continuous maintenance and repair to keep them in first-class condition. Heavy use on many golf courses, sports grounds and recreational areas causes wear and damage to the turf. Grounds staff work all year round with a range of equipment to keep turf areas well groomed. Some of the machinery suitable for turf care is also used in agriculture and horticulture.

SPIKERS, SLITTERS AND AERATORS

This group of machines is designed to improve turf and minimise the effects of compaction

PLATE 15.1 *Top dressing a sports ground.* (Charterhouse)

PLATE 15.2 *Turf slitter on a compact tractor.* (Kubota)

PLATE 15.4 *Fine slitting drum for trailed sports turf spiker.* (Charterhouse)

caused by constant use in all weathers. Aeration of the root zone increases growth by improving surface drainage and allowing rapid movement of oxygen to the plant roots.

Spikers and Slitters

Slitters and spikers may be tractor mounted, trailed or pedestrian controlled. The smaller self-propelled pedestrian operated machines have a working width of about 1 m. They are powered by an air cooled petrol or diesel engine which also drives the slitter mechanism.

Tractor mounted machines may be up to 2.5 m wide. The weight of the machine and forward movement of the tractor combine to rotate the slitter mechanism.

A slitter consists of a horizontal shaft, supported by heavy duty bearings, with a series of knives bolted on to lugs set at right angles to the shaft. The knives are arranged in a spiral pattern around the main shaft. As the shaft rotates, the knives cut slits in the turf. A well designed slitter with carefully positioned knives has a large underground slitting arc with very little surface disturbance.

PLATE 15.3 *Trailed sports turf spiker.* (Charterhouse)

For the deep slitting action needed on sports fields and similar areas, knives up to 230 mm long, spaced 100 mm apart, are used. On fine turf areas like golf greens, smaller knives 100 mm long, spaced at 40 mm, are widely used. Lightweight machines are frequently used during the summer to prevent the surface sealing and capping. For heavy duty work, large trailed spikers, with a working width of 2.5 m requiring at least 30 kW (40 hp), are used to cover large areas of turf very quickly.

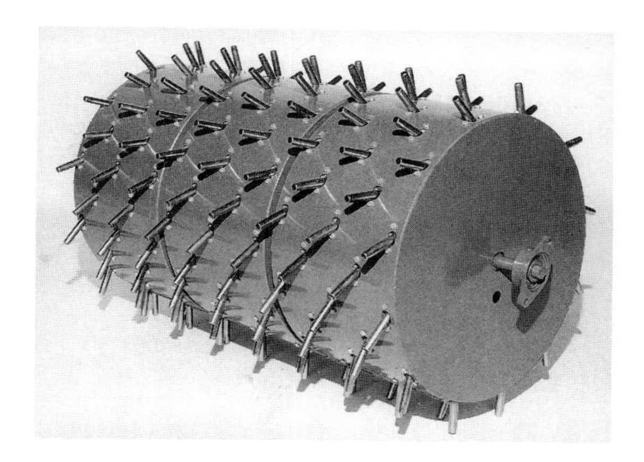

PLATE 15.5 *Hollow tine corer drum. The cores are collected in the drum, and a hinged trapdoor is used to empty the cores from the drum.* (Charterhouse)

Aerators

A type of aerator specially designed for use with fine turf produces round vertical holes. Sometimes the holes are left to close naturally, but it is more common for the aeration process to be followed by top dressing and brushing, which fills the holes with a free draining and open top dressing material. This dramatically improves surface drainage and allows air to reach the root zone.

The holes are produced by pushing a series of solid or hollow tines vertically into the surface of the turf. To minimise surface damage, most aerators are designed so that the tines enter and withdraw vertically.

Pedestrian machines usually have a 5–7 kW engine to propel the aerator and power a hydraulic pump, which supplies oil to a ram which forces the tines into the ground and withdraws them again. Some models have a special linkage which allows the aerator to move forward while the tines are still in the ground. After the ram has withdrawn the tines, the tine frame moves forward before the tines re-enter the turf. Another design has intermittent drive to the wheels so that it only moves forward when the tines have been withdrawn.

Most pedestrian controlled aerators have a working width of about 1 m. The tines are spaced at 50–100 mm (2–4 in), giving approximately 75–100 holes per square metre of turf. The depth of the holes varies from 75–100 mm.

There are three common types of tine:

- *Hollow coring tines* remove a plug of turf as they push into the surface. The cores are usually left on the surface in windrows behind the aerator. A separate pedestrian operated machine can then be used to pick up and elevate the cores into a small trailer or they can be collected by hand.
- *Solid tines* are used to improve drainage where surface smearing is unlikely to occur.
- *Flat bladed tines* are used for general turf improvement and root pruning.

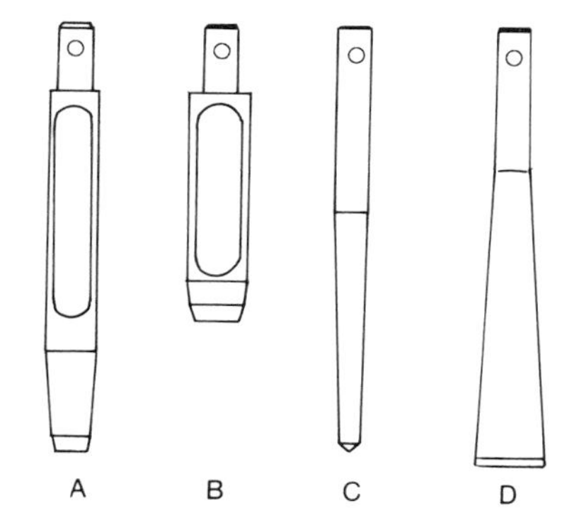

FIGURE 15.1 Types of tine for a turf aerator: A. long hollow tine; B. short hollow tine; C. solid tine; D. flat bladed tine. *(Sisis Equipment)*

Tractor mounted aerators with a power requirement of about 12–20 kW (16–28 hp) are suitable for compact tractors fitted with grassland tyres. Many have pivoted tines attached to a horizontally mounted drum which rolls forward when pulled by the tractor. The drum, about 600 mm in diameter, is hollow. When hollow coring tines are used, the cores fall into the drum and leave a clean, aerated surface. A trap door in the drum is used to empty out the cores at a convenient point away from the area of turf under treatment. The coring tines can work to a depth of 75 mm.

Heavy duty tractor mounted aerators with a power requirement of up to 45 kW, depending on working width, are used to treat large areas of turf. These machines are designed for severely compacted and panned turf, and penetration depths of up to 400 mm can be achieved. The action of heavy duty aerators is not only to pierce round holes or remove cores from the turf, but also to heave the subsoil and break it up to improve drainage and root growth. The heaving action will raise the surface uniformly by 50 mm or more; this is achieved by moving the tines through an arc whilst they are buried in the ground. Having entered the ground vertically, the tines are retracted at an angle of approximately 30 degrees from the vertical.

Heavy duty aerators are power take-off driven, tine movement being controlled by a crank mechanism. Working widths vary from 1 m for an aerator suitable for use with a compact tractor, up to 2.5 m for use with more powerful tractors of 45 kW (60 hp) or more. Tine spacings vary from 50–100 mm.

PLATE 15.6 *Engine driven aerator for spot treatment.* (Sachs-Dolmar)

Aerators for spot treatment

Where small areas of turf are seriously compacted, as at approaches to golf greens and well used grass paths, specialist aeration equipment can be used to deal with the problem.

One type of spot aerator is mounted on a compact tractor and driven by the power take-off shaft. It has a series of drills, similar to metal cutting drills, which make holes in the turf to a depth of 300 mm at 100 mm spacing.

An alternative machine, similar in appearance to the pneumatic road drill, comprises a small, two stroke engine driving a compressor attached to the aerator handles. A single hollow spike is forced into the turf by vibration to a maximum depth of 500 mm. Air is injected from the spike as it penetrates the ground to aerate the loosened subsoil.

Mole Ploughs

The mole plough can be used to deal with severely compacted playing areas, and as an aid to renovating existing turf areas, especially on clay soils. Heavy duty mole

PLATE 15.7 *Turf drainer. The spring loaded rear roller consolidates the turf lifted by the drainer.* (McConnel)

ploughs are used by contractors as part of the drainage process for heavy soils. These machines may either be trailed or mounted on the three point linkage.

A mole plough consists of a very strong frame which slides along the ground when the machine is in work. A heavy leg, similar to one used on a subsoiler, is attached to the frame. A circular section share is attached to the bottom of the mole plough leg. A slightly larger diameter expander bullet is attached by a short length of chain to the back of the circular share. A disc coulter is usually fitted in front of the share to reduce the heaving effect of the mole plough leg on the surface of the soil.

When the mole plough is pulled through the ground, the leg makes a vertical cut to improve drainage, and the circular share forms a tunnel about 75 mm in diameter, rather like a mole run. The expander bullet presses the soil outwards to form a long-lasting drainage channel.

Mole ploughs are normally used in conjunction with a plastic or clay pipe drainage system in clay soils. The working depth of the mole plough will depend on the depth of the plastic or tile drains which can be as deep as 900 mm.

Turf Drainers

When using a mole plough on turf, great care is necessary to avoid surface upheaval and damage. For this reason a turf drainer, a lighter type of mole plough without an expander, is normally used.

Surface damage is minimised by a large diameter disc mounted immediately in front of the drainer leg. This makes a clean vertical cut through the turf so that the drainer leg does not tear the surface of the turf. A

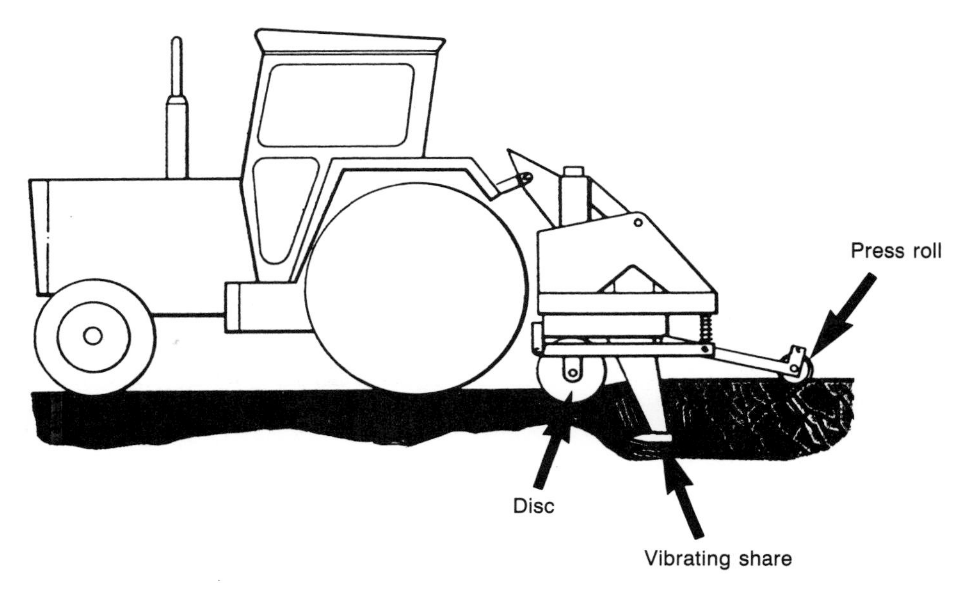

FIGURE 15.2 Tractor mounted turf drainer. *(Twose)*

spring loaded roll is normally fitted behind the drainer to push the cut turf back together and leave the area ready for use.

Turf drainers work at a much shallower depth, normally not more than 200 mm (8 in). Single leg models are made for the smaller compact tractor, and drainers with two or more legs are made for larger tractors, which have a power requirement of 30 kW or more, depending on the number of legs.

A turf drainer is often used in conjunction with other equipment, such as a roll and a slotter, as part of a turf renovation programme.

SLOTTERS

A slotter is used to cut a narrow drainage trench in the turf, which is then backfilled with a free draining material. The slotter unit and backfill hopper are attached to a frame mounted on the three point linkage of a tractor.

A pair of vertical discs, about 50 mm apart, make two parallel cuts in the turf. The discs are followed by a chisel share which lifts out a strip of turf 50 mm wide to a depth of 150 mm or more and then turns it to one side. The slot is filled with a free draining material, usually calcified clay, which falls by gravity from a hopper mounted above the slotter unit.

The turf soon grows back over the slot, leaving permanent drainage channels beneath the surface.

RAKES

Various types of pedestrian controlled and tractor mounted rotary rake are used for turf maintenance and renovation. Rotary rakes scarify or scratch the surface of the turf to remove thatch, which is the dead grass, lateral growths, etc. usually found in well established turf.

Pedestrian operated rotary rakes look like cylinder mowers, but the cutting cylinder is replaced by a thatch removal reel consisting of a series of knives or spring steel tines (fingers) which cut into the surface of the turf. The depth of tine penetration can be adjusted to suit the condition of the turf and the quantity of thatch to be removed.

Rotary rakes are powered by a small air cooled petrol engine. Many are self-propelled but some of the smaller models are pushed by hand. A typical heavy duty rotary rake has a working width of 460 mm (18 in) driven by a 3.7 kW (5 hp) petrol engine. An example of a rotary rake designed for lighter work has a 460 mm reel powered by a 2.2 kW (3 hp) engine.

On some of the smaller machines the reel is driven in the same direction as the forward travel. The removed thatch is then left on the surface behind the rake. Most professional rotary rakes have a contra-rotating reel which gives more efficient thatch removal, but which requires a more powerful engine. Some have a collection box or bag similar to a cylinder type lawn mower which catches the thatch.

Some pedestrian operated rakes designed for fine turf have interchangeable reels that make it possible to vary the severity of treatment. The choice includes a thatch reel with thin section knives, a spring wire rotary brush or a nylon bristle rotary brush.

Three point linkage rotary rakes are designed for use with a compact tractor or specialist turf maintenance power unit of at least 12 kW. A typical machine has a working width of 1.1 m, though wider models are available for larger tractors.

Brushes

Polypropylene, nylon and plastic bristle brushes are widely used to brush in top dressings, remove dew from playing surfaces and renovate hard play surfaces.

Turf brushes are mounted on a three point linkage frame or to a turf care power unit. Most of them used with compact tractors are about 2 m wide. They can also be used on a combination turf care frame where they have the added benefit of providing a striping effect.

Drag Mats and Harrows

Drag mats of very light, chain link mesh are dragged over turf to level the surface and work in top dressings or to remove dew from playing surfaces.

Drag mats may be trailed behind a small tractor or mounted on a combination turf care frame. The frame mounted version is simple to transport from site to site.

PLATE 15.8 *Tractor mounted turf brush. Brackets on the frame can be used to attach other items of turf care equipment.* (Charterhouse)

Lightweight chain harrows are sometimes used for turf maintenance work. Chain harrows without projecting tines have a gentle scarifying action on the turf. The more traditional spiked grass harrows are used on outfields and coarse grass recreation areas where they will remove thatch and generally rejuvenate the grass. However, their action is too aggressive for fine turf.

Rollers

Heavy flat rollers are regularly used to level and firm turf and to stop small stones and worm casts from damaging mowing machinery.

Rollers, usually between 600 mm and 1 m wide, may be hand pulled, pedestrian controlled or ride-on machines. Wider tractor models from at least 2 m wide may either be mounted on the hydraulic linkage or trailed. Weights vary from 250 kg for a small hand roller to 3,500 kg for a trailed model.

Some large rollers are mounted on a tool frame for ease of transport. One example

PLATE 15.9 *Drag mat attached to the same tool frame as the turf brush shown in plate 15.8.* (Charterhouse)

PLATE 15.10 *Heavy ballast roller with rear transport wheels. The roll is lifted from the ground with hydraulic rams and then turned through 90 degrees to give a narrow transport width.* (Charterhouse)

has the roller pivoted centrally under a tool frame with pneumatic tyred wheels at the rear which are raised and lowered with a hydraulic ram. When the wheels are lowered for transport, the roll is lifted clear of the ground and hydraulically turned through 90 degrees to allow it to pass through narrow openings.

Smaller diameter rollers are one of the implements often used on a combination turf care frame particularly to level and firm hard play areas, running tracks, etc.

Single Pass Turf Machines

Specialist toolbar frames, either mounted or trailed, are in common use with compact tractors and turf care power units. The tool frames can have up to four separate implements working in tandem. Numerous combinations of turf care implements can be attached to the frame as the situation demands.

For example, a mounted tool frame with a working width of either 1.8 or 2.45 m could be fitted with a slitter at the front followed by a spring tined finger rake, roller and a nylon bristle brush. A platform above the tool frame is used to carry hand tools and materials.

Much heavier trailed outfits are carried on pneumatic tyred wheels for rapid road transport, and they have a hydraulic lift ram to lift and lower the frame. The working height or depth of each piece of equipment can be adjusted independently with screw jacks. Implement combinations can be varied but are usually similar to those for tractor mounted tool frames.

Turf Edgers

A variety of machines exist for edging turf around borders and along paths, roadways, etc.

For domestic use, there is a hand pushed, roller driven scissor cutter, which has a crank mechanism that moves a blade up and down against a second fixed blade. These mechanical edging shears are only suitable for small gardens. Various electric trimmers

PLATE 15.11 *Trailed tool frame with a slotter, spring tined rake, brush and roller. Each tool can be adjusted independently.* (Sisis Equipment)

PLATE 15.12a *Front-mounted disc edger. The 400 mm diameter disc is hydraulically raised and lowered, the pneumatic tyred wheel providing height control.* (Ransomes, Sims and Jefferies)

are also available to maintain grass edges around borders. One type has a small nylon cord trimmer with a cutting head which can be turned to cut vertically along grass edges. Another design of electric turf edger has vertical rotating plastic blades supported by a roller. The machine is pushed along on its roller close to the edge of the lawn so that the blades can leave a neatly trimmed edge.

Professional turf edgers include pedestrian operated and tractor mounted models. One type has a rotating blade, driven by a small, two stroke engine or battery electric motor. It edges the turf by cutting the grass against a fixed guide blade as the machine is pushed along by a single handle mounted between two rollers. The cutting rotor is mounted at one side of the rollers and is suspended over the edge of the grass.

Disc edgers are used to cut grass which grows over the edge of hard paths and roadways. They can be front-, mid- or rear-mounted on a compact tractor or a turf care power unit.

One design of front-mounted disc edger has a frame with a vertical disc attached to the nearside; the offside of the frame is carried on a rubber tyred wheel. The disc is set at a slight angle to the direction of travel. A share is used to move the material cut by the disc away from the edge of the turf and leave it on the roadway for later collection.

One example of a mid-mounted disc edger has a pivoting arm with an angled disc at one side of an underslung frame. A hydraulic ram which lifts and lowers the disc also transfers some of the weight of the tractor on to the disc. A heavy spring allows the disc to rise and fall as changes occur in the level

PLATE 15.12b *Mid-mounted disc edger. The disc arm linkage allows the disc to move sideways allowing the disc to follow the edge of the kerb.* (McConnel)

of the roadway or path. The angled cutting disc is attached to the frame by a linkage designed to allow some sideways movement. This compensates for steering errors by the driver to ensure that the disc can accurately follow the line of a raised or level kerb. A share turns the cut material away from the edge ready for collection at a later time.

Another disc edger is rear-mounted on the three point linkage with a vertical cutting disc followed by a share which lifts and turns over the cut material.

Top Dressers

Fine turf areas need regular top dressing to maintain healthy growth and good drainage.

PLATE 15.13 *Edging a lawn with an electric nylon cord trimmer.* (Flymo)

Bulky top dressings may be applied to the surface of the turf or after hollow core aerating.

Most top dressers are trailed or semi-mounted with large diameter turf tyres, but some smaller models are mounted on the three point linkage.

Trailed top dressers have a hopper with a hydraulic motor driven floor conveyor, which carries the material to a rotary spreading brush extending the full width of the hopper. The brush which has polypropylene bristles is driven by a hydraulic motor. Application rate is regulated by an adjustable shutter at the back of the hopper where the material is fed to the rotary brush.

PLATE 15.14 *Top dresser mounted on a compact tractor, with a brush and roller feed and a hydraulically driven vibrating unit to stop bridging. It has a spreading width of 1.5 metres.* (McConnel)

PLATE 15.15 *The hydraulic motors which drive the rotary brush and floor conveyor are at the left-hand side of this top dresser. The adjusters for the shutter which controls flow of material from the hopper are above the rotary brush.* (Charterhouse)

A typical top dresser with a spreading width of 1 m has a hopper capacity of about 0.5 cubic metres, and needs a tractor of at least 12 kW (16 hp) and a minimum hydraulic pump output of 14 litres per minute.

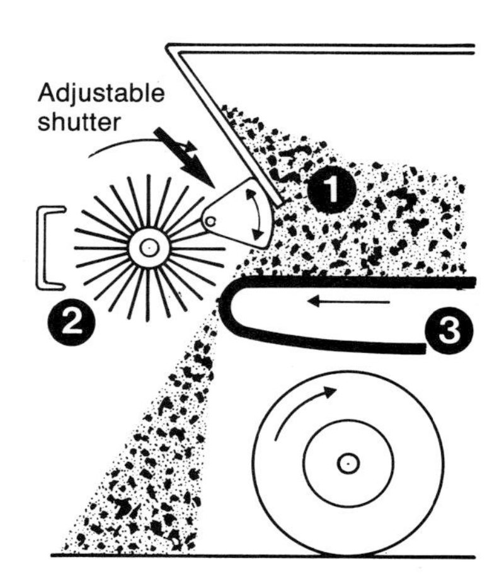

FIGURE 15.3a Moving floor conveyor and rotary brush feed mechanism on trailed top dresser. *(Charterhouse)*

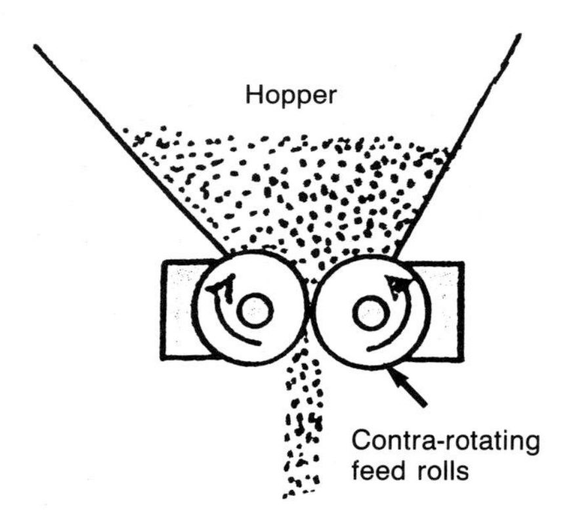

FIGURE 15.3b Roller feed mechanism on small mounted top dresser. *(Sisis Equipment)*

Some small mounted top dressers have a tapering hopper with a land wheel driven rotary brush. An adjustable shutter in the hopper controls the application rate by restricting the flow of the material to the rotary brush. Another design has a roller feed mechanism in the bottom of the hopper which distributes the material across the full width of the machine. Adjustable shutters in the hopper control the application rate.

PLATE 15.16 *This top dresser hopper has been tilted backwards and filled by reversing it into a sand heap.* (Charterhouse)

PLATE 15.17 *The top dresser shown in plate 15.16 spreading sand on a sports ground. The roller feed mechanism in the bottom of the hopper is land wheel driven.* (Charterhouse)

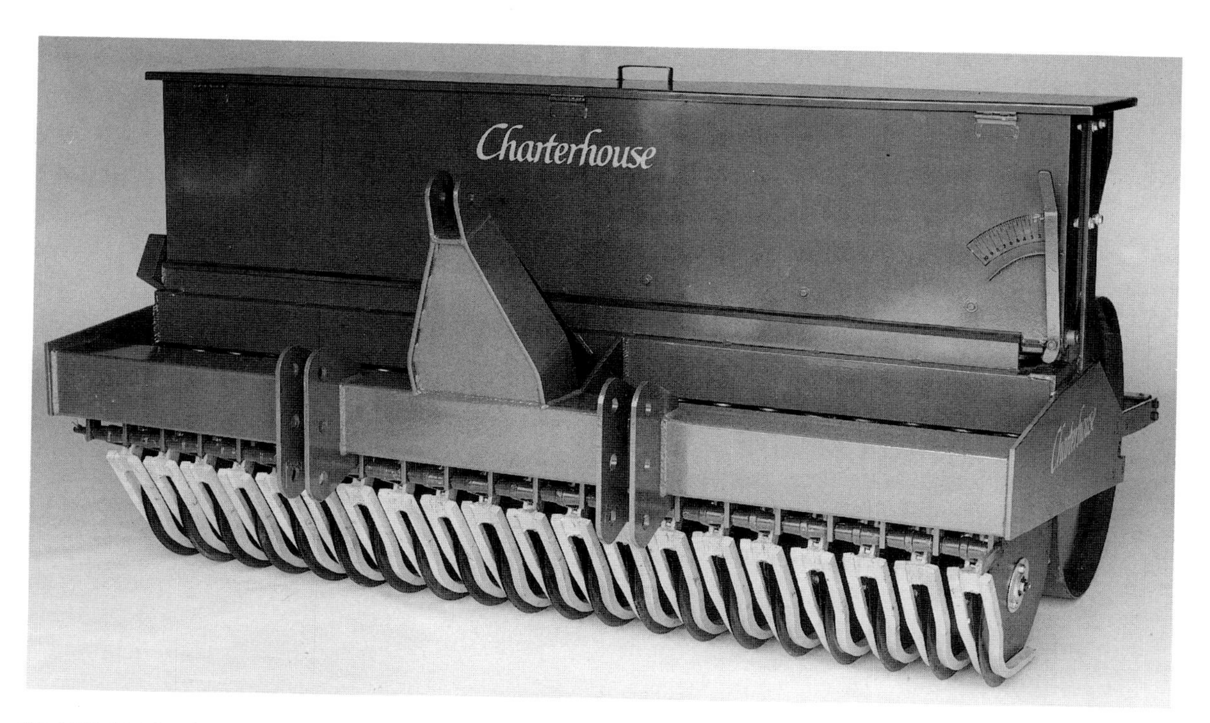

PLATE 15.18 *An overseeder used to sow seed directly into turf. The discs, which can float individually, cut slots for the seed. The roller at the back of the overseeder firms the soil. The roller can be filled with water when working in hard conditions to ensure the discs penetrate the ground to the required depth.* (Charterhouse)

Machines used by contractors and local authorities for large scale top dressing have hopper capacities of up to 5 tonnes. In the bottom of the hopper there is a moving floor conveyor which carries the material to twin contra-rotating spinning discs driven by hydraulic motors. Spreading widths of 10 m can be achieved, and a tractor of 30 kW is required.

Overseeders

Where very severe turf wear has occurred, overseeding may be carried out with either a pedestrian controlled or a tractor mounted machine. Some overseeders scarify and loosen the surface and then deposit seed into scratches or depressions left in the surface. Another type of overseeder has discs which cut slots for the seed in the surface of the turf.

Pedestrian controlled overseeders cut slots into the turf with a series of vertical knives attached to a horizontal rotor. A simple seeder unit with a hopper and a grooved roller to distribute the seed into the slots is mounted above the knife rotor. Some models have a rear roller or other type of coverer to level the soil after seeding. A typical pedestrian operated overseeder has a 3.7 kW (5 hp) air cooled petrol engine. It is propelled by a pair of rubber tyred wheels and has a working width of 460 mm.

Tractor mounted overseeders usually have a series of discs which cut grooves in the turf for the seed. The machine illustrated has discs spaced at 100 mm which can float individually over any unlevel spots on the turf. Sowing depth is limited by the fork shaped skids around each disc. The seed metering unit distributes small quantities of seed to the grooves in the soil, after which a large diameter flat roll at the back of the machine consolidates and levels the surface. Overseeders may be up to 2 m wide, and in good conditions they can operate at speeds of up to 14 kmh (9 mph).

Turf Lifters

Turf lifters may be pedestrian controlled or tractor mounted. The turf is cut by either a pair of vertical discs and an undercutting blade, or by a 'U' shaped blade. A knife swings down at regular intervals to cut each turf to the required length; this is normally 1 m but can be varied on some models. It is usual for the undercutting blade to vibrate as this improves cutting performance in hard ground conditions. Depth of undercutting can be adjusted to suit soil conditions and the type of turf being lifted.

Many pedestrian machines have a front driving roller with an engine above it. The engine, usually about 4.5 kW, also drives the blade vibrator and the knife which cuts the turves to length.

Where small areas of turf are to be lifted, a simple tractor mounted undercutter can be used. Vertical cuts are made by discs with an undercutting blade running beneath the turf. This machine lifts continuous strips of turf, usually about 300 mm wide, which are then cut to length by hand. A typical tractor mounted turf cutter can cut a strip of turf up to 80 m long in one minute.

PLATE 15.19 *Tractor mounted turf cutter with a blade which undercuts a strip of turf 310 mm wide. It is possible to cut a strip 80 metres long in one minute.* (Sisis Equipment)

Line Markers

White lines can be marked with dry powders, with marking powder mixed with water or with an aerosol paint.

Dry powder markers have a wheel driven feed mechanism beneath the powder hopper which delivers a band of powder close to the surface to avoid drifting in the wind. Different line widths, typically 25–100 mm, can be achieved by adjusting a shutter in the bottom of the hopper.

Roller markers are the traditional machine for line marking, using powder mixed with water to form a stiff solution which is transferred to the ground by a roller.

The machine consists of an open topped container or tank supported by two small wheels at the back and a larger diameter marker roller at the front. The marker roller is in contact with an idler wheel which runs in the marking solution in the tank. When the machine is pushed, the idler wheel turns and transfers marking solution to the rim of the marker roller which marks turf. Different line widths can be achieved by using different width marker rollers. An alternative design has a continuous belt which picks up the white line solution from the tank and transfers it to the turf.

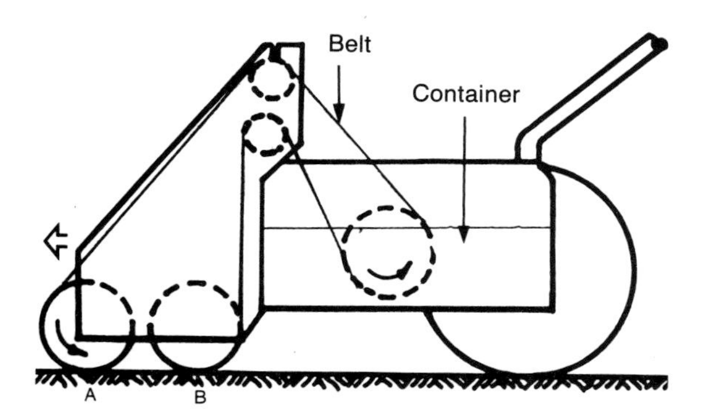

FIGURE 15.4 Belt type line marker. The marker belt is in contact with the surface to be marked between points A and B.

Aerosol markers in many different paint colours are widely used on artificial playing areas, car parks, etc.

A pressurised aerosol canister is attached to the line marker frame on its own wheels. When the marker is pushed along, the aerosol, controlled by a trigger switch on the handlebars, sprays a line on the ground through a special nozzle. Empty aerosol cans must be carefully disposed of and never thrown into a fire.

Spray-on markers can be used with a range of marking materials. Pre-mixed marking solution is put in a small tank and pressurised to 1–3 bar with a hand pump. A simple trigger switch controls the flow of liquid to the nozzle. Spray-on markers may be mounted on a small hand trolley or carried like a knapsack sprayer.

PLATE 15.20 *White line marker. The container is pressurised with the hand pump lever (top) after filling it with marking solution. A trigger on the handlebars is used to release the solution from the nozzle between the front wheels.* (Supaturf Products)

MAINTENANCE OF TURF CARE EQUIPMENT

- Regular engine maintenance must be completed according to the instruction book for all pedestrian controlled turf equipment. Power take-off shafts and guards must be checked for damage and wear, and broken guards should be replaced immediately.
- Mechanical linkages should be checked for correct operation and lubricated where necessary. Knives and blades should be inspected for tightness and damage.
- Check hydraulic systems and hoses for leaks and inspect oil reservoir levels.
- Worn blades and knives should be replaced in complete sets. It is not advisable to replace an odd blade as this may upset the balance of the rotor and result in severe vibration.
- When turf care machinery is used on hard ground, nuts and bolts may work loose through vibration. For this reason, it is important to check them for tightness at frequent intervals.
- Machines should be carefully cleaned after use. Soil working parts should be protected with a rust preventing solution during long term storage.
- It is usual practice to give turf care machinery an annual overhaul during the winter months. This involves a complete check of all working parts, belt, chain and gear drives, blades and bearings. Engines should be serviced and damaged guards renewed. This work is best done as soon as the season is over because it may take some time to obtain the necessary replacement parts.

Considerable mechanical damage can result if care is not taken when using turf machinery, especially near roadways, pavements and kerbing. The equipment should be set in the transport position when it is moved from one site to another. Many of the wider machines have pneumatic tyred transport wheels which make it possible to transport the equipment along the road at high speed.

SUGGESTED STUDENT ACTIVITIES

1. Look for and identify as many turf care machines as possible when you visit sports grounds and amenity areas.
2. Study the effect of the different types of aeration tines and blades on sports turf.
3. Find out how to adjust the feed mechanism on various types of top dresser.

SAFETY CHECKS

Always disconnect the drive and stop the engine before clearing blockages in turf maintenance machinery. Remember that some machines will continue to run after the drive has been disengaged.

Never work on or under a raised machine, especially if it is operated by hydraulics; always use an axle stand or similar support.

Chapter 16

GLASSHOUSE EQUIPMENT

The glasshouse is the only area of horticultural production in which it is possible to control most aspects of the environment. Modern computerised systems provide the nurseryman with the ability to control temperature, humidity, light intensity and irrigation programmes.

Glasshouse Ventilation

Ventilation of glasshouses and plastic structures is necessary to:

- control temperature and avoid overheating.
- regulate the humidity.
- restore levels of carbon dioxide.

Where possible, natural ventilation is used by fitting ventilators near the ridge of the glasshouse which can be opened and closed when required. As warm air rises through convection, some of it will escape through the opened ventilators, and at the same time cooler air is drawn inside.

Where ridge ventilation is not possible, for example in plastic covered tunnels, forced ventilation using electric fans may be necessary. However, fan ventilation systems are costly to run and tend to be noisy in operation.

There are two main systems used for roof ventilation. The traditional ridge system employs a continuous run of glazed panels hinged at the ridge. The opening panels extend the full length of the glasshouse and are often found on both sides of the ridge. The ventilators are opened and closed by a series of toothed racks fitted to the ventilator panels, which engage with pinion gears fitted to a shaft running the full length of the building. In the vast majority of glasshouses, the pinion shaft is driven by a reversible electric motor, though some are chain operated.

The advantage of a continuous run of ventilator panels on each side of the ridge is that ventilation is usually possible no matter which direction the wind is blowing. Normally, the run of ventilators on the leeward side—away from the wind—would be opened.

A second type of ventilator is often found on Venlo or Dutch Light glasshouses, with glass panels extending from the ridge to the gutter, are used to provide ventilation. The ventilators are hinged at the ridge and can be lifted open at gutter level. It is common for one panel in four to be hinged. On some houses, the ventilators are opened by a series of cables with either large springs or in some cases counter balance weights to help close

PLATE 16.1 *An oil fired boiler installation.* (John Blyth)

PLATE 16.2a *Rack and pinion mechanism for opening glasshouse ventilators. The straight row of teeth (rack) is moved up and down, to open and close the ventilators, by a small gear (pinion) on a shaft driven by an electric motor.* (John Blyth)

PLATE 16.2b *Electric motor and gearing for opening the glasshouse ventilator system shown in plate 16.2a. The shaft (top right) carries the pinions which raise and lower the ventilator racks. The pulley at the right can have an endless chain around it to allow manual operation at ground level.* (John Blyth)

them. A system of rods and levers is an alternative method of controlling the ventilators.

Most modern ventilator systems are automatically controlled using electric motors and gearing. For efficient ventilation, the total area of ventilating panels in a glasshouse should be approximately 20 per cent of its floor area.

Aspirated Screens

Most nurseries have automatically controlled glasshouse ventilation systems. A simple system involves a sensitive thermostat set in an aspirated screen which sends accurate temperature information to the control unit. This switches on an electric motor which opens or closes the ventilators in small steps or stages. These small movements ensure that there are no sudden violent changes of temperature inside the glasshouse.

The aspirated screen is an important part of a glasshouse control system. It consists of an insulated box or tube open at both ends. The screen is often made of plastic, or sometimes aluminium, and lined with polystyrene for insulation. This ensures that the thermostat sensor placed in the screen is not affected by the heat of the sun and cannot receive heat from the metal parts of the glasshouse. To further increase the sensitivity of the aspirated screen, it has a small fan which draws a continuous stream of air through the tube. This constant flow of air over the sensor allows it to identify very slight changes in temperature.

Fan Ventilation

In some circumstances when, for example, there is insufficient provision for natural ventilation, or in plastic structures without roof ventilators, fan ventilators can be used. The normal arrangement is to have a series of large, low speed fans, up to one metre in diameter, fixed in either the sides or the gable ends of the structure. The fan outlets are covered by louvres; they only open when the fans are running, thus avoiding draughts.

Fans are installed to extract air from the inside of the glasshouse or tunnel. This is replaced by fresh air which enters through a series of hinged ventilators on the opposite side of the structure to the fan units. The

PLATE 16.3 *Aspirated screen showing the sensors which are connected to the glasshouse control systems.* (John Blyth)

ventilators usually have automatic controls similar to those for rack and pinion roof ventilators. The air speed in the glasshouse must be limited to a maximum of about 1.5 metres per second to avoid draughts.

Fan ventilation systems are expensive to run, and a substantial electricity supply backed up by a standby generator is essential. A typical fan ventilation system will need between 30 and 60 kW per hectare, depending on the shape, layout and planned cropping in the structure.

Ventilator Maintenance

- Electric motors used to operate the ventilator gear or fans will be the enclosed type with sealed bearings. The external surfaces should be kept clean and cooling air inlets dust free.
- The control unit requires no maintenance. When servicing is required, a specialist engineer should be called.
- The ventilator operating gear requires some maintenance as it works in a rather corrosive atmosphere. Gearboxes should be checked and lubricated annually. Chain drives should be lubricated with graphite grease. Rack and pinion mechanisms should have the old grease scraped off once a year, as the lubricant tends to harden at high temperatures. After cleaning, the moving parts should be coated with a high melting point grease.

Glasshouse Heating Equipment

Many glasshouses are heated to allow continuous production throughout the year. Heating systems developed in the middle of the nineteenth century relied on supplies of cheap coal and a simple boiler which heated and circulated hot water through large diameter cast iron pipes. Water circulation relied on the thermo-syphon effect of hot water rising and passing through the top run of pipes and returning to the boiler through the lower ones.

Heating Systems

Glasshouses can be heated by warming the *air*, by warming the *soil* or by a combination of both methods. Three heating systems are in common use:
Warm air heaters These burn fuel to provide hot gases which are passed to a heat exchanger. This allows the hot gases from the burnt fuel to transfer much of their heat into the glasshouse without actually coming into contact with the glasshouse atmosphere. The waste gases from the heat exchanger pass through a chimney and into the atmosphere. The air warmed by the heat exchanger is circulated within the glasshouse by a fan unit in the heater cabinet. To aid even distribution, the warm air is sometimes directed through metal or polythene ducts at ground level or overhead.

Hot water heating Modern hot water heating systems use a variety of fuels to heat the water, which is circulated with the assistance of a pump through pipework installed in the glasshouse. The pipes are usually fixed to the structure at a height of between 300 mm and 1 m above the surface of the soil. The majority of heat transfer is through radiation, but careful siting of the pipework can also produce beneficial convection currents which help to produce an even distribution of heat. Hot water heating is generally used where higher temperatures are necessary.

Under floor heating Two basic systems are in use. Soil warming cables powered by electricity have the advantage of a thermostatic unit which gives accurate temperature control. The second system employs small diameter hot water pipes placed either in the soil or in a special design of polystyrene tray and covered with felt matting.

Under floor systems are designed to warm up the growing medium and promote rapid root growth, and they may be used with or without another method of heating. Under floor heating is mainly used for propagation of plant material and the production of house plants in pots.

Fuels for Heating

Oil, gas, coal and electricity are the most commonly used fuels for glasshouse heating. Each has its own particular advantages.

Coal Many large nurseries, especially those near major coalfields, use coal since they benefit from low transport costs for this bulky fuel. Coal can provide an economic form of heating provided the area to be heated is large enough to justify investment in automatic coal handling equipment.

Coal fired boilers can be fully automated. They use specially prepared graded fuels which can either be fed into the boiler by a conveyor or small grains of coal can be blown into the boiler by a fan unit. The coal is delivered in bulk and stored in hoppers or silos. A considerable storage capacity should be provided to allow for irregular deliveries, transport difficulties, etc.

Oil The most widely used grade of oil is gas oil or 35 second oil, commonly known as diesel oil. It is convenient to store and can be burnt cleanly with relatively simple pressure jet burners. Storage tanks and fuel pipes do not have to be heated in temperate climates, but as this type of oil is liable to thicken in very cold weather, many companies supply a winter grade oil which has an anti-thickening additive.

Oil supplies are generally reliable so only limited storage is necessary; most growers aim for at least two to four weeks' storage capacity. Gas oil is used for a wide range of heating equipment including both warm air and hot water boilers.

Gas Mains gas, where available, provides a very economic form of heating. It cannot be stored by horticulturalists but piped supplies are very reliable. It is a clean fuel, generally producing less soot (carbon) and sulphur dioxide than oil or coal.

Gas burners are more complex than those used for burning liquids like oil, because it is necessary to purge (remove) any traces of gas from the combustion chamber when the boiler cuts off or is about to start up. This is done by an electric fan which blows air through the chamber. Some gas boilers can also burn oil if the gas supply should fail, and these are known as dual fuel boilers.

Liquid petroleum gas (LPG) is an alternative where mains gas is not available. Propane gas can be purchased in cylinders, or for larger installations storage tanks are filled by road tanker deliveries. Propane is often used to provide an inexpensive mobile heating system for glasshouses and plastic tunnel structures. Direct fired mobile convector heaters can also be used for frost protection, avoiding the need for expensive permanent heating installations. LPG is also used for temperature controlled growing with warm air heaters or overhead radiant tube heaters.

A major advantage of propane is that it introduces carbon dioxide when burnt efficiently. Many growers use propane heaters as carbon dioxide generators in their glasshouses.

Electricity is clean and efficient but often rather costly compared with fossil fuels.

Electricity is used in horticulture chiefly in the propagation of plant material. Electric soil heating cables are in common use to warm propagating beds, mist and fog propagation units and beds for the growing-on of young plants. In some glasshouses, heating cables are used to warm both soil and air in closed propagating cases. Electric fan heaters are used in potting and work areas and will be found in some small leisure glasshouses.

Particular care must be taken when installing and using electricity in horticulture. The installation must be able to withstand high humidity and water splashing, and all components must be properly earthed. Many electric heating cables and heating foil pads now available are designed to operate on a 24 volt supply, which considerably reduces the risk of electric shock. A transformer is used to reduce mains voltage down to 24 volts.

Other fuels are used to a limited extent for heating purposes. Most have a lower heat value than coal, gas or gas oil but may well be cheaper to buy. Such fuels as straw, heavy bunker oils, 28 second oil (kerosene), wood chippings and wood off-cuts have limited horticultural use. Before installing a system which uses one of these less common fuels, there needs to be a guaranteed continuity of supply and an acceptance that boiler settings and maintenance are more complex.

Heat Requirement

A very large proportion of the surface area of modern glasshouses is covered by glass, even more so than earlier structures which have a wooden framework and dwarf walls. This means that a glasshouse with narrow glazing bars will lose more heat than one with a thicker timber framework.

The heat loss from different surface materials is denoted by its 'U' value or heat transmission value. This is measured in watts per square metre of surface area per degree C difference between the outside temperature and that required inside the greenhouse (W/sq m/degree C). Glass has a much higher 'U' value than timber. The heat requirement in kilowatts to lift the temperature inside a structure by a stated amount can be calculated, provided that the 'U' value is known. In the case of glass and polythene sheet, this is approximately 5.7 watts per square metre of surface area per degree C temperature lift, usually shown as 5.7 W/sq m/degree C.

To find the heat requirement for a glasshouse:

- Calculate the total surface area of the building, side walls, gable ends and roof in square metres. For calculation purposes, assume that the glasshouse is made entirely from glass. Losses through walls and floors are less significant.
- Multiply the surface area by the 'U' value to find the total heat loss.
- Make an allowance of an additional 30 per cent for heat losses caused by leakages, broken glass, ill-fitting doors, etc.
- Multiply the total figure by the temperature lift required. This is the difference between the minimum temperature required inside the glasshouse and the

coldest expected ambient (outdoor) temperature. For fully heated indoor crop production, an average recommended temperature lift is 22 degrees C.

The formula to use is:

Maximum anticipated heat needed (watts)

$$= \text{surface area of glass (sq m)} \times 30\% \times \text{'U' value} \times \text{temperature lift (°C)}$$

Example: For a glasshouse 30 metres long, 3 metres wide, 2 metres to the eaves and a ridge height of 2.6 metres (total glass area of 354 square metres), calculate the maximum heat requirement to lift the temperature by 22 degrees C.

Maximum heat required

$$= \text{surface area} \times 30\% \times \text{'U' value} \times \text{temperature lift}$$

$$= \frac{354 \text{ sq m} \times 130 \times 5.7 \times 22}{100}$$

$$= 57{,}709 \text{ watts}$$

$$= 57.7 \text{ kilowatts}$$

Warm Air Heaters

In most circumstances, it is only possible to use the heat exchanger type of warm air heater in a glasshouse. This ensures that none of the products of combustion can come into contact with glasshouse atmosphere as most food and ornamental crops are very sensitive to sulphur compounds, carbon monoxide and carbon products found in a boiler chimney or flue. Small quantities of these substances can prove fatal to plants.

Warm air heaters are usually free standing cabinets or cylinder shaped units. The burner unit and combustion chamber is in the central part of the heater cabinet and is surrounded by an air space. The flame from the burning fuel heats the combustion chamber and the hot gases are passed through a series of tubes within the heat exchanger. After passing through the heat exchanger tubes, the gases will have lost most of their heat and they are exhausted into the atmosphere through a flue or chimney.

Cool glasshouse air is drawn into the heater cabinet by a fan which directs it over the hot combustion chamber and heat exchanger. The air takes up most of the heat from these metal components in the boiler and transfers it through the top of the heater cabinet into the glasshouse as a continuous stream of warm air.

The warm air may be distributed throughout the glasshouse by a series of perforated metal or clear polythene ducts, or it may be free to circulate directly from the heater cabinet. The warm air tends to rise initially, then falls as it cools creating a circulation of air within the glasshouse. Foliage should not be placed near the heater outlets as it may be

PLATE 16.4 *Oil fired, warm air heater in a glasshouse.* (John Blyth)

scorched by the hot, dry air. Warm air heaters are supplied as complete units and only require connection to a fuel supply, electricity and a flue before they can be used.

The boiler thermostat must be carefully situated in the glasshouse to ensure effective temperature control. The thermostat switches the boiler on when the temperature falls below a pre-set level, and turns it off when the temperature rises above the required setting. A second thermostat in the heat exchanger unit controls the circulating fan, which is not allowed to operate until the exchanger is thoroughly warm. In a similar way, the fan is not switched off by the thermostat until the heat exchanger has cooled down.

A disadvantage of warm air heaters is that they have no reserve of heat. As soon as the boiler stops, the air in the glasshouse begins to cool down, often quite rapidly. The thermostat may not be able to respond quickly to such rapid variations in temperature, and this can result in a fluctuating temperature pattern in the glasshouse with the peaks and troughs giving a range of up to 4 degrees C either side of the required temperature.

Hot Water Systems

Modern hot water heating systems are suitable for use with a wide range of fuels. The most popular are gas oil (35 seconds), mains gas and coal. Modern hot water boilers are normally of welded steel construction and sold as a package by specialist boiler makers.

Boiler output is measured in kW or B.Th.U. (British Thermal Units) per hour. For example, a boiler would need an output of approximately 2,100 kW to be capable of lifting the temperature by 22 degrees C in a glasshouse with a floor area of 10,000 square metres (1 hectare).

Package boilers are made in standard sizes and supplied with most of the equipment necessary for easy installation. The package includes a boiler feed pump to top up the water in the system when necessary, water treatment equipment, a control panel, etc.

Boilers are usually of two or three pass construction. This indicates the number of times the hot gases pass backwards and forwards through the boiler to transfer heat to the water.

Fuel is burnt in a combustion chamber and the resulting flue gases pass through a series of fire tubes or smoke tubes. There may be as many as one hundred tubes in the boiler, each of which is surrounded by water. The heat from the hot gases in the tubes is transferred into the water by conduction. After most of the heat has been removed, the flue gases are collected and directed to the boiler chimney. Not all of the heat is removed from the flue gases as some is required to make the gases rise rapidly up the chimney and, in so doing, avoid low temperature corrosion in the flue. This is caused by various pollutants in the gases which mix with water, which is always present when fossil fuels are burnt, to form very corrosive acids and alkalis.

Most glasshouses heated by hot water have a pressurised heating system. The boiler and glasshouse pipework is pressurised to about 3 atmospheres (44 psi) in order to raise the boiling point of the water, in the same way that an engine radiator is pressurised to raise the boiling point of the cooling system.

PLATE 16.5 *Hot water heating pipes in a glasshouse. The pot plant bed also has under floor hot water heating.* (John Blyth)

By raising the water temperature in a pressurised system and circulating the water at high speed with a circulator pump, rapid transfer of heat from the water to the glasshouse air can be achieved with considerably reduced runs of heating pipe.

Control Systems

Modern heating systems are rather complex. They have a thermostat to keep the water in the boiler manifold at a constant temperature. Mixing valves connected to the manifold distribute hot water to individual glasshouse heating circuits. The amount of hot water fed to each circuit is controlled by individual thermostats in each glasshouse. When the temperature in a particular glasshouse falls well below the required level, the mixing valve supplies large quantities of hot water to that unit. When the difference between the desired temperature and the actual temperature is small, much less hot water will be supplied to the heating circuit.

Many nurseries have a computer control system, which integrates information from both inside and outside the glasshouses to give accurate and efficient control of the heating and ventilating equipment, which in turn saves energy and reduces costs.

Under Floor Heating Systems

Under floor heating systems, using either hot water or electric heating cables, are used to provide even warmth to bed areas used to grow plants in pots and containers. The compost in the containers absorbs the warmth which results in significantly increased root growth. A further benefit is that it often allows glasshouse air temperatures to be lowered, saving considerably on fuel.

Hot Water Systems

The water temperature required for bed heating is much lower than that for pipework heating systems. Bed temperatures are normally in the region of 15–24 degrees C (60–75 degrees F). Even heating is achieved by placing a series of pipe loops below bed level.

PLATE 16.6 *Under floor heating system. Plastic hot water pipes are laid in shaped polystyrene trays. Foil sheeting and absorbent matting (folded back) are placed over the hot water pipes.* (John Blyth)

There are two systems. Hot water piping lies in the soil at a depth of about 500 mm (18–20 ins) in one, whereas in a more up to date system, flexible plastic piping is placed on a specially moulded polystyrene base which reduces heat loss to the surrounding soil. The heating pipes are covered with sand or metal foil to give an even spread of heat. A similar bed heating system using warm water passing through flexible plastic pipework can be used for heating glasshouse benches.

The hot water for under floor heating is often provided by the main glasshouse boilers. However, as the water temperatures required are lower than that for pipework heating, it is more usual to pass the hot boiler water through a heat exchanger or calorifier. This allows the water in the under floor heating circuit to absorb some of the heat from the main boiler pipework. The warmed water then circulates through the system before returning to the calorifier for reheating.

Electric Heating Cables

Electric heating cables are an alternative method of providing under floor heating. The heating cables are supplied as complete waterproof units which are laid beneath the surface of the bed in a series of loops, and are normally covered with moist sand. The resistance to current flow in the cables warms them and the surrounding sand. It is usual to have a thermostat to control soil temperature.

Many electric soil heating units operate at mains voltage (240 volts), but as high voltages can be dangerous, especially in damp conditions, modern units operate at a lower voltage, frequently at 24 volts.

A wide range of heating cables and flexible heating blankets are available. Because it tends to be more expensive to run than pipework systems, electric heating is normally confined to propagation units or areas of the nursery where hot water is not readily available.

MAINTENANCE

Regular maintenance of boilers and heating installations is essential to ensure long, trouble-free service. Neglect may result in a breakdown in cold weather when loss of heat in the glasshouse can prove very expensive.

Heating engineers offer maintenance contracts for boilers and ancillary equipment. They are a good insurance against breakdowns which are bound to occur without a proper preventative maintenance programme.

Solid fuel boilers need to be serviced once or twice a year according to type and use. All inspection covers must be removed so that soot can be cleaned from all internal surfaces and the smoke tubes can be rodded out. The internal surfaces should be examined for cracks and corrosion, taking special care to ensure that any fire brick linings are not breaking up. There may also be a soot trap to clean out at the base of the chimney. The ignition equipment, electric motor and drive systems should be serviced by a competent engineer.

Steam boilers must have an annual inspection for insurance purposes. It is necessary to drain down the boiler, remove inspection covers from water passages and mud hole doors, and then flush the boiler to remove scale and other deposits.

Gas fired boilers Because of their specialist nature and the possible dangers of operating gas fired boilers, a qualified service engineer should carry out major servicing twice a year.

Oil fired boilers need regular servicing, usually at intervals of six months. This is a task which can be completed by nursery staff provided that training has been given; otherwise it is best left to an engineer.

Carbon should be gently scraped away from all internal surfaces. The lining material should be checked for deterioration and replaced if damaged. Soot should be removed from the flue passages with an industrial vacuum cleaner. The burner nozzle should be checked, cleaned and replaced if necessary. Igniter electrodes will need to be removed and checked. If reusable they should be cleaned and the gap reset, or it may be necessary to replace them. The 'magic eye detector', which cuts out the boiler if a fault occurs, needs cleaning and the fuel line filter should be replaced.

Finally the boiler is reassembled and tested. It may be necessary to adjust the oil burner operating pressure and primary air supply to ensure efficient combustion.

If an oil fired boiler will not fire, a few simple checks can be made before calling out a service engineer.

- Has the boiler run out of fuel? If there is an adequate supply, is the fuel tap on?
- Can the ignition system be heard buzzing as the sparks jump between the electrodes? The electrodes will need either replacing or cleaning and resetting if they are not operating correctly. Ignition failure is the main cause of boilers not starting.
- Has the boiler locked out?

Boiler lock out Modern boilers are fully automated and cut in and out according to need. However, if a boiler fails to start when it is switched on it is necessary to have a safety device in the control system to avoid the risk of explosion. Boilers have a start up sequence and if any part of that sequence fails, for example the fuel does not ignite, the boiler will lock out and a warning light shows or an alarm will sound.

The lock out will also operate if the flame goes out while the boiler is running. A magic eye or light detector in the burner monitors the flame while the boiler is running. Should the flame fail, the control unit will activate the ignition system for a short time, usually about thirty seconds. If the fuel fails to re-ignite, the boiler will shut down and lock out. The fault must be located and repaired before the boiler is restarted by hand.

Warm air heaters also need the six monthly check required for oil fired boilers. In addition, the heater cabinet case should be removed to allow inspection of the outside of the heat exchanger for cracking and corrosion. Cracks must be repaired or the heat exchanger replaced; otherwise toxic fumes may contaminate the glasshouse.

Hot water pipework should be checked for leaks and corrosion, and severely corroded sections of pipework may need replacement.

Control equipment, aspirated screens, switch gear, etc., must be kept as clean and dry as possible. Any damaged equipment should be replaced. Make sure that air inlets to glasshouse control equipment are not obstructed by foliage.

Propagation Equipment

Mist Propagation Systems

Mist propagation was developed in the United States during the 1950s to improve the speed of rooting and the quality of rooted cuttings. Mist propagation is only suitable for cuttings with leaves.

There are four main requirements for efficient mist propagation:

- A heated bed or bench to warm the growing medium and encourage rooting. The beds are usually heated with hot water pipes or electric heating cables, which are described earlier in this chapter.
- A zone of high humidity surrounding the stem and leaves.
- A cool, well ventilated glasshouse.
- High light intensity.

The high humidity needed for optimum rooting is provided by a series of misting nozzles mounted on or suspended 500 mm–1 m above the propagating bed or bench. The nozzles are usually spaced at intervals of about 1 m, but this may vary with different bed widths. Anvil type nozzles are in common use; they have an output

PLATE 16.7 *Mist propagation unit with over head spray mist line.* (John Blyth)

of 30–70 l/hr (7–15 gal/hr) and an operating pressure of 2–4 bar (30–60 psi).

An electric solenoid valve is fitted to the water supply to switch the system on and off. A filter unit in the pipeline prevents blocked nozzles provided that it is serviced at regular intervals. Mist nozzles produce water droplets of approximately 100–120 microns in diameter (one micron is a thousandth of a millimetre).

The amount of misting required from the nozzles is determined by the rate of evaporation from the surface of the leaves. Some common methods of detecting evaporation rate and controlling misting include an electronic or artificial leaf, a solar detector and a time interval controller.

The *electronic leaf* is the most widely used system. A small block of plastic material has a pair of electrodes usually about 12 mm apart set in its flat upper surface. The unit often has a spike to fix it into a tray or bed among the cuttings. For efficient operation, the upper surface of the electronic leaf must be level. The leaf is linked to a control box by a low voltage electric cable, usually 24 volts.

When mist from the nozzles falls on to the cutting leaves, it also settles on the surface of the electronic leaf. The moisture builds up to form a continuous film of water which connects the two electrodes in the plastic block. The water acts as a conductor to complete the electric circuit from the leaf to the control unit. The flow of current through the circuit shuts off the solenoid valve in the water supply and the nozzles stop misting.

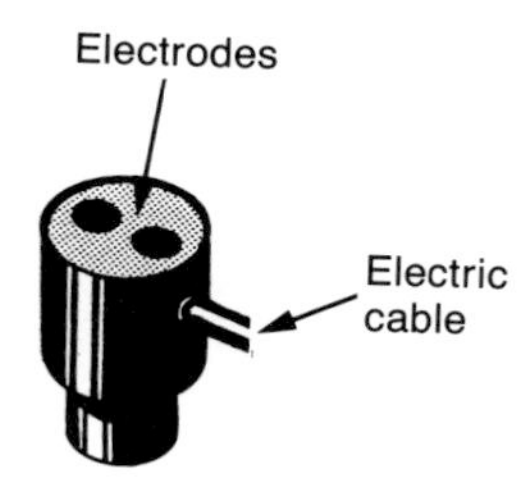

FIGURE 16.1 An electronic leaf. When the top of the unit is wet, the two electrodes are connected by the moisture and the pump is switched off. When the top dries out, the pump delivers water to the misting nozzles.

As moisture is lost from the surface of the electronic leaf through evaporation, a point occurs when the water no longer connects the two electrodes. This breaks the electric circuit, causing the solenoid valve to open and misting to restart.

Many mist controllers have an additional feature which alters the sensitivity of the leaf to give some control over the amount of water provided at each misting. To ensure that the leaves of the cuttings are kept thoroughly wet, it is necessary to give extra misting so that some water runs off the leaves into the rooting medium.

Solar energy detectors have a magic eye light detector or solar sensor which collects energy from sunlight. The amount of solar energy detected by the sensor is used to control the misting unit, and the unit should be suspended above the propagating bed or bench in an unshaded position. The system works on the principle that water loss from leaves is directly proportional to solar energy received by the detector. As there are many factors involved in water loss, additional adjustment is provided in the control box to give satisfactory operation under a range of conditions.

Time interval control is provided by using a time clock which gives periods of misting at set intervals. The length of each misting can be varied as required. This system is only suitable in areas with long periods of uninterrupted sunlight, and is often used in conjunction with one of the other control systems.

Maintenance of Mist Control Systems

Most mist controllers and solenoid valves are maintenance-free, and any control system faults will require attention by a qualified electrician.

Mist nozzles require occasional dismantling for cleaning, especially when hard water is used, and it may be necessary to soak the nozzles in a descaling solution to remove lime deposits. After cleaning, the nozzles must be reset according to the manufacturer's recommendations. Many misting nozzles have filters to help avoid blockage.

For efficient operation, the upper surface of the electronic leaf must remain flat and smooth. It should also be grease-free, as limescale and grease deposits severely interrupt its operation. The leaf can be cleaned by gently rubbing the top surface on a pad of fine 00 grade sandpaper, ensuring that the surface is kept perfectly flat. The surface should not be wiped with the hand, as this leaves a greasy film.

Fog Propagation

The problem of overwetting, associated with mist propagation, can be overcome by fog propagation. This system uses much finer water droplets to provide a 'fog' which remains suspended in the glasshouse for some time. To help maintain the fog, it is usual to cover the inside of the glasshouse with a fabric screening material which stops the fine droplets coming into contact with the glass.

Most of the water droplets should be about 10 microns in diameter, which is only one-tenth the size of those used for mist propagation. The suspended fog maintains the glasshouse atmosphere at up to 98 per cent relative humidity. (Relative humidity relates to the amount of water present in the air. Air with 100 per cent relative humidity is saturated. By heating air, moisture is removed and the relative humidity will be reduced.)

PLATE 16.8 *Fogging head installed in a propagation unit. The right-hand connection supplies water to the nozzle and the left-hand connection is the main compressed air feed. The centre pipe is the pilot air feed used to operate the fogging head control mechanism.* (John Blyth)

The most popular fog propagation system in use has a low pressure water supply, normally 1.3–3 bar (20–45 psi), piped through small bore nylon pipework to the fogging heads. A supply of compressed air at 5–7 bar (75–105 psi) is also piped to each fogging head. The high pressure compressed air strikes the jet of water leaving the nozzle, atomising it into a fine mist.

The fogging heads are spaced at intervals of 2–3 m throughout the propagating area. The production of fog is controlled by a humidistat which can be set to give the desired relative humidity within the propagating unit. Additional controls are provided to stop or reduce fogging at night and to wean material which has already rooted.

The compressor unit which provides compressed air must be installed in a dry, well ventilated position outside the propagating unit. Compressors tend to be noisy so they should be sited well away from working areas. The propagating bed area is heated with warm water or electric heating cables in the same way as a mist propagation system.

Maintenance

The filters in the water supply line require frequent cleaning, and the water which collects in the air compressor reservoir or tank should be drained at weekly intervals throughout the season. The air supply water trap should be checked daily and drained as necessary.

The fogging heads should be dismantled, cleaned and readjusted annually by a trained service engineer.

Carbon Dioxide Generators

Enriching the carbon dioxide level of the atmosphere in a glasshouse gives significant improvements in crop growth, yield and

PLATE 16.9 *Compressor unit for supplying compressed air to the fogging heads in a propagation unit.* (John Blyth)

quality. Enrichment is only required when the plants are in vigorous stages of growth, and enrichment of between two and six times that found naturally has been used. The normal level of carbon dioxide in the atmosphere is 300 parts per million (0.033 per cent). The most common enrichment level is in the region of a threefold increase, to approximately 1,000 parts per million (0.1 per cent).

Three hydrocarbon sources—propane gas, natural gas and paraffin (kerosene)—are widely used to provide carbon dioxide. Stored liquid carbon dioxide piped to nozzles throughout the glasshouse can also be used, but cost and problems of installation have limited its use.

There are various types of carbon dioxide generators. These include:

Suspended open burners Propane or natural gas is piped to a series of burners suspended at intervals throughout the glasshouse. Depending on size, one burner can enrich up to 600 sq m of glasshouse floor area. The carbon dioxide produced by burning the gas is distributed by natural air movement within the glasshouse. A pilot light burns continuously in each burner to allow automatic control of the unit. Propane and natural gas burnt in purpose-made units provide near-perfect combustion with no harmful pollutants and virtually no smell.

Combined direct fired space heaters and carbon dioxide burners are constructed in a similar way to industrial direct fired warm air heaters. The carbon dioxide produced is distributed throughout the glasshouse by an electric fan. The burner control and electric ignition system is the same as that used for glasshouse heating boilers. A single unit can enrich up to 2,000 sq m of glasshouse floor area.

Paraffin fired carbon dioxide generators are normally of the direct fired space heater type. High quality paraffin with a very low sulphur content of 0.03 per cent or less must be used to avoid damage to the crops. A sophisticated direct fired burner unit is necessary to ensure that the fuel is completely burnt. Some models have a two stage combustion process to ensure that all poisonous carbon monoxide produced is converted to carbon dioxide. Even small concentrations of carbon monoxide can be fatal to plants.

The paraffin is burnt in a pressure jet burner, and the hot gases are directed on to a cone or plate within the combustion chamber which soon reaches red heat. At this temperature the fuel is completely burnt and the carbon monoxide produced is converted to carbon dioxide.

A side effect of burning fuel for carbon dioxide enrichment is that a considerable amount of heat is produced. A threefold increase in carbon dioxide content can raise the glasshouse temperature by approximately 2 degrees C.

Control Systems

Any carbon dioxide enrichment is lost as soon as the glasshouse ventilators are opened, so an automatic control system for the generator is desirable to avoid wastage of gas or paraffin. Where a computerised system is not available, partial automation can be achieved by using one or more of the following controls:

A solar time switch will start up the unit at or soon after sunrise and turn it off at the end of the day. This type of control automatically compensates for time of year and changes in day length.

Micro-switches attached to the ventilators can be used to shut off or start up the carbon dioxide generator when the ventilators start to open or close.

Photocells offer another method of automatic control by switching on the carbon dioxide generator when light levels are high, and switching the unit off during prolonged dull periods.

Maintenance

Open flame burners require no regular servicing, but air intake screens should be brushed clean occasionally and any soot removed from the pilot jet.

Direct fired heaters should be serviced in the same way as warm air heaters.

Control units are splash-proof, but they should be kept as clean as possible and never allowed to become soaked with water.

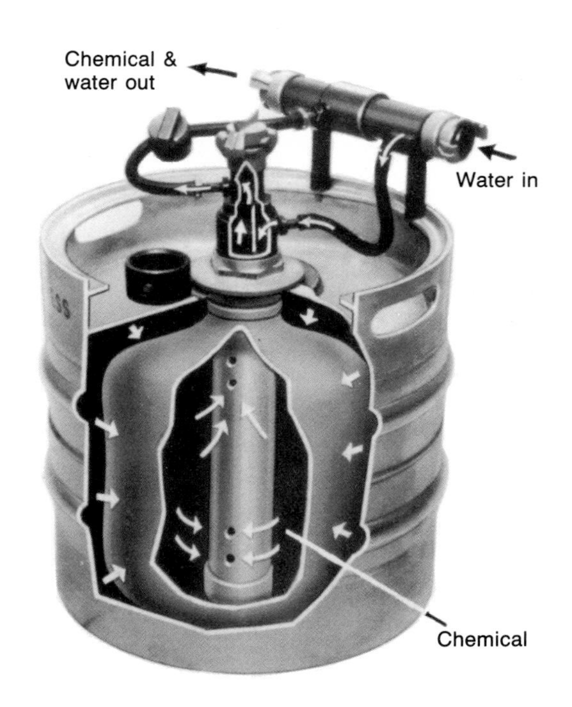

PLATE 16.10 A displacement dilutor. The concentrated chemical in the bag is surrounded by the incoming water. The water pressure slowly forces the chemical through a nozzle into the irrigating water which carries the feed to the crop.

LIQUID FEEDING AND INJECTION EQUIPMENT

Intensive horticultural cropping requires vigorous plant growth. To help achieve this, regular applications of liquid feed are used, especially on glasshouse crops. A wide selection of fully soluble liquid feeds are available and it is common practice to apply them to individual plants through an irrigation system.

Displacement Dilutors

On many holdings, liquid feeding is achieved with a displacement dilutor incorporated into the watering system. Dilutors are designed for medium or low pressure systems with an operating pressure of 2–3 bar.

A displacement dilutor consists of a concentrate container, usually made from polythene or aluminium, and a dilutor head. Various sizes of dilutor are available, with the larger models having a container capacity of up to 160 litres. The container has a sight gauge to indicate the level of concentrate solution in the tank. The irrigating water flows through the dilutor head, which can be fitted with a range of nozzles to give a range of liquid feed outputs.

Some types of displacement dilutors are fitted with a strong PVC inner bag, which separates the incoming water from the concentrate solution and thus avoids any possibility of the two solutions mixing. It also enables a wider range of materials to be used, especially those with a specific gravity (density) very similar to that of water, such as dilute acid.

Concentrated liquid feed solution is placed in the container up to an indicated level. As the irrigating water passes through the dilutor head, a small proportion is fed down a tube into the concentrate container. This displaces some of the concentrate which is forced through the dilutor head nozzle into the irrigating water to give a dilute feed solution. There is no significant mixing of incoming water with the concentrate solution in the container because of the difference in specific gravity. Water is less dense, so it floats to the top of the heavier concentrate rather than mixing with it.

Dilution will continue until all the concentrate is used, at which point the container will be full of water. A dye is often added to the liquid feed to indicate whether feeding is taking place and also to show when the concentrate solution has run out.

By using different dilutor heads, it is possible to vary dilution rates from 1 part of concentrate in 25 parts of water to a rate of about 1 in 2,000. Dilutors are accurate enough for most liquid feeding situations, but dilution rate can be affected by sudden changes in water pressure or flow rate.

The spare dilutor head nozzles and dilution charts supplied with the equipment should be carefully stored. The dilutor should be flushed with plenty of clean water and stored dry when not in use.

Proportional Injectors

As glasshouse holdings become larger and more specialised, more accurate automated liquid feed systems have become necessary. Proportional injection systems which give a precise dilution ratio, irrespective of water pressure or flow rate, meet this need. Many of these systems operate by using the pressure of the irrigation water; others are electrically operated. Proportional injection is mainly used to apply liquid feed, though it can also be used for pest and disease control, pH correction, etc.

Proportional injectors are generally made from plastic, which is unaffected by most chemicals used for liquid feeding. The main components are a container for the concentrate and an injector pump. A double acting piston driven hydraulic motor linked to the injector pump unit is a widely used design. Pressure from the irrigating water drives the hydraulic motor piston backwards and forwards in its cylinder. A connecting rod links the piston to the pump plunger, which sucks concentrate from the storage tank. On the opposite stroke, the plunger forces the

PLATE 16.11 *A proportional injector. The water inlet is on the left-hand side and concentrated chemical enters through the bottom of the injector body. Diluted chemical leaves the unit from the outlet on the right-hand side of the injector body.* (Hingerose)

concentrate into the irrigating water. Mixing of concentrate and water takes place in a mixing chamber before the diluted feed is passed to the irrigation lines.

A range of models is available to suit different water flow rates. Many are linked to a computer controlled feeding system covering the whole nursery. Dilution rates can normally be altered by a simple external control. The equipment is virtually maintenance-free but, as with all irrigation equipment, the water must be clean.

Electrically operated models often use a similar piston driven hydraulic motor, but injection is controlled by an electrical flow meter in the irrigation water main. This is connected to one or more solenoid valves which control the supply of water to the hydraulic motor in order to regulate the injection of concentrate into the water supply.

Soil Sterilisers

Soil sterilisation should more correctly be described as partial sterilisation, because the aim is to kill weed seeds and harmful organisms within the soil or compost. After sterilisation, the growing medium still contains organic matter and nutrients, but should be weed-, pest- and disease-free. Soils can be sterilised with either chemicals or heat, and both methods can be used to treat batches of pots and containers or large areas within a glasshouse or tunnel.

Although more expensive, sterilisation by heat is more effective in the long term, although both systems are sometimes used in rotation. Chemical sterilisation is usually done by a licensed contractor, as the materials used are highly toxic. After chemical sterilisation, no person should be allowed to enter the structure until it is declared safe to do so.

Steam Sterilisation

Batches of material or complete glasshouse floor areas can be sterilised with the use of a steam boiler or mobile steam generator.

Sheet steaming is a common method. Low pressure steam (0.1–0.3 bar) is injected under heavy covers or sheets laid over well cultivated, moist soil. Perforated metal pipes, laid on the surface of the soil under the sheets, distribute the steam over the entire area. The steam percolates into the soil where it condenses, and the resulting hot water drains further down into the soil. The soil

PLATE 16.12 *Steam generator suitable for steam sterilisation of soil in a glasshouse.* (John Blyth)

temperature gradually rises, and it may take from two to six hours to achieve a temperature of approximately 80 degrees C, which must be maintained to a depth of 150 mm for at least twenty minutes. Sterilisation can be effective from 150–225 mm below the surface, depending on soil type.

Many holdings have a modern steam generator which uses heated coils surrounding a boiler combustion chamber. Water is instantly turned to steam when it comes into contact with the heated coils. There is never a large quantity of water to be heated at any one time and there is no long warming-up period. The coils are heated by a pressure jet burner which runs on either diesel fuel or kerosene.

Electric Batch Sterilisers

An electric steriliser can be used when small batches of sterilised soil or compost are required. The growing medium is placed in a container. Heat is applied by using an electric heating element either in a water bath below the container or within the moist compost. The heat converts the water or moisture in the compost into steam which sterilises the soil or compost. Electric sterilisers with capacities of up to one cubic metre are available for both domestic and commercial use.

Hot Gas Batch Sterilisers

Soil and compost can be sterilised with portable equipment which uses heat from a kerosene burner. The growing medium is fed into the upper end of a rotating drum which is set at an angle so that the material falls towards the lower end as the drum rotates. The drum runs at about 40 rpm and may be driven by an electric motor or small petrol engine. A burner unit is situated at the lower end of the rotating drum. Hot gases from the burner pass up into the drum in the opposite direction to the soil or compost. The degree of heating depends on the heat output from the burner and the time taken for the material to pass through the drum; the longer it takes, the hotter it will become. Moisture in the soil or compost will be turned to steam by heat from the burner, and this will raise the temperature sufficiently to sterilise the soil. Outputs of up to about 6 cubic metres per hour can be achieved, depending on the size of the machine. Kerosene consumption varies, depending on machine size, from 13–30 litres per hour.

Suggested Student Activities

1. Look for different types of glasshouse ventilation system and find out how each type is controlled.
2. Measure the dimensions of a commercial glasshouse, and then calculate the heat requirement in kilowatts to lift the temperature by a maximum of 22 degrees C.
3. Find out how mist propagation controls can be adjusted to suit different plant species and conditions.
4. Investigate the different methods of applying liquid feed to glasshouse crops.

Safety Check

Electricity is an important source of power in glasshouses. Moisture is always present in a glasshouse, so make sure the equipment used is designed for damp conditions.

Chapter 17

IRRIGATION

Many horticultural enterprises, sports grounds and amenity areas need some form of irrigation system. Glasshouse crops *must* be irrigated. The controlled application of water will, for example, improve the yield of fruit and vegetable crops, promote good quality turf on sports grounds, golf greens, etc., and make the regular watering of container beds, amenity areas and plant borders a simple task.

Sprinkler systems can be used to give frost protection as well as for irrigation purposes. Frost damage can be reduced by covering the flowers and developing fruit with water. As the temperature drops towards freezing point, the sprinklers are turned on. As freezing temperatures occur, heat is removed from the water rather than from the flowers or fruit. Sufficient water must be used to avoid water previously applied from freezing, and application rates of about 3.2 mm per hour are necessary to give good protection. Frost damage can usually be avoided in temperatures as low as about minus 5 degrees C.

Frost protection is usually limited to the spring when sudden frosts can occur. Excessive use of this technique can cause waterlogging of the soil. As many sprinkler systems are semi-permanent, the same equipment can be used for irrigation during the summer.

Sources of Water

Mains water supply is a reliable and economical source of water for irrigation in many areas. However, water authorities are increasingly placing restrictions on the amount available for irrigation in order to avoid large pressure drops in the mains supply system. Many water companies will only permit 12 or 19 mm piped and metered supply, which is inadequate for many types of commercial irrigation equipment.

For this reason, many growers have their

PLATE 17.1 *Rotary irrigator on a sports field. The booms are made to rotate by the water moving through the nozzles. The rotary boom also drives the cable winding drum which pulls the irrigator along on a cable which is anchored at one end of the field.* (Briggs Irrigation)

own water storage, either in large tanks above ground or in a reservoir which is usually lined with butyl rubber sheet to avoid loss from leakage. The tank or reservoir is filled from the mains and pumped around the holding when required. Mains water is clean and reliable, although in areas with hard water, lime deposits usually furr up nozzles and other items of equipment. Mains water is mainly used for irrigation within glasshouses, for propagation and for diluting spray chemicals such as herbicides, pesticides and fungicides. Where adequate supplies of mains water are not available, other supplies must be found.

Boreholes can be sunk in many areas of the country. This is an expensive process with no guarantee of success, and a licence to abstract water is required. The water could be saline, even away from the coast, or it may contain fine sand particles which cause premature wear of pumps and equipment.

Only rarely is there enough artesian pressure to bring the water to the surface naturally. A submersible pump is usually required, and it may just bring the water to the surface or, as in many installations, lift the water and distribute it under pressure to all parts of the holding.

Rivers are another source of irrigation water, but levels vary and maximum demand is in the summer months when rivers are at their lowest level. For this reason, the authorities restrict the quantity which may be abstracted and may even stop irrigation equipment being used in periods of drought. Pollution of the supply can be a hazard, especially near industrial areas. However, rivers and water courses are widely used to irrigate field vegetable crops.

Ponds and lakes, usually supplied by springs or streams, have a variable water level according to the season. It may therefore only be possible to abstract small quantities of water at any one time. The main problems with pond and lake water are the possibilities of pollution and the build-up of algae and of surface rubbish such as leaves, especially on large areas of slow-moving water. Good quality filtration is essential to reduce the risk of blockages and damage to the equipment.

Roof water collected from glasshouses and other buildings is a useful supplement to water stored on the holding in reservoirs and large tanks. It is generally pollution-free but care should be taken with surface water collected from roadways, especially if there is a risk that spray chemicals or diesel fuel could pollute the water.

Reservoirs will hold large quantities of water for field irrigation purposes. It may be necessary to excavate a large reservoir which is fed from a water course from which abstraction may only take place when large volumes of water are flowing. Much of the stored or impounded water would fill the reservoir during the winter and spring but little, if any will be available during the summer period. Reservoir construction is a major undertaking requiring suitable ground, and permission must be obtained, both to build the reservoir and to abstract the water.

Glasshouse Irrigation

Glasshouse irrigation systems are designed either to give complete coverage of the crop grown in the structure or to water individual plants or groups of plants grown on benches, in pots or in containers. Overall coverage is ideal when the crop is grown in beds or in the glasshouse soil. Irrigation may be carried out from above the crop with, for example, a sprayline system, or within the crop using a drip or trickle system. There are many forms of irrigation equipment available to suit a wide range of crop needs and to meet the personal preference of the grower.

Low Level Irrigation

Trickle irrigation Lengths of flexible or 'lay flat' tube are placed alongside rows of plants. The tubes are perforated, stitched or porous, so that there is a continuous run of drips along their length, thus producing strips of watered soil. Because of the low operating pressure of the system, usually 0.5–1 bar, even distribu-

PLATE 17.2 *Trickle irrigation system is used for pot plants.* (John Blyth)

tion of water can only be achieved if the tube is level, and sloping surfaces seriously affect drip output. This system is mainly used for irrigating cash crops under glasshouses and plastic structures.

Trickle or drip irrigation systems are used to water individual plants. They use a series of low density polythene or PVC lateral tubes usually 12–15 mm in diameter. Outlet nozzles are fitted at intervals along the laterals, either directly into them or in small diameter tubes (spaghetti tubes) leading off from them. The spaghetti system allows one lateral to trickle irrigate two rows of plants or groups of plants in pots.

A number of different trickle irrigation systems are available. A typical example has an operating pressure of 0.3–1 bar with an output of 2–7 litres per hour. Trickle irrigation is normally used for high value crops such as heated tomatoes, cucumbers and pot plants. As each plant or pot has its own watering nozzle, the system is ideal for applying liquid feed, as it ensures that each plant gets the same amount.

As most low level irrigation systems operate at very low water pressure, they are subject to a number of problems:

- *Air locks* result in some nozzles not operating. This problem is usually avoided by careful design and installation of the system.
- *Blockages* can occur, as the nozzles are very small and are easily blocked by any grit or algae in the water supply. Good filtration of the water supply is essential, with frequent cleaning of the small in-line gauze filters which are installed in the water feed pipes supplying each section of the trickle system.
- *Variations in output* are usually caused by changes in level and by pipe friction. For efficient operation, the system must be as level as possible. A height rise of 1 metre will reduce the pressure in the system by

0.1 bar which has a major effect on the water output. However, careful design can minimise the problem.

Maintenance

The system should be flushed with clean water after a period of use, and several tube end plugs and joints should be pulled apart to allow rapid drainage. Some manufacturers supply an acidic solution to help remove lime scale and algae when flushing the system. It is possible to dismantle individual nozzles for cleaning on some trickle irrigation systems.

After flushing the system, the pipework should be drained and coiled up for storage, taking care to avoid damage through kinking. It should be stored in a cool dark place, well away from sunlight. It is good practice to label the equipment so that it can be laid in the same place when reassembling the system in preparation for the next season.

Sprayline Irrigation

Overhead spraylines, fitted with equally spaced nozzles or jets, are normally suspended above the crop. They produce an overall coverage of fine droplets—like rain. The spraylines are made from PVC, extruded aluminium or galvanised steel. The pipe lengths vary but are usually 4.5 m long with a diameter of 25 or 32 mm.

PVC spraylines are usually joined with a solvent cement to make a permanent joint. Aluminium spraylines generally have special couplings which allow the joints to be taken apart when necessary. Galvanised pipe is threaded. The nozzles are either threaded directly into the spraylines or into threaded sockets spaced at intervals along the sprayline. Nozzle spacing may vary from 750 mm up to 1.3 m. Closely spaced nozzles have small apertures designed to irrigate a narrower area of soil, whereas more widely spaced nozzles with larger apertures and at wider spacings will irrigate a wider area.

Spraylines can be designed to cover areas from approximately 1–5 m in width, depending on nozzle size, nozzle spacing and water pressure. Typical sprayline operating pressures are from 1–3 bar with individual nozzle outputs varying from 15–1,000 litres or more per hour, depending on nozzle size. Sprayline application rates are generally high, 25 mm of water in 6–8 hours being common.

PLATE 17.3 *Overhead irrigation sprayline.* (John Blyth)

PLATE 17.4 *Overhead sprayline nozzle.* (John Blyth)

Calculating Irrigation Application Rates

A typical irrigation installation in a glasshouse 14.6 m wide and 33 m long would consist of four spraylines, each covering a width of 3.8 m, with nozzles spaced 2.3 m apart, giving 14 nozzles on each run of line. Typical output from each nozzle is about 325 litres per hour at a pressure of 2 bar. The application rate for the irrigation system described can be calculated in the following way:

Total output per sprayline =
325 l/hr × 14 nozzles = 4,550 l/hr

Total area covered by one sprayline =
33 m long × 3.8 m wide = 1,254 sq m

Application rate =
4,550 l ÷ 1,254 sq m = 3.6 l/sq m/hr

Next, convert to cc/sq cm/hr =
3,600 cc/10,000 sq cm/hr

This is equivalent to 0.36 cm/hr =
3.6 mm/hr

In seven hours the application rate =
3.6 mm × 7 hrs = 25 mm of water

The application rate in this example of 25 mm in seven hours is well within the standard rate for a sprayline system.

Sprayline Supports

Spraylines need adequate support to operate efficiently. It is usual to suspend the sprayline from a strained wire running along the length of the glasshouse. Support from the wire is necessary at intervals of 500 mm for PVC spraylines and one metre for aluminium and steel pipework.

It is desirable to install spraylines with a slight fall in one direction in order to improve drainage when not in use. Many installations have self-sealing stop ends which seal when under pressure, but open when sprayline pressure drops.

Care of spraylines
No routine maintenance is necessary, but it is important that the water is filtered to avoid nozzle blockages. To prevent the build up of algae inside the pipework, the spraylines should be completely drained when not in use. Blocked or damaged nozzles should be cleaned or replaced immediately, as failure to do this will result in very uneven irrigation.

Other Glasshouse Irrigation Systems

Although most glasshouses and polythene structures have sprayline irrigation, other systems are occasionally used. Rotating nozzles which produce larger droplets are sometimes used to dampen foliage and help set fruit. A number of growers have found that sprayline nozzles mounted on short risers and joined together by short lengths of hose are convenient for short term cropping. They can be moved with ease from crop to crop.

Outdoor Irrigation

Amenity Beds and Containerised Plants

A wide range of irrigation systems can be used for plants in amenity beds, garden centres and for hardy plants in containers.

Trickle or drip irrigation is a development of the system used in glasshouses. Laterals are laid on the ground or across the top of pots and containers, with trickle nozzles spaced at intervals on the beds or in each plant container. For large containers like those with standard trees, it is common practice to place two drippers in each container to ensure the plants receive sufficient water.

In amenity plantings, where it is not desirable for the irrigation system to be visible, low level watering systems can be buried below the surface of the soil.

Rotary sprinklers and rotating nozzles can be fitted on risers, usually 500 mm–1 m high, which are usually fitted to 12 mm galvanised steel piping. The nozzles and sprinklers are quite small and are designed to give coarser droplets than those from glasshouse spraylines, making them more suitable for outside use. The larger droplets are less likely to drift and can also penetrate dense foliage more effectively.

Individual sprinkler units or nozzles will irrigate an area of between 5 and 9 m in diameter, with outputs of 500–800 litres per hour at a pressure of 1–3 bar.

The sprinklers or nozzle risers are linked by flexible pipework to a pumped water supply. Filters are usually included in the system to prevent blockages.

PLATE 17.5 *Rotating head sprinkler irrigation unit installed in a plant container bed.*
(John Blyth)

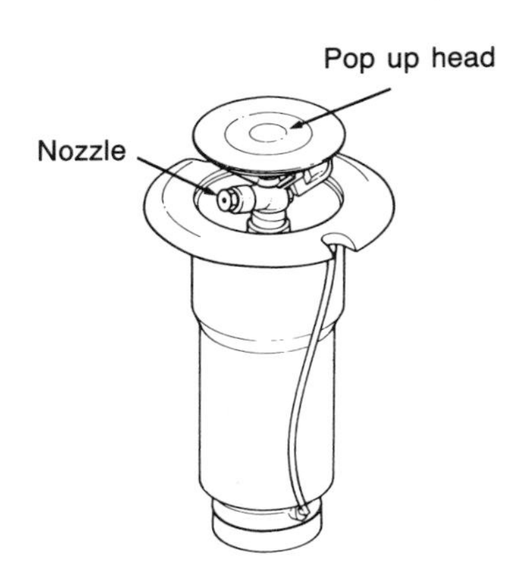

FIGURE 17.1 Pop up type, full circle sprinkler nozzle. The sprinkler head retracts below ground level when the unit is not in use. *(Cameron)*

Turf and Amenity Areas

Pop up sprinklers are widely used to irrigate amenity areas, golf courses, etc. Traditionally, agricultural field irrigation equipment has been used for this purpose, but the less obtrusive pop up sprinkler system has become popular for watering turf and amenity areas.

Pop up sprinklers have a retractable head which is lifted above ground level by the pressure of the irrigating water when the system starts up. Pop up sprinklers automatically return flush with the surface when not in use. They cause a minimum amount of obstruction, can be driven over by mowing machinery and are vandal-proof.

Most pop up sprinklers are impact driven in a similar way to those used for field scale irrigation. The sprinklers are linked by a system of underground pipes to the water supply which is pumped at a pressure of about 3 bar to the nozzles. To avoid the need for large diameter pipes and high output pumps, most systems are automatically controlled with an electric solenoid valve on each sprinkler unit. When the system is started up from a central control panel, the sprinklers are operated in sequence, one at a time over a period until the whole area has been irrigated. In many situations, such as golf courses, irrigation takes place at night when the area is not being used. Output varies considerably ranging from 15–160 litres per minute, with irrigated circle diameters of 20–50 m.

Alternative permanent irrigation systems for amenity areas where the system must be unobtrusive and not readily damaged include fixed nozzle spray heads and pop up rotating nozzle sprinklers which are impeller or gear driven.

Self travelling sprinklers, which are compact versions of field scale mobile irrigators, are also used to water turf and amenity areas. Some models have a reeling drum mounted on a small irrigator trolley, which pulls itself along on a cable to give about 120 m of travel before the machine needs to be reset. Before irrigation starts, the trolley cable is pulled out and anchored. The reeling drum, rotated by the irrigation water, gradually rewinds the

PLATE 17.6 *A small reel-in irrigator. The sprinkler head is pulled away from the unit as the hose pipe is unwound from the reel. Water flowing to the sprinkler head operates the piston arrangement on the side of the hose reel. This slowly winds in the hose, pulling the skid mounted sprinkler head along with it. The irrigator is anchored with a spike pushed into the turf.* (Andrews Sykes)

cable and in so doing pulls the trolley along as irrigation takes place.

Field Irrigation Equipment

Sprinkler irrigation, using portable aluminium mains and lateral pipes fitted with impact driven sprinklers on short risers, has been common practice for many years. Water, which may be supplied from various sources, is distributed to fields by underground mains or a semi-permanent network of surface mains.

Irrigation installations for field scale production of some fruit and vegetable crops often consist of two sets of spraylines and sprinklers. While one is in operation, the other is moved to the next site. With some systems only the risers and sprinklers are moved. They are plugged into self-sealing sockets in the sprayline laterals which are left in position for the season.

Impact sprinkler nozzles are used for most field scale pipeline irrigation systems. The nozzle has a spring loaded arm which is pushed sideways by the pressure of the water as it leaves the nozzle. Spring pressure returns the arm to its rest position and, in so doing, slightly rotates the complete nozzle unit. This cycle is repeated continuously, causing the nozzle to rotate through 360 degrees, to irrigate a complete circle of ground.

Portable aluminium spraylines, normally either 50 or 100 mm in diameter and 6 or 9 m long, are in common use. The sprayline sec-

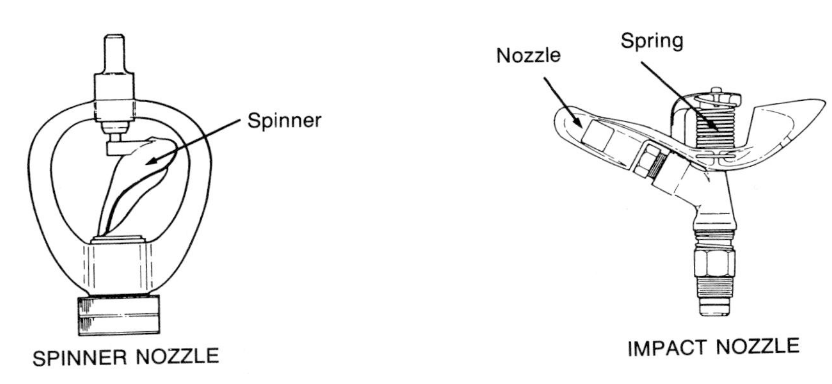

FIGURE 17.2 Examples of full circle irrigation nozzles used in horticulture. *(Cameron)*

tions are connected with quick release couplings which allow speedy assembly and dismantling. The normal working height of the sprinkler units is between 600 and 1,000 mm above ground level, and the sprinklers may be spaced at 9 or 12 m intervals along the spraylines, depending on nozzle output. Longer risers may be used when irrigating

PLATE 17.7 *Automatic reel irrigator. The rain gun trolley is pulled across the field as the hose is unwound by a tractor. The hose reel, rotated by the water flowing to the rain gun, re-winds the hose pipe and drags the trolley towards the anchored irrigator unit. The hose re-winding speed can be altered to give a range of irrigation rates.* (Andrews Sykes)

or frost protecting standard trees and soft fruit. Sprinkler systems designed to irrigate small areas of young plants and seedbeds are also available. They are ideal where there is a limited water supply, and they also meet the needs of the grower who cannot justify the expense of installing field scale sprinkler irrigation.

Oscillating spraylines provide an alternative method of irrigating small areas of land. A single row of jets on an aluminium sprayline is oscillated by the irrigating water. The pressure of the water (usually about 1.5 bar is required) causes the sprayline to rotate backwards and forwards through an arc of about 80 degrees so that the ground on both sides of the sprayline is watered. A number of spraylines can be joined to give runs of up to about 200 m. A typical oscillating sprayline unit will apply approximately 25 mm of water in a 6 hour period.

Mobile Irrigators

A range of self-propelled mobile irrigators has been developed to overcome the high cost of installing portable aluminium sprayline equipment.

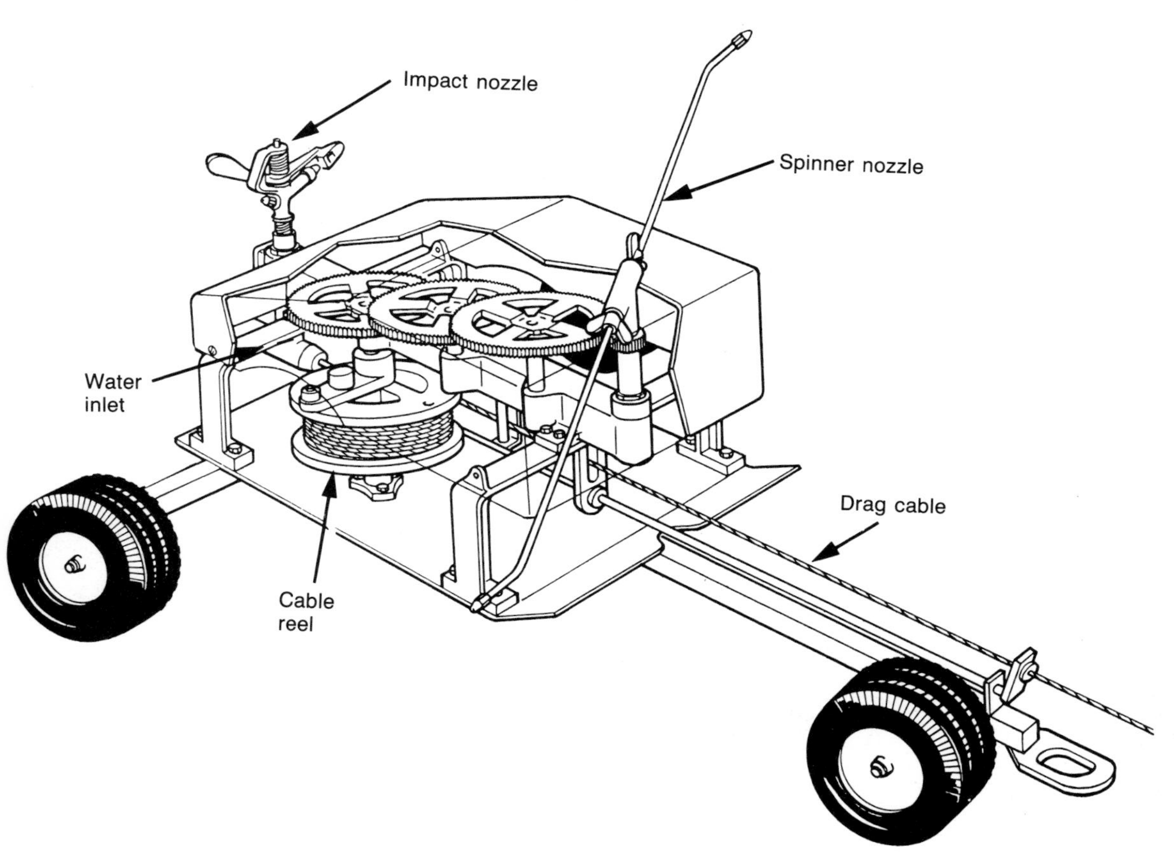

FIGURE 17.3 Mobile hose drag irrigator. The spinner nozzle provides the drive to the cable drum and the impact nozzle delivers most of the water to the crop. The irrigator is moved to one end of the plot to be watered, with the cable unwound and staked at the end of the run. The spinner turns the cable drum to slowly wind in the cable and pull the irrigator along the row. At the same time, the hose is dragged along behind the machine as it is unwound from the hose reel. An automatic stop brings the irrigator to rest and cuts off the water. *(Wright Rain)*

Mobile irrigators use either a rain gun or an impact driven rotary sprinkler carried on a wheeled trolley. With the irrigator at one side of the field, the trolley is pulled across to the opposite side as the water hose is unwound from the reeling drum. The hose is rewound by turning the drum in reverse using a pump or a water turbine operated by the irrigating water. The pump or turbine and the drum are mounted on a trailer making it a simple task to move the irrigator from field to field.

When irrigation starts, the reeling drum turns very slowly and the water hose is gradually rewound. As the rain gun or sprinkler trolley is attached to the water hose, it will be slowly pulled back to the irrigator unit as watering takes place. Most mobile irrigators have a variable hose rewind speed so that the irrigation time can be set for between 5 and 24 hours. This makes it possible to operate the irrigator on a 12 or 24 hour cycle.

There are various models of mobile irrigator. Some have up to 400 m of hose; there is, however, considerable water pressure loss in a hose of this length so a water supply with a pressure of between 4.5 and 10 bar is necessary for efficient operation. A mobile irrigator can cover up to 1.5 hectares on each run.

Suggested Student Activities

1. Examine the different types of jet used for glasshouse irrigation systems and find out the particular use of each type.
2. Investigate the range of crops being irrigated on a nursery or holding you know, and find out which of these crops take priority in the irrigation programme.
3. Study a mobile field crop irrigator to find out how the machine is propelled across the field.

Safety Checks

Numerous accidents, including some fatalities, have occurred when irrigation equipment has been used near overhead power lines. Important points to remember are:

- When working near overhead cables, always carry lengths of irrigation pipe in a horizontal position and keep them as near to the ground as practicable.
- Do not direct rain guns at an overhead cable; the wires are not normally insulated and water is a good conductor of electricity.
- Plan the irrigation procedure so that the irrigator does not pass below overhead cables.

Chapter 18

VEGETABLE AND FRUIT MACHINERY

As well as the various types of drills and planters for potatoes and many other vegetable crops (see Chapter 12) there are numerous machines for harvesting, cleaning and grading these crops.

VEGETABLE HARVESTERS

Some of the machinery primarily designed for harvesting potatoes can be modified for lifting such crops as carrots, parsnips and onions. In many cases, the tops are removed either mechanically with a flail topper, similar to a flail grass cutter, or by spraying a desiccant chemical to kill off the foliage.

Correctly adjusted harvesting machinery will:

- Lift and recover as much of the crop as possible.
- Separate the crop from soil, stones, tops and weed rubbish to give a clean, damage-free sample.
- Remove unmarketable material such as chats (small potatoes) to avoid unnecessary handling and transport.

Elevator Diggers

Small scale growers will find that an elevator digger, along with a certain amount of hand work, is suitable for their needs. Elevator

PLATE 18.1 *Harvesting potatoes with a single row machine. After hand sorting, the potatoes are bagged up on the harvester ready for market.* (Standen)

diggers are power take-off driven, and may be mounted on the three point linkage or semi-mounted with a pair of rear castor wheels. They have a working width of either 1 or 2 metres and lift either one or two rows of potatoes at a time.

The machine consists of a broad horizontal share that undercuts the crop, which is then passed on to an open link elevator. The elevator is supported by a combination of plain rollers and agitators which shake the crop as it passes up the conveyor. This action causes the soil, small stones and undersized crop to fall between the links back to the ground. The partly sorted crop then passes to a second open link elevator which completes the separation process before the crop is returned to the ground in a windrow for hand picking.

Elevator diggers used mainly for potato lifting have a steel link rod conveyor with a space of about 50 mm between the links. Some models have rubber covered links to reduce crop damage. Machines used for root vegetables have closer spaced rubber covered links attached to endless rubber belts, and the degree of agitation is less than that for potatoes.

Working depth is controlled hydraulically or by wheels at the front of the digger. The depth should be set so that all the crop is lifted without damage, but excessive depth must be avoided to prevent too much soil passing on to the link conveyor.

Complete Potato Harvesters

Large scale potato producers lift two or more rows at a time with a complete harvester. After the crop is lifted and the haulm removed mechanically, the crop passes onto a sorting table where hand pickers remove stones, clods and damaged tubers. The sorted crop is elevated to a trailer towed alongside by a

PLATE 18.2 *Two row elevator digger with a narrow spaced link conveyor.* (Terry Johnson)

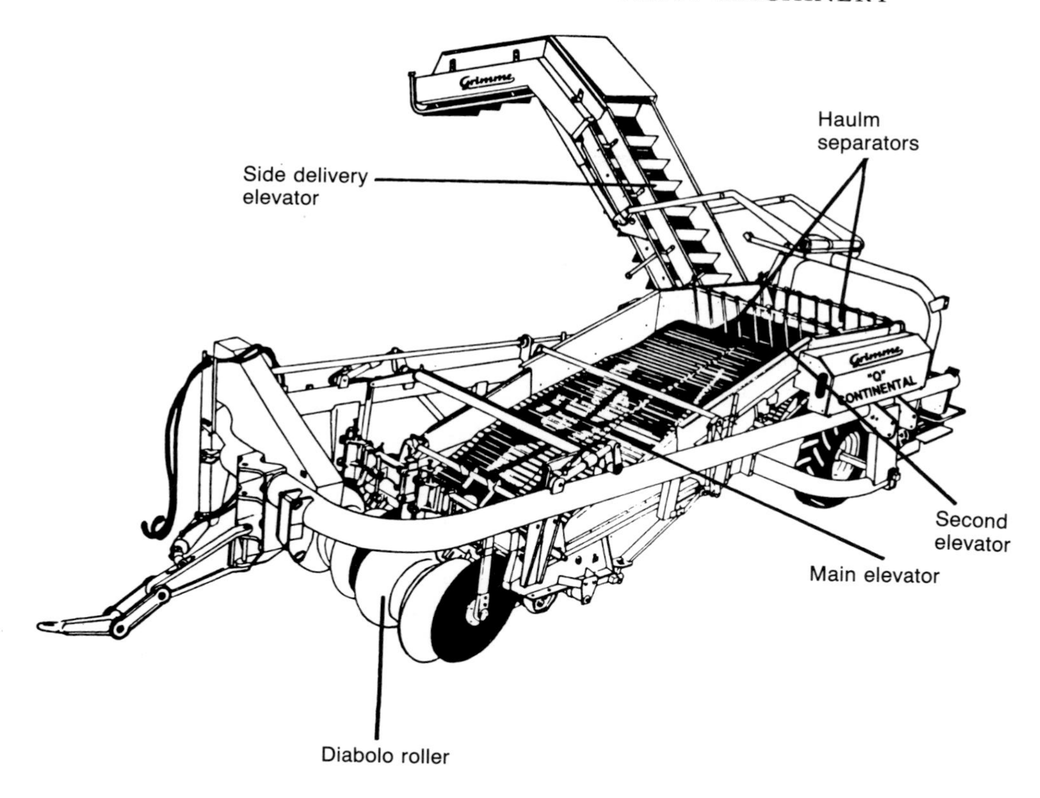

FIGURE 18.1 Un-manned two-row potato harvester. *(Grimme)*

second tractor or bagged up on the sorting platform.

Green Vegetable Harvesters

Most of these trailed machines are not complete harvesters; they are really mobile grading and packing units which rely on a gang of people to select and cut the crop. The hand cut crop is placed on a conveyor which carries it up to a grading table on the harvester for further sorting and hand packing.

Numerous attempts have been made to mechanise the harvesting of such crops as hard cabbage and celery with some success, but this equipment is very specialised and therefore too expensive except for contractors and very large scale producers. The main problem with this equipment is keeping mechanical damage and bruising within acceptable limits.

Green vegetable harvesters need a tractor in the power range of 20–60 kW. Most are semi-mounted, the front being supported by the tractor hydraulic linkage, and a pair of castor wheels at the back carry much of the harvester weight. Working depth, or height, is controlled by the tractor hydraulic system. The elevating and sorting mechanism is driven by hydraulic motors or the power take-off shaft. The tractor driver has the controls in the cab so that changes in elevator and belt speeds can be made as necessary to suit changes in harvesting conditions.

Fruit Harvesters

Soft Fruit

The most common soft fruit harvester is the one for black currants used by cordial and jam makers. A black currant harvester straddles the row of bushes and mechanically strips off the fruit and much of the foliage. The fruit is removed by a combination of brushing with rotating fingers and shaking the bushes. The mixture of fruit and foliage drops onto a sorting belt where separation is normally achieved with an air blast which removes most of the leaves and other light material. The crop is usually boxed up on the machine, and the filled trays are stacked at intervals across the field ready for collection.

Tree Fruit

Although they have met with little success in Great Britain, mechanical tree fruit harvesters are available.

A pair of hydraulically operated arms fitted to a modified tractor are clamped around the trunk. A mechanism is used to vibrate the arms, which shake off the fruit. The harvested fruit is collected in a fabric funnel fitted around the tree, and then placed in boxes or bulk bins ready for sorting and grading in the packhouse.

Washers

A number of specialist manufacturers produce a variety of washers for a full range of horticultural crops.

Vegetable Washers

Root vegetables are normally washed before packaging for market. Primary cleaning takes place in the field and a carefully adjusted harvester will considerably reduce the amount of cleaning required at the packhouse. However, soil will often cling to the roots and some stones will be mixed in with the crop.

Root washers agitate the crop while it

PLATE 18.3 *Vegetable washer. The crop is elevated to the cylinder shaped washing drum. The rubber conveyor carries the clean produce away from the washer for grading and packing. Soil and crop debris is removed from a trapdoor in the bottom of the hopper under the washer drum.* (Stanhay Webb)

is partly or completely immersed in water. Most washers have an open ended drum or barrel rotated by an electric motor. The washer barrel is placed in a large tank of water and it is set with a slight fall towards the outlet. The crop is elevated into the rotating barrel where it is washed as it passes through. Soil and stones drop into a trough beneath the washer or into a settling tank nearby. Some washers have nylon brushes inside the barrel to improve cleaning, and water jets may also be used to give the crop a final rinse as it leaves the barrel. The cleaned crop is usually carried from the cleaning barrel on a slatted conveyor.

Another type of washer consists of a narrow cleaning trough with rotating brushes. The crop is often propelled along the trough by water, and spray jets provide a final rinse.

Vegetable washers require a great deal of water. Many vegetable processing plants have settling tanks and filtering equipment so that the water can be re-used. Water consumption can be in the region of 3,000 litres (650 gallons) for each tonne of washed crop.

GRADING MACHINERY

Vegetables and fruit can be graded by size or by weight. For the majority of horticultural produce, size grading is widely used. There are two types of size grader: constant aperture and increasing aperture.

Constant Aperture Graders

The produce is graded by passing it over a series of grids or link chains which have different size apertures (holes). The crop is placed on the grader, where stones, chats and other small produce are removed as it passes over the smallest grid. It then moves on to grids with progressively larger apertures, and any vegetable smaller than the grid falls through to a padded collecting area or into a container. The very large items move across all the grids and pass out from the end of the grader. The crop is moved across the grids by a shaking or vibrating movement created by the grader mechanism.

Some graders have a chain conveyor to carry the produce through.

As with most graders, a roller inspection area is provided for the operators to remove damaged or ungradable material to ensure a high quality final product. This type of grader is widely used for potatoes, onions and similar crops.

Increasing Aperture Graders

This type of grader is capable of a much more gentle action, and is therefore preferable for grading easily damaged produce, such as apples. Increasing aperture graders can also be constructed to handle long, thin root crops, such as carrots, which would be difficult to grade with a constant aperture machine.

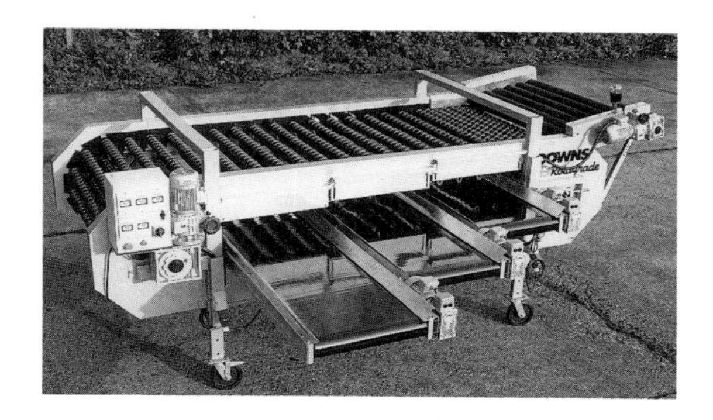

PLATE 18.4 *Increasing aperture grader. As the space between the rows of rollers becomes wider, larger produce can pass through to the cross conveyors. These carry the different sizes of fruit or vegetables to sacks, boxes or the packing area.* (Downs)

The grading table consists of a series of round belts or rollers, arranged so that they become wider apart as the produce passes along the table. The crop is fed onto the grader at the end where the rollers or belts are closest, allowing only the smallest items to fall through. As the space becomes wider, larger produce can fall through, usually onto a padded tray, where it is collected by hand and packed according to grade.

PLATE 18.5 *Single row potato harvester with an automatic weigher and bagger unit on the platform at the front of the machine.* (Grimme)

Weight Graders

Weight graders are used for crops which are sold by weight or where a minimum weight of the product is required. This has become more important since the introduction of European Economic Community grading standards. Examples of weight graded produce include lettuce and cucumber.

Individual items are placed into moving pans which are counterbalanced and arranged so that particular weights tip the pans at a certain point along the grading table. The produce is then deposited in the appropriate collecting area ready for packing.

Weighers

Semi-automatic weighers are widely used to help reduce labour costs. Produce, such as potatoes, is fed into a hopper by a conveyor. Discharge chutes at the base of the hopper feed the produce into sacks, bags or boxes until the preset weight is achieved when the flow from the hopper is automatically cut off. To make the weight as accurate as possible and to avoid too many overweight bags or boxes, some weighers have two stage filling. The container is first filled to an approximate desired weight, followed by a final gentle and slow topping up to achieve an accurate weight.

SUGGESTED STUDENT ACTIVITIES

1. Look at the different types of potato and vegetable harvester and follow the crop flow through the machine.
2. Find out how vegetable washers handle

the crop and how the soil is removed from the machine.

3. Study the working principles of two types of fruit or vegetable grader, and find out how the crop is packed after it has been graded.

Safety Check

Stack pallets and boxes on firm level ground. Damaged storage boxes should not be stacked when filled; potato and vegetable boxes can weigh as much as one tonne and there is a risk that damaged boxes could collapse with serious results if stacked.

Chapter 19

MACHINERY FOR ESTATE AND GROUNDS MAINTENANCE

The equipment needed for general maintenance of private estates, woodland areas and public parks extends far beyond that used for turf care; it includes hedge cutters, diggers, sweepers, trailers and chain saws.

HEDGE CUTTERS

There are three designs of hedge cutter used to maintain park and estate hedges. One type has a high speed flail rotor at the end of a hydraulically controlled arm, which can be set at various angles to cut the sides and top of hedges. Another type has a reciprocating knife cutter bar which can also be angled hydraulically. A third design, used by some contractors, has a circular saw blade driven by a hydraulic motor at the end of an arm which can also be set at various angles by a hydraulic ram.

Flail hedgers are a popular and versatile type of hedge cutter. They can deal with quite thick hedge material and can also be used to cut grass verges and banks. Reciprocating knife hedge trimmers give a very neat finish to a hedge, and are preferred by those who do not like to see the rather untidy finish sometimes left by a flail hedger. Used correctly as a trimmer, a flail hedger will give a very neat finish to a hedge.

PLATE 19.1 *Fencing in progress.* (Kawasaki)

PLATE 19.2 *Flail hedge cutter. The hydraulic rams used to angle the arm are supplied with oil from the reservoir on the machine and a power take-off driven pump.* (Bomford Turner)

Flail hedge cutters have a hydraulic motor driven rotor, with free swinging flails, carried on a tractor mounted arm. Hydraulic rams are used to alter the angle of the arm and the position of the cutting head in relation to the hedge or bank. Many models have a two speed rotor; a typical machine has a rotor speed of 2,400 rpm for cutting grass and 3,000 rpm for hedge cutting. Some machines have their own self contained oil reservoir and hydraulic system complete with a power take-off driven pump. Others use the main tractor hydraulic system for the rams and flail motor.

Most flail hedgers have a rotor with a cutting width of about 1 m, while others have a 1.5 m rotor head. The maximum reach of the rotor head—the furthest distance the cutter can extend from the tractor—depends on the length of the rotor arm. Standard models have a two part arm which gives a maximum reach of about 5 m, and some with a three part arm can reach further than 7 m. A break-back mechanism is built into the frame to protect the rotor from damage, i.e. the arm and rotor will swing backwards if the head hits an obstruction. Some hedge cutters have a break-back mechanism which operates in both forward and reverse directions.

Reciprocating knife hedge trimmers may be hand held or tractor mounted. Hand held trimmers are powered by either a small, two stroke petrol engine or an electric motor, and are used extensively for trimming hedges in public amenity areas and private gardens.

Hand held trimmers of the type illustrated in Plate 19.3 have a double sided cutting blade which is used to cut on both upward and downward strokes. Smaller hedge trimmers, usually electric motor driven, have a blade length of 300–450 mm. A typical heavy duty professional hedge trimmer has a 2.3 kW (3.1 hp) two stroke engine very similar to that used for a chain saw, with a 600 mm double edged cutting blade.

A second design of hand held hedge trimmer has a single sided blade usually driven by a small two stroke petrol engine. Blade length varies from 600–750 mm. The trimmer is held two handed, with one hand on the power unit and a second at the back of the cutter bar.

Mounted hedge trimmers with a reciprocating knife are attached to the three point linkage, and driven by a hydraulic motor or a mechanical system of gears and belts from the power take-off shaft. The machine is very similar in design to a flail hedge cutter (see

PLATE 19.3 *Hand held hedge trimmer. The double sided knife is driven by a two stroke petrol engine.* (Sachs-Dolmar)

Plate 19.2) with a reciprocating knife cutter bar instead of a flail cutting head.

Mounted hedge trimmers can be used on compact tractors. A typical machine has a three section cutter bar arm angled by hydraulic rams, so that the knife can be set to cut the sides and top of a hedge. A safety break-back unit in the arm allows the cutter bar to swing backwards if it hits an obstruction. The break-back unit is automatically reset by reversing the tractor. Some models have a cutter bar arm with a swing over arrangement which makes it possible to cut on either side of the tractor.

Maintenance

Flails must be kept in good condition and if one is lost, an opposite pair must be fitted to maintain rotor balance. A missing flail will cause rotor vibration.

The condition of the reciprocating knife blades should also be checked and they should be replaced when necessary. Most machines have serrated edge blades which do not require sharpening.

Tractor mounted hedge cutters have guards to protect the operator from flying debris, and wire mesh guards are sometimes fitted around the cab to give extra protection. It is important to keep these guards in good condition.

Check the condition of cables to electric hedge trimmers and test the safety circuit breaker before starting work.

Diggers

Heavy industrial diggers with a backhoe and front loader are a familiar sight, but they are rather large for horticultural work such as landscaping and ditch maintenance.

Two basic types of digger are used in horticulture: one is built around a tractor with a front loader and backhoe; the other is a digging unit fitted to the tractor three point linkage. By fitting different sized buckets, it can be used for digging, trenching and ditch cleaning.

The digger illustrated is built around a 15 kW (20 hp) compact tractor. It has a maximum digging depth of 2,060 mm and the front bucket has a maximum lifting capacity of about 420 kg. When in work, the legs at the back are lowered hydraulically to take the weight of the digger off the rear wheels. The front of the tractor is supported by lifting it off the ground with the loader bucket. Hydraulic rams are used to move the digger boom up and down and swing the bucket through a working arc of about 180 degrees.

The type of work possible with the backhoe depends on the size of the bucket. A typical range includes 200, 325, 400 and 600 mm wide trenching buckets and a 750 mm ditch cleaning bucket. The front bucket is used for scraping, levelling and loading soil and other material.

Some three point linkage mounted backhoe diggers use the tractor hydraulic system, while others use the independent hydraulics on the digger with a power take-off driven pump. These machines work on the same principles and are used in the same way as tractor diggers. The main limitation of linkage mounted diggers on compact tractors is their work rate, but they are ideal for trenching, etc., in confined spaces. Three point linkage backhoe diggers are made to fit compact and farm tractors.

Grass Trimmers

Hand held trimmers are widely used to cut around the base of trees, fence posts and walls, and also sloping bank sides and other areas of grass in confined or inaccessible places. They may be driven by a small two stroke petrol engine or an electric motor. The cutting head consists of a high speed rotor with a single or twin nylon line cutters. Depending on the size of trimmer, the rotor head has a cutting width of 200–450 mm (8–18 in), and an engine or motor of 0.5–2.5 kW or more.

The nylon cutting line is carried on a replaceable spool in the rotor head. The line wears quite quickly and it must be returned to its correct length to ensure maximum cutting efficiency. This normally occurs automatically when the trimmer is

PLATE 19.4 *Compact loader and digger.* (Kubota)

PLATE 19.5 *Nylon cord grass trimmers; the heavy duty model (right) has four cutting cords. Both have two stroke petrol engines.* (Allen Power Equipment)

in use. Only the tip of the line actually cuts, and as it wears, the spool in the rotor housing releases sufficient line to restore it to the correct length. If the rotor fails to release more line, the problem can be solved by tapping the rotor head on the ground. The guard over the trimmer head has a line cutter which automatically trims the line to the correct length each time it is extended from the spool.

Some trimmers have a release button or other mechanism on the rotor which allows the operator to pull more line from the spool to restore it to the correct length. An alternative trimmer head consists of a small rotating disc with either two or three swinging plastic blades equally spaced around the circumference.

The angle of the rotor head can be adjusted on certain models of grass trimmer so that the cutter can also be used as a lawn edger.

On many professional grass trimmers, the nylon cord cutting rotor can be exchanged

PLATE 19.6 *Brush cutter with a four bladed disc.* (Stihl)

for other types of cutting head. These attachments include rotors with swinging plastic blades, a miniature grass cutting rotor similar to that of a rotary lawn mower, steel brush cutting blades and small circular saw blades up to about 250 mm in diameter.

With the exception of the smaller domestic models, usually with an electric motor, grass trimmers have a shoulder harness and two handles to give the user full control of the cutting head.

Drive from the two stroke petrol engine is transmitted to the rotor through a centrifugal clutch and bevel gears. The cutting head does not turn until the engine reaches a certain speed. The centrifugal clutch provides a method of stopping the trimmer head while the engine is running.

Using Grass Trimmers

- For safe operation, eye protection must be worn when using a strimmer. Stout footwear, if possible with steel toecaps, thick trousers or leggings and gloves should be worn. Shorts and trainers are not suitable clothing when using a nylon line or other types of powered grass trimmer.
- Before working on the cutting rotor, stop the engine or electric motor and isolate it, so that it cannot be started accidentally.
- Place the trimmer in a safe position on the ground before starting the engine. Always check the residual current circuit breaker and make sure the cable is safe before using a mains operated electric trimmer.
- For trimming around trees, walls, fences, etc., and for scalping, the head should be held at an angle to the ground with the tip of the line making contact and the back of the rotor about 75 mm above the ground.

PLATE 19.7 *Using a brush cutter with full protective clothing – face shield, ear defenders, gloves and safety boots.* (Sachs-Dolmar)

Trimming and scalping

Mowing

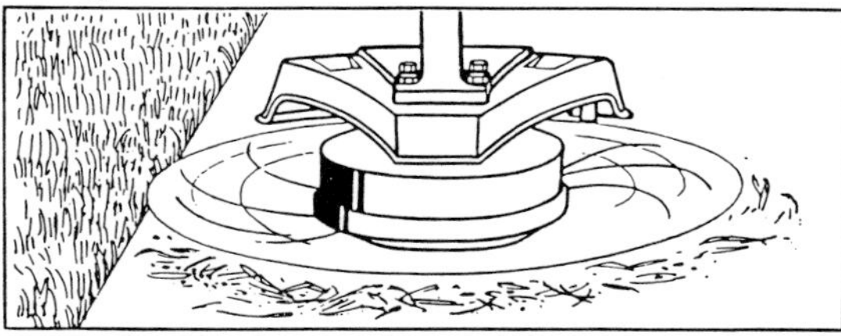

Sweeping

FIGURE 19.1 Working positions for a nylon cord grass trimmer head. *(Flymo)*

To avoid line breakage, it is advisable to use less than full throttle for this type of work. The rotor should be held level and at full throttle when mowing small areas of grass.

- The fanning action of the rotor can also be used for sweeping cut grass from paths, etc. This is done by holding the head level with and above the surface being cleared, and moving the head from side to side.

All Terrain Vehicles (ATVs)

ATVs, as they are commonly known, are useful for transporting equipment at high speed and for working on ground where a heavier machine may damage the surface. Their low ground pressure makes them suitable for applying spray chemicals and fertilisers with small spreaders and sprayers, either carried on the vehicle or towed behind. Grass cutting with a trailed engine driven flail or rotary mower is another task suitable for the ATV. It can achieve quite high outputs on level areas of grass.

ATVs have a single cylinder air cooled engine. Power output varies from 7–18 kW (9.3–24 hp) or more, depending on the model. There are two wheel and four wheel drive versions, and gearboxes provide a range of forward speeds and reverse. A twelve volt socket, connected to the battery, can be used to drive small electric motors on trailed implements, for example a spinner broadcaster with an electrically driven spinning disc.

A typical four wheel drive ATV has a single cylinder, four stroke, air cooled, 15 kW (20 hp) engine. It has a centrifugal clutch and the gearbox provides ten forward speeds and one reverse. The vehicle has rear drum brakes and disc brakes at the front.

Forklift Trucks

Industrial forklift trucks are widely used in horticulture for handling produce and materials. They are invaluable in packing sheds, cold stores and warehouses for stacking pallets and boxes and for loading lorries. Because of their weight and small wheels, industrial forklift trucks are limited to working on hard surfaces. Rough terrain forklifts are used to handle pallets, boxes and bags in fields and on similar uneven sites.

Forklift trucks may be powered by a diesel or propane gas engine. Where engine fumes are unacceptable, forklift trucks are usually powered by electricity. Electric forklifts have a set of large batteries which need recharging at regular intervals with a mains operated unit, usually overnight.

Industrial forklifts are arranged with the mast and forks at the front. The driver's seat is centrally mounted for maximum visibility of the load. The batteries and motor or engine are at the back, surrounded by heavy weights to counterbalance the weight lifted by the forks when the machine is in use. Industrial forklifts have front wheel drive and rear wheel steering.

Diesel and gas powered trucks often have a hydrostatic or similar automatic transmission

PLATE 19.8a *Trailer work in a plant nursery.* (Ford New Holland)

PLATE 19.8b *Tipping trailer with extended wire mesh sides.* (Wessex)

to give smooth operation together with a wide range of speeds. A rapid reversing system is usually incorporated into the gearbox. The power unit also drives the hydraulic pump which operates the pallet fork ram, the smaller mast tilt rams and, on some models, the side shifting rams or grab.

Two section masts are in common use; they allow the machine to pass through low doorways and openings when the mast is fully lowered, but still enable it to achieve a maximum lift height of 4 metres or more.

Great care is necessary when driving a forklift truck. All forklift operators are required to undergo formal training, and in many cases must hold a certificate of competence. Many serious accidents have occurred because forklift trucks have been overloaded. It is essential for the operator to be aware of the maximum safe load and other limitations of the machine.

TRAILERS

Every horticultural enterprise needs trailers and barrows. Trailers are available in many shapes and sizes; some are designed for special purposes such as bulk transport of potatoes, flat beds for hauling boxes and pallets, and low loaders for moving machinery from site to site. There is a huge variation in the size of trailers, from small two wheelers for lawn and garden tractors up to models

PLATE 19.8c *Garden tractor with mid-mounted mower deck hitched to a small trailer.* (John Deere)

PLATE 19.9 *Self-propelled material transporter powered by a 7.5 kW (10 hp) engine. The load platform can be tipped when required.* (John Deere)

with a capacity of many tonnes for general haulage work behind large tractors.

Trailers have a steel frame with steel or timber bodies. The smaller models are hitched to the tractor with a drawbar pin, which must have a securing clip to stop the pin jumping out when the trailer is being used. Larger trailers may be hitched to the drawbar or connected with an automatic hitch—a hydraulically operated hook which passes through a ring on the drawbar and then lifts and secures the trailer to the tractor. Many trailers have a tipping body with a hydraulic ram supplied with oil by the tractor hydraulic system. As well as rear tipping trailers, there are models which can tip sideways, and others with a high lift system which raises the trailer body about 2 metres before it is tipped.

The law requires that trailers with a jaw-type drawbar have a screw jack to raise and lower the drawbar when hitching it to the tractor. This does not apply to very small trailers. There are also regulations concerning the provision of brakes on trailers over a certain size when used on the public road. Full lighting must also be fitted to trailers used on the highway after the official lighting up time or in poor daylight conditions.

BRUSHES AND SWEEPERS

Hard recreation areas, roadways and similar surfaces require sweeping periodically, and during the autumn there are many leaves to remove from turf and amenity areas. Various mechanical aids are alternatives to labour intensive hand sweeping.

Rotary brushes speed up the job of cleaning roadways, car parks, etc. Power take-off or hydraulic motor driven rotary brushes may be front- or rear-mounted on a tractor. The brushes have a horizontal or vertical sweeping action, depending on the model. Some sweepers have a collecting bucket mounted behind the brush to collect the swept up material, which is then transported to a compost heap or storage area where it is hydraulically or mechanically tipped.

The rotary brush can be angled on some models to sweep the rubbish to one side. The bristles may be polypropylene or wire or a mixture of both materials. Sweeping height is usually controlled with a pair of adjustable castor wheels, but some rotary brushes have hydraulic height control.

A typical power driven rotary brush for a compact tractor has a sweeping width of 1.45 m and a brush diameter of 500 mm. The brush can be turned with hydraulic rams by 25 degrees to right or left for angle sweeping.

Leaf sweepers may be towed by a tractor or pedestrian controlled, and can be used on both turf and hard surfaces. The brush is driven by the power take-off shaft, a small petrol engine on the machine or the sweeper wheels. The leaves are collected in a large capacity container, in front of or behind the brush, which may be emptied mechanically or tipped by hand. Brush height is adjusted with screw handles on the castor wheels.

A typical pedestrian operated leaf sweeper has a 200 mm diameter nylon bristle brush with a sweeping width of 900 mm. It has a 2.25 kW (3 hp) four stroke petrol engine to drive the brush and land wheels. The drive to the brush and a dead man's handle on the frame is used to control forward travel. The wheel drive can be disengaged with the brush still running when sweeping in confined spaces.

A similar design of sweeper for use with lawn tractors has a wheel driven brush which sweeps the leaves into the container. Much larger leaf sweepers with a power requirement of 19 kW (25 hp) have a sweeping width of 2.1 m. They are semi-mounted, power take-off driven and have a large capacity leaf hopper which can be hydraulically or mechanically emptied from the driving seat.

SHREDDERS AND CHIPPERS

Shredders are used to turn shrubby material into a mulch. They range from small, petrol or electric motor driven models for garden waste, to machines for the professional which will shred small brushwood and need

PLATE 19.10 *Sweeping a road with a rotary brush.* (Massey-Ferguson)

PLATE 19.11 *Trailed leaf sweeper behind a lawn tractor. The wheel driven sweeper brush throws the leaves forward into the collector.* (Massey-Ferguson)

PLATE 19.12 *Power take-off driven wood chipper blowing finely chopped material into a trailer.* (Massey-Ferguson)

a compact tractor power take-off shaft to operate them.

Chippers are even more power hungry. A typical machine requires a minimum of 12 kW (16 hp) to convert brushwood up to 100 mm thick into finely chopped wood chips suitable for mulching shrub beds or as a dressing for pathways.

Chippers and shredders have a series of hardened steel blades on a rotor which pulverise the material against a perforated metal grid. The chipped or shredded material passes through the grid. Shredders usually return the pulverised material through a chute to the ground. It is more usual for chippers to blow the chips through a discharge spout which can be rotated through 360 degrees. An adjustable flap at the top of the spout is used to control the direction of the chippings as they are blown into a trailer.

Chippers and high capacity shredders can be mounted on the three point linkage or trailed. Both types use the power take-off to drive the knife rotor at a speed of approximately 1,300 rpm. It is possible to vary the size of chip by changing the arrangement of the blades on the rotor.

Chain Saws

Chain saws, which are widely used in parks and for estate maintenance work, are very dangerous tools and must be handled with great care. They can only be used right-handed, and nobody should attempt to use one without first receiving adequate training. The health and safety regulations require that operators wear full protective clothing.

Most chain saws have a two stroke petrol engine with a diaphragm carburettor which allows the engine to work in any position. Some of the smaller chain saws have an electric motor. When at full speed, the engine runs at 7,000–10,000 rpm, depending on the model, and the cutting chain reaches a speed of approximately 60 mph.

Various lengths of chain guide bar are used, ranging from 300 mm on electric saws up to 600 mm or more on large petrol engined saws. Much longer guide bars are made, but they are not in common use except by tree specialists. A typical chain saw for general estate work will have a guide bar from 380–540 mm in length and an engine in the range of 2.1–3.6 kW (2.8–4.8 hp).

PLATE 19.13 *Chain saw showing the recoil starter handle, chain brake lever and engine controls.* (Stihl)

Controls

The *recoil starter* should be pulled slowly until compression resistance is felt, and then sharply pulled with the right hand. Do not pull the starter cord to its full extent, or suddenly release it; the cord should be allowed to rewind slowly.

The safe way to start a chain saw is to hold it firmly on the ground with the left hand on the front handle and the right foot through the rear handle.

The *throttle trigger* controls the speed of the cutting chain, and it cannot be used until the throttle trigger lock is depressed. When the engine reaches a preset speed, the centrifugal clutch engages the drive to the chain. At tick-over the cutting chain will be stationary.

The *on/off switch* isolates the ignition system so that the engine cannot start or run.

The *chain brake* is a safety device which will stop the cutting chain instantly when pushed forward. This is done by moving the back of the left hand, which is on the front handle, against the chain brake lever. The chain brake should be tested before using the saw. The chain brake lever also serves as a guard to prevent the hand coming into contact with the chain if it accidentally slips off the front handle.

Kick back is the uncontrolled and violent upward movement of the end of the guide bar when it comes into sudden contact with an obstacle, such as a branch, or when the wood closes in and pinches the chain in the cut.

When kick back occurs, upward movement of the saw brings the chain brake lever into contact with the back of the left hand to stop the cutting chain instantly.

Some saws have an inertia chain brake which will engage automatically without the hand striking the chain brake lever.

Safety Devices

In addition to the chain brake, other chain saw safety devices to protect the operator include:

- The *rear handle guard* protects the right hand if the chain breaks.
- The *chain catcher* is a pin which catches the chain if it breaks; it works in conjunction with the rear handle guard.

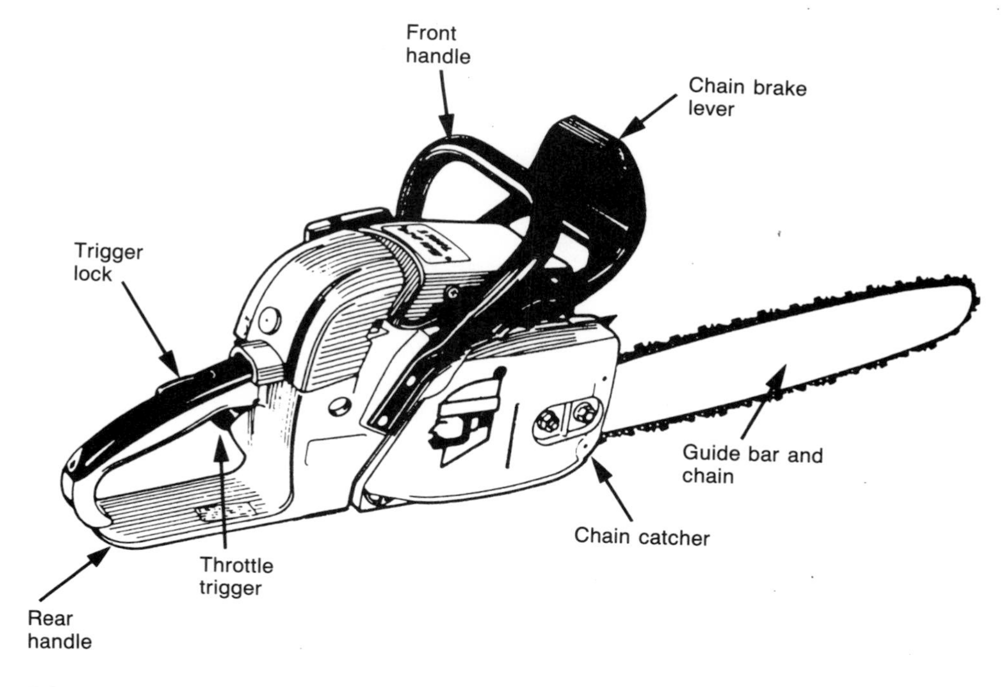

FIGURE 19.2 The parts of a chain saw. *(Stihl)*

- The *anti-vibration system* dampens saw vibration to reduce operator fatigue and improve cutting accuracy.
- A *safety lock-out switch* isolates the throttle trigger so that the chain cannot be accidentally engaged.

Chain saw users are required by law to wear a safety helmet, ear defenders, eye protection, protective gloves and leggings (or trousers made from a special material) and safety boots. Clothing should be close fitting.

Maintenance

The most important factor in chain saw operation is keeping the chain sharp. It is cutting correctly when the chain throws out chips rather than dust, which is a sure sign that the saw needs sharpening.

Frequent sharpening is required, but only two or three strokes of the file are necessary on a well maintained chain to return it to full efficiency. Chain tension on the guide bar is also important. The chain might come off if it is too slack, and a tight chain will cause excessive wear of the guide bar and drive sprocket.

An automatic chain lubrication system is standard on all but inexpensive electric chain saws, which have a hand operated oiler for lubrication. For both systems, the oil reservoir must be topped up to ensure the chain does not run dry.

The engine will need regular maintenance with particular attention paid to the air cleaner and sparking plug. A blocked filter will cause excessive exhaust smoke and choke the engine. A dirty sparking plug is a frequent cause of non-starting. It is a good idea to have a spare plug carefully stored in the tool kit.

Using Chain Saws

Check the saw for condition, making sure the chain is correctly tensioned and the safety devices are working.

To start the saw, place it on firm ground,

PLATE 19.14 *A lightweight chain saw with a 1.3 kW two stroke engine and 350 mm guide bar. This saw is suitable for pruning, cross cutting and felling small trees.* (Sachs-Dolmar)

holding the right foot on the rear handle plate and the left hand on the front handle. Engage half throttle, set the choke if the engine is cold, switch on and pull the starter cord firmly. When the engine starts, run the saw and check chain brake operation.

Never start work until the immediate area is cleared of obstructions which could restrict escape in an emergency, and keep bystanders well away.

The saw should be running at full speed before starting to make a cut. Keep the saw body close to the timber whenever possible.

Before attempting to use a chain saw for felling, attend a training course or get a skilled person to show you the various techniques for felling and logging. Only experienced chain saw users should tackle hung and wind blown trees; this can be a dangerous operation, especially for those with little experience.

Suggested Student Activities

1. Look for the different types of hedge trimmer and trace the drive from the tractor to the cutting unit.
2. Use a nylon cord trimmer to carry out the three cutting and sweeping operations described in this chapter. Make sure you wear suitable protective clothing.
3. Study the regulations concerning the safety clothing required when using a chain saw.
4. Attend a practical training course on maintenance and the safe use of chain saws.

Safety Check

Do not allow anyone to ride on a tractor drawbar when towing a trailer or other wheeled implement. This practice is against the law and is extremely dangerous. The only safe place for a passenger is on the floor of a trailer or in the tractor cab which has a passenger seat.

Chapter 20

THE WORKSHOP

Every horticultural holding has a workshop. This may be a building set aside for the purpose or a bench at the back of a shed. Even the smallest workshop needs a basic set of hand tools, a power drill and a grinder to help keep a keen edge on a variety of cutting tools. Larger holdings often have a well equipped workshop with enough hand and power tools to carry out all manner of construction and repair work.

WORKSHOP HAND TOOLS

Spanners

There are several easily recognisable types of spanner, but spanner sizes are a mystery to many occasional mechanics. This often results in taking the easy way out with an adjustable spanner—a simple solution, provided that the jaws are not stretched and the thread is not worn, in which case it will soon remove the corners from every nut and bolt head.

Open ended or flat spanners are part of every toolkit. However, they are not always convenient for undoing nuts in confined spaces, and they can also slip off the nut or bolt head if used carelessly.

Ring Spanners fit around the nut or bolt head and are less likely to slip off. They are more suitable for nuts in confined spaces as only one-twelfth of a turn is necessary before the spanner can be repositioned on the nut. Ring spanners may be flat or have cranked ends, which make them suitable for nuts in a small recess. Ring and open ended spanners have a different size at each end.

Combination spanners have an open jaw at one end and a ring at the other; the open jaw and the ring are both the same size on the spanner.

Socket spanners are more expensive but have many advantages, especially speed and the ability to reach nuts in recessed holes. They can be used with a range of handles, including a speed brace, ratchet, tee-handle and long bar which can be attached to the socket or used with extension bars to reach nuts inside a component and other awkward places. A universal coupling allows the handle to be used at an angle to the socket. The standard range of sockets has a 12 mm square drive, but larger and smaller square drive sockets are also made.

Box spanners, often called plug spanners, are a cheap and effective form of socket spanner. They have six faces while sockets have twelve, and are turned with a steel rod or tommy bar.

Adjustable spanners may be a mechanic's friend but they will also take the skin from knuckles very quickly. New adjustable spanners have parallel jaws and close fitting adjuster thread. They are made in various sizes, according to handle length and maximum jaw opening width. When the jaws of an adjustable spanner become stretched and worn, it should be thrown away because it will take the corners off nuts or slip off completely, often with painful consequences.

Torque spanners are used to tighten a nut or bolt to a precise setting or torque. Used with sockets, they are a mechanics tool needed to

PLATE 20.1 *Pressure washing.*

tighten, for example, cylinder head studs after an engine overhaul.

Allen keys are 'L' shaped and hexagonal in section. They are used for Allen screws, which have a hexagonal recess in the head.

Spanner sizes

Spanner size is related to the thread on the nut or bolt which may be coarse or fine. A coarse thread has more strength, while a fine thread is less likely to vibrate loose in service.

The main thread types in use are:

Metric threads have been used on continental equipment for many years. Metric thread is now standard on most horticultural equipment. Spanners suitable for metric nuts and bolts are marked with the width across opposite faces or flats of the nut in mm. A metric spanner marked 22 fits a nut which is 22 mm across the flats.

American National Coarse and Fine threads (ANC and ANF) will be found on some horticultural equipment of American origin. The spanners are usually marked A/F; the size refers to the distance across the flats. For example, a ½ inch A/F spanner fits a nut ½ inch across the flats.

British Standard Whitworth and Fine threads (BSW and BSF) are the coarse and fine threads found on some older horticultural machines. Spanners for these thread rates are marked with the bolt diameter; a spanner marked ⅜ inch has jaws much wider than

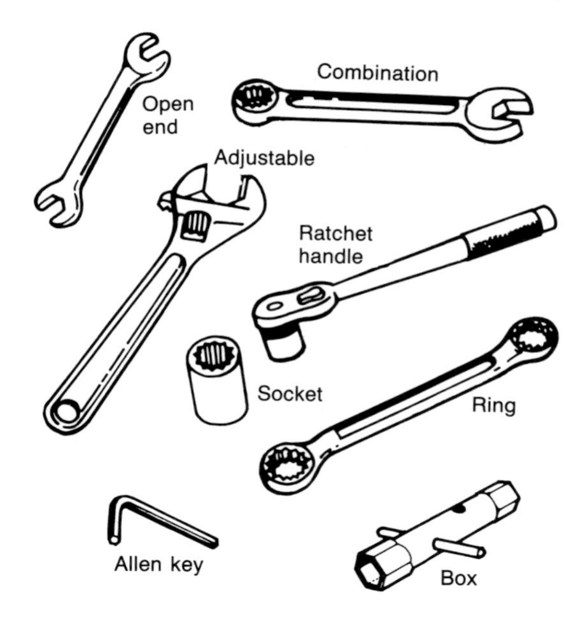

FIGURE 20.1 Types of spanner.

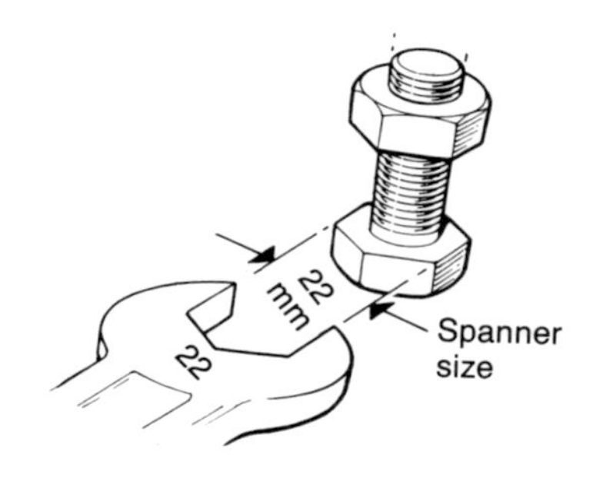

FIGURE 20.2 Spanner size.

3/8 inch. A 3/8 BSW spanner also fits a 7/16 BSF nut.

Unified National threads, both coarse and fine (UNC and UNF), were used before metric thread rates were adopted. The nuts and bolt heads, usually marked with a line of small joined circles, fit A/F spanners.

British Association threads (BA) are a very fine thread rate used for small diameter nuts and bolts, mainly found on electrical equipment. The very small spanners are marked with the BA size. The most common sizes are 0 BA (6 mm diameter) to 10 BA (1.7 mm diameter). Much electrical equipment has small diameter metric thread nuts and bolts.

Screwdrivers

There are several types of screwdriver with short, medium or long blade lengths and various blade widths. They are meant to be used for turning screws and small bolts with straight or cross slotted heads. The blade will soon be bent or damaged if used as a lever, and paint stirring is more effective with a stick!

Cabinet makers' screwdrivers have wooden handles. They are used for carpentry and the handle should not be hit with a hammer as such treatment will eventually split it.

Engineers' screwdrivers have square or round metal shanks in tough plastic handles. They can be used when it is necessary to give the screwdriver handle a sharp tap to help free a rusted screw.

Electricians' screwdrivers have tough, insulated handles made of bakelite or plastic.

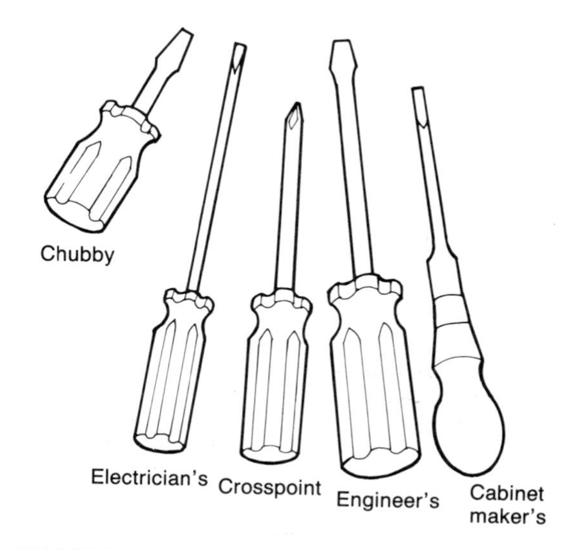

FIGURE 20.3 Types of screwdriver.

They are designed for use with electrical installations (after the supply has been turned off), vehicle electrics and light engineering work.

Cross point screwdrivers also known as Phillips and Posidrive screwdrivers are made for turning screw and small bolt heads with star shaped slots. Four point sizes are available.

Ratchet handle screwdrivers are useful for high speed work. Most have a set of different size flat and crosspoint blades.

Sharpening screwdrivers is best done with a file. A grindstone can be used, but the tip of the blade must not be overheated as this will make it very brittle. The end of the blade should be flat and square, not sharpened to a knife edge. A blade with rounded corners will not grip in the screw slot properly and will damage the slot. Choose a screwdriver blade of similar width to the slot in the screw. Using the wrong size blade will damage the screw head and the blade.

Cold Chisels

These are used for cutting small pieces of metal, splitting seized nuts and various other jobs. They are also used by some to cut brick and concrete, though a brick bolster with a wide blade is better for this purpose. The best brick bolsters are those with a hand guard on the handle.

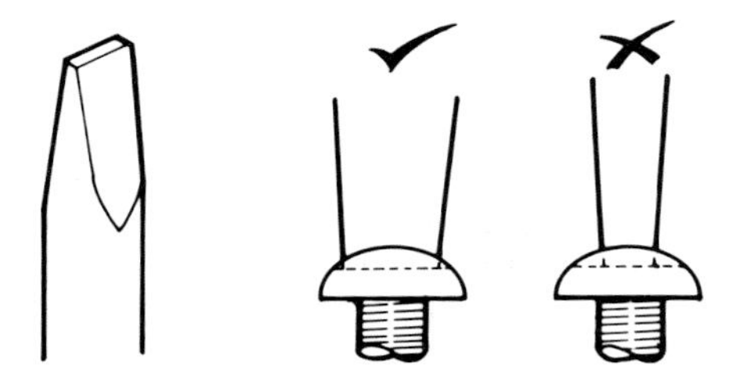

The correct shape for a screwdriver blade.

Always use the correct sized screwdriver; one which is too narrow will damage the slot in the screw head.

FIGURE 20.4 Correct and incorrect sized screwdrivers.

The end of a cold chisel handle will become mushroom shaped with use. In this condition it is dangerous, as a glancing hammer blow may cause small pieces of metal to fly off the handle which could injure you or a bystander. Another risk is that the hammer may slip off the chisel head with painful results. For safety's sake, grind off any burrs which form on a cold chisel handle and wear goggles when chiselling at metal or masonry.

Some small cold chisels can be sharpened with a file, but a grindstone is normally used. Prevent overheating at the cutting edge by frequently dipping it in cold water. A cold chisel has two taper angles with the cutting edge ground to an angle of about 60 degrees.

Pliers

Engineers' or combination pliers should be part of every toolkit. The jaws have flat faces for holding flat material and a rounded section for gripping bolts and other round material. The inner part of the jaws has a knife edge for cutting wire. Some engineers' pliers have slots on the outside of the handles, close to the hinge, which can be used to cut wire.

Side cutting pliers are used to cut wire, split pins and similar items. Side cutters are also very useful for removing stubborn split pins when repairing machinery.

Pointed nose pliers have rather limited use but are very handy for holding small parts in confined spaces. They should be used with care to avoid breaking off the tips of the jaws.

Circlip pliers are used to remove circlips from shafts, bearings and other parts. Pliers are available for both internal and external circlips.

Electricians' pliers may be of the engineers' side cutting or long nosed pattern, but with heavily insulated handles. They should only be used for electrical work after the power has been turned off. Electricians also use

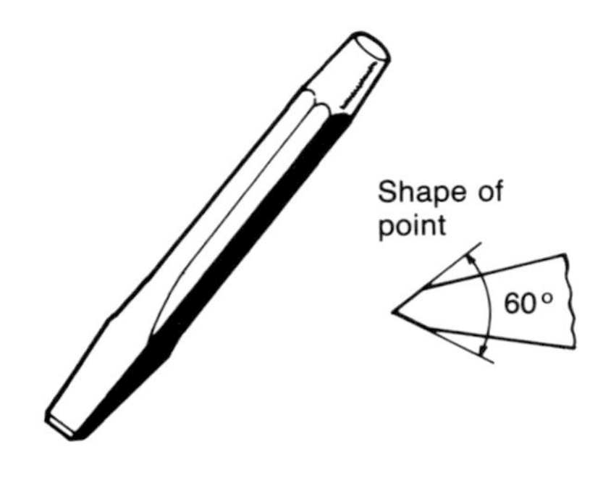

FIGURE 20.5 Cold chisel.

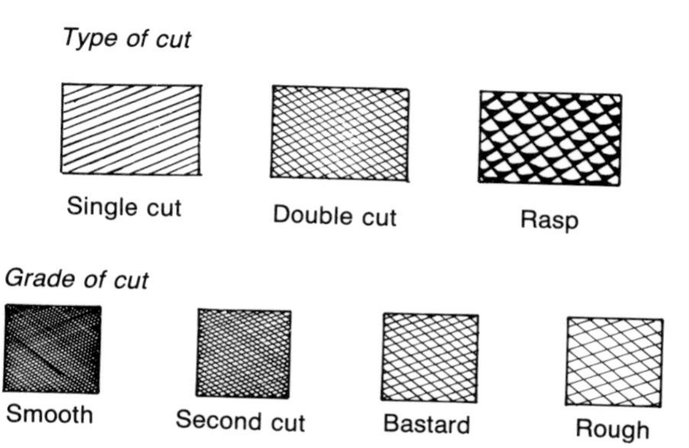

cable stripping pliers to remove the insulation from wire before connecting it to sockets, etc. The pliers can be adjusted to suit various wire sizes and are an efficient way of stripping off the insulation when wiring a plug.

Files

Files are made in a range of sizes, shapes and types of cut. A small selection of files is needed in any workshop, and careful storage and use will extend their working life.

Single cut files have rows of teeth running across the file in one direction only. Sometimes called reaper files, they are useful for sharpening cutting edges such as the blades on a reciprocating knife grass cutter.

Double cut files have rows of teeth running diagonally across the file in two directions to form small pyramid shaped teeth. They are used for general filing work.

The number of teeth on a file will determine how much metal it will remove. The following table is a guide to the main file cuts:

Type of file	*Teeth per cm*	*Teeth per inch*
Smooth	24–26	60–65
Second cut	12–16	30–40
Bastard	8–12	20–30
Rough	6–8	15–20

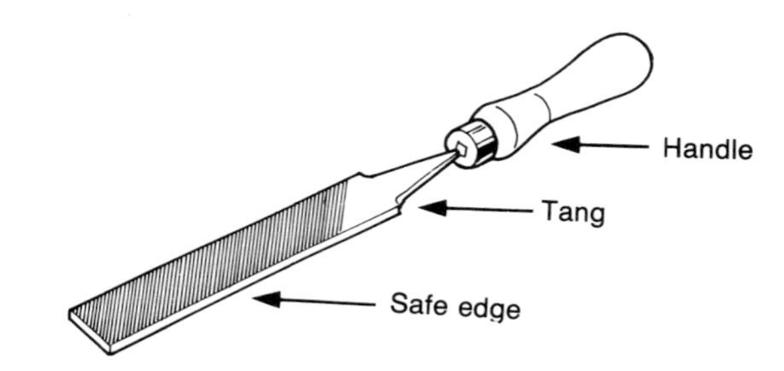

FIGURE 20.6 Files—grades and types of cut.

There are a number of different file shapes and sections. Common file shapes are round, half round, triangular, flat and flat with a safe edge. A safe edge file, which has one smooth edge, can be used to file a square corner without undercutting it with the file edge while using the teeth on the face of the file.

Using files

File teeth are meant to cut on the forward stroke only; hand pressure applied to the file on the cutting stroke should be released on the return stroke. Use the full length of the file and do not work too fast; one cutting stroke per second is best.

The teeth will soon become clogged if a file is used on dirty or rusty metal. Files should be cleaned with a file card, which is similar to a wire brush but has short, stiff wire bristles on a flexible backing.

Do not store files so that they can rub against each other as this will ruin the teeth.

Keep files wrapped separately in cloth or keep them upright in a stand. Always put a handle on the tang before using a file as the sharp point can cause a nasty injury to your hand.

Hammers

The best quality hammers are made of high quality carbon steel with a hickory wood handle. Less expensive hammers may have an ash handle. The head is secured to the handle with hardwood or steel wedges.

Hammers are made in various sizes, denoted by the weight of the hammer head. A small 0.23 kg (½ lb) hammer is useful to cut paper gaskets and other light work. A 1.1 kg (2½ lb) engineers' hammer is a basic piece of equipment for a mechanic, and plenty of work will be found around the holding for a 3.2 or 4.5 kg (7 or 10 lb) sledge hammer.

Engineers' or ball pein hammers are the most useful type of hammer in the mechanic's toolkit. They have a ball at one end of the head and a flat face at the other. The ball pein is used for riveting.

Cross pein hammers have a blunt, chisel shaped head instead of a ball. They are really a carpenters' hammer.

Soft faced hammers with copper, hard rubber or hide faces should be used when an ordinary hammer would damage the work, for instance when hammering items made from brittle or soft materials such as cast iron, copper and brass.

A *claw hammer* with a nail drawing claw forming part of the head is another tool for the carpenter. With a good quality steel shaft, it makes a useful general purpose hammer.

The *club hammer* or lump hammer is used mainly by bricklayers, but its short handle and heavy head make it a handy general purpose tool.

A *sledge hammer* is a must for most holdings. It is ideal for straightening bent components

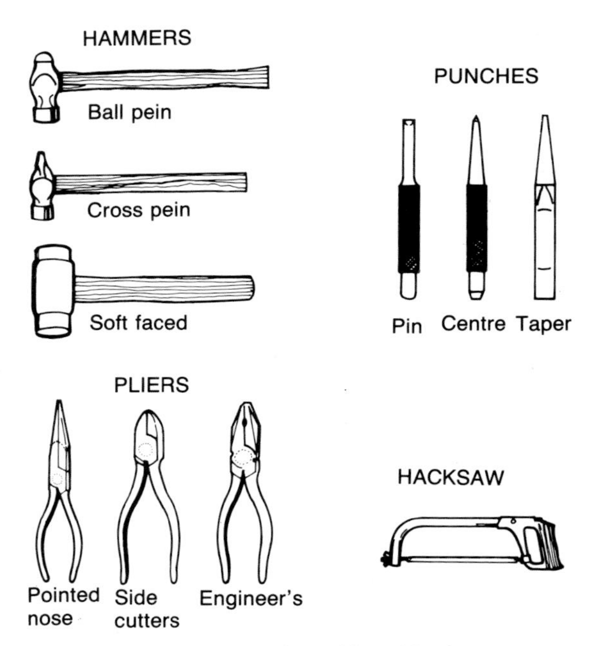

FIGURE 20.7 A selection of hand tools.

and driving steel posts into the ground. A large wooden mallet or beetle should be used for wooden stakes.

Using hammers
Always hold a hammer close to the end of the handle or shaft and use your wrist as the pivot point. When it makes contact, the hammer face should be parallel with the part being struck.

Do not use a hammer with a loose or damaged handle. It is usually possible to tighten the head by driving an extra wedge into the end of the handle. Hammer faces are very hard; do not strike two faces together as this could crack or damage them.

Hacksaws

Metal cutting hacksaws usually have adjustable frames which will take blade lengths of 250 or 300 mm (10 or 12 in).

The blade must be fitted with teeth pointing away from the handle. It is tensioned by taking up all the slack and then tightening the wing nut two or three more turns. The blade used must have the correct number of teeth for the job in hand, and at least two teeth must always be in contact with the metal. There is a choice of flexible or hard blades. The flexible blade is best for the handyman as it will not break with careless use. A hard blade will give an accurate cut, but soon snaps if the saw is twisted while on a cutting stroke.

- A coarse blade with 7 teeth per cm (18 teeth per in) is best for thick sections and soft metals.
- For general work, use a blade with 9.4 teeth per cm (24 teeth per in).
- A fine blade with 12.6 teeth per cm (32 teeth per inch) is required when cutting very thin section material.

Using hacksaws
Always cut with the full length of the blade, making about one cutting stroke per second. Apply steady pressure on the saw for the forward cutting stroke and release that pressure on the return. Very little saw pressure is needed when cutting thin section materials. When possible, secure the work in a vice before starting to saw. Slacken the blade tension when the saw is not being used.

Punches

Taper punches are mainly used to line up pin and bolt holes when assembling equipment. They are made in various lengths and diameters.

Parallel or pin punches as the name suggests are the same diameter throughout their length. They are useful for driving out bolts and pins from various components such as gears, pulleys and shafts. A taper punch is not really suitable for this job as it would probably become stuck in the pin hole before the pin is removed. A pin punch is likely to break if used as a lever to align holes.

A *centre punch* is used to mark the centre spot for drilling a hole in metal. The impression made by the punch guides the point of the bit when starting to drill.

Workshop Power Tools

Every workshop, no matter how small, needs an electric drill and probably a small bench grinder. Most power tools run at mains voltage (240 volts), but it is possible to buy tools which run at a lower voltage for added safety. For even safer operation, there is a range of cordless power tools which can be used where no mains power is available. These tools have rechargeable batteries from 4–24 volts depending on the size of the tool and the power required.

Some workshops have an air compressor. Apart from its obvious use for pumping up tyres, there are many types of compressed air operated tools, including small hand grinders and power drills, which eliminate all risks of an electric shock.

Drills

An electric hand drill is an essential workshop power tool. With suitable attachments it can be used for sawing, sanding, wire brushing and polishing as well as boring holes in wood, plastic and metal. There are many sizes of electric drill, from the handyman's 8 mm capacity model to a heavy duty professional model with the ability to drill holes of 19 mm diameter. Larger drills, fixed in a drill stand, can be used to drill holes of 25 mm diameter or more.

Most drills have two speeds and many have a hammer setting which is useful when drilling holes in masonry. Use the correct drilling speed for the job in hand. It is better to use the slower speed for holes over 5 mm in steel and other hard materials. The chuck capacity limits the size of parallel shank drill bit which can be used. However, large diameter drills with a reduced diameter shank to fit the chuck are available. These drill bits are very suitable for making quite large holes in timber, plastic and other soft materials; it is better to use a pillar or heavy bench drill for drilling large holes in steel and similar hard metals.

Twist drills
These are used for drilling holes in metal. There are two types:

- Low speed drill bits, which are not very expensive, are only suitable for hand drills or for power drills when working in very soft materials.
- High speed twist drills made of good quality carbon steel are much more expensive. They are suitable for electric drills and with careful use will give long service.

Twist drills for use with hand held drills have parallel shanks which must be tight in the drill chuck. Drill sharpening is a skilled job which can be learned with practice. The point should have a cutting angle of 60 degrees. It can be sharpened with a grinder but care must be taken to avoid overheating the tip. Use a new twist drill as a guide to get the correct angle.

Using drills

- Check that the drill bit is fully inserted into the chuck and tighten it firmly using the chuck key in all three positions.
- Keep your drill bits sharp. There is little point in trying to force a drill through metal if it is blunt.
- Clamp the work securely, especially when using a pedestal drill or a hand drill in a bench stand.
- A centre punch dot mark will help to start the drill bit into work.
- Large holes are easier to drill if a small pilot hole is drilled first. Use a slower drilling speed for large diameter holes.
- Care is needed when drilling holes in thin sheet metal. Clamp the sheet so that it does not pick up with the drill bit when the hole is almost complete. For best results, it helps if the sheet metal is placed on a piece of wood before starting to drill.
- The drill bit will cut better and give a good finish if it has suitable lubrication. It should be lubricated with water soluble oil when drilling steel. This is a white fluid which can be bought from a tool shop and diluted with water. Cutting oil, as it is usually known, has good lubricating properties and helps to cool the drill bit. Cast iron should be drilled dry, and soft metals such as aluminium and copper can either be drilled dry or lubricated with paraffin.

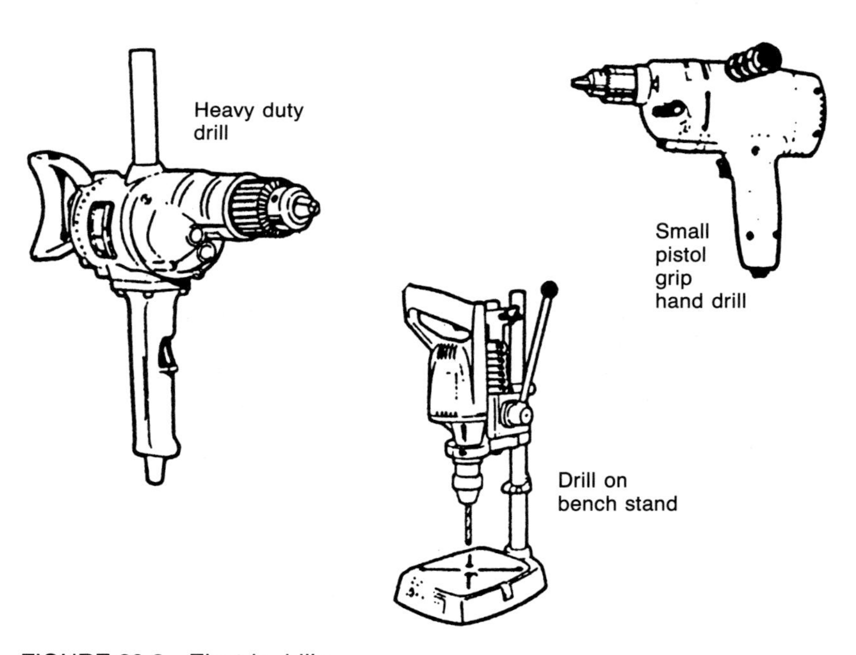

FIGURE 20.8 Electric drills.

Bench Grinders

A small bench grinder is another useful item of workshop equipment. There are various models of bench and floor standing grinder, and the model selected will usually depend on the type of work to be done. For most horticultural workshops a 150 or 200 mm (6 or 8 in) bench grinder should be adequate.

Most grinders have two abrasive wheels, one fine and one coarse. The fine wheel should be reserved for sharpening drills, cold chisels and cutting tools such as bill hooks and axes. Some grinders have a wire brush at one side and an abrasive wheel at the other.

Using bench grinders

- Always use goggles to protect your eyes from flying sparks and stray metal chips.
- Try to use the full face of the grinding wheel and do not exert any pressure on the side of the wheel.
- Set the work rests close to the grinding wheel; the metal being ground may be pulled between the rest and the wheel if the gap is to wide.
- Keep the grinding face wheels in good condition. They will become uneven with use and can be restored to the correct shape with a stone dresser; one type has a set of hard rotary cutters which is held against the face of the grinding wheel until the edge is restored to its correct shape.

Angle Grinders

Angle grinders are similar to an electric drill but have a sanding or grinding disc running at right angles to the motor drive shaft. Sizes vary from a small handyman model with a 115 mm (4½ in) diameter disc up to a heavy duty 300 mm disc. Many workshops have a 180 mm (7 in) angle grinder. An angle grinder is useful for grinding work in situ, e.g. a piece of machinery or a large piece of metal. The grinding disc can be replaced with a cutting disc for masonry, roof sheets and thin section steel. The smaller models may be fitted with sanding discs and polishing mops.

Using angle grinders

- Always wear goggles. A particle mask is also needed when using a grinding disc.
- Angle grinders are very noisy so wear a pair of ear defenders to protect your hearing.
- Check that no other person is at risk from flying sparks when you are angle grinding.
- Clamp small items before using the angle grinder.
- Make sure that the electric cable is in a safe position before starting work.

Compressors

Most air compressors are driven by an electric motor. Some are small portable units, while others are larger, permanent installations with a fixed air line around the workshop.

A portable compressor with a small air receiver is useful in the workshop for inflating tyres, spray painting, cleaning down equipment and driving compressed air power tools.

Using compressors

- Wear goggles when blowing dirt from machinery and equipment with compressed air.
- Never direct a jet of compressed air at the body as this can cause serious injury.
- Water from the air will build up in an air compressor tank. This must be drained away at frequent intervals to prevent damage to the tank.

WORKSHOP MATERIALS

All workshops have at least a small stock of metal, nuts, bolts and other hardware items. Many will carry a selection of replacement parts such as filters, fan belts, rotary mower blades and similar items.

Numerous metals and other materials are used in the construction of horticultural machinery. Metals have various physical properties which make them ideal for some applications and totally useless for others. Cast iron, which is brittle, is excellent for making intricate engine castings, but quite useless for such items as a trailer drawbar.

A basic knowledge of the more important metals is needed by anyone using a workshop to make or mend large or small pieces of equipment for the nursery, glasshouse, sports ground, etc.

Ferrous Metals

Ferrous metals contain iron. There are many types including cast iron, carbon steel and alloy steels used in the construction of horticultural machinery and equipment. Ferrous metals contain carbon in varying quantities according to the type of iron or steel.

Cast iron

Cast iron is relatively cheap and is easy to drill, machine and weld. It is rather brittle but has great strength in compression (i.e. it resists crushing forces). It is used to make castings, as its name suggests, sometimes of quite intricate shape. Molten cast iron is poured into moulds made from sand to produce the required shape of casting.

Most types of cast iron contain between 3 and 4 per cent carbon plus small amounts of manganese, phosphorus, silicon and sulphur. The actual amounts of these materials varies with the grade and type of iron. When molten metal is allowed to cool naturally in a casting box, grey cast iron is produced. Rapid cooling, usually achieved with a chilling block in the casting box, results in white cast iron which is harder and more brittle.

Chilled cast iron is in fact white cast iron, and is sometimes used to make soil wearing parts such as plough shares. The underside of the share is chilled to give a self sharpening effect because the softer top of the share wears faster than the hard underside.

Cast iron has many uses including gear box housings, engine cylinder blocks and cylinder heads (small engines often have aluminium alloy cylinder blocks and heads).

Malleable cast iron castings with a 0.5 to 1 per cent carbon content start off as ordinary iron castings. Special heat treatment removes most of the carbon to remove the brittleness. This rather expensive process is used to produce castings which must be able to withstand shock loads.

Wrought iron, with a minute 0.1 per cent carbon content, is almost pure iron. It is very expensive and rarely used. Old ornamental wrought iron gates are probably made from wrought iron, but almost all modern ones are made from mild steel, which is much cheaper.

Steel

Low carbon steel (mild steel) is the cheapest and most common form of ferrous metal used to make horticultural machinery and equipment. Low carbon, or mild steel, contains from 0.1–0.3 per cent, and it is easy to saw, drill, machine, bend and weld. It has no great strength and rusts very quickly if it is not protected from the weather. Painting, galvanising (zinc coating), tin plating and plastic coating are some of the methods used to prevent corrosion.

Uses of mild steel include garden tractor bonnets and mudguards, machine frames, sheet metal work and guards on many horticultural machines.

Medium carbon steel is harder and tougher than mild steel, and contains between 0.3 and 0.7 per cent carbon. It can be forged, welded, drilled, machined and sawn, but its increased hardness makes these processes more difficult. It can be annealed (softened) to make the machining processes easier, but it will then need further heat treatment to restore its hardness.

Uses of medium carbon steel include shafts and axles, plough discs and beams, cultivator tines, etc.

High carbon steel has a carbon content ranging from 0.7–1.4 per cent. The higher carbon content indicates that this is a hard though rather brittle type of steel. It is used to make good quality tools including chisels, punches, files, and high speed twist drills.

Alloy steel contains one or more non-ferrous metals to give the steel additional properties. Some examples of alloy steels include:

- *Stainless steel* contains chromium to make a non-corrosive material used for some spraying equipment.
- *Tool steel* has a small quantity of tungsten which gives a good cutting edge.
- *Nickel steel* is very tough and has the necessary strength for making high quality steel axles and shafts.

Heat treatment of steel

Various heat treatments are required when manufacturing tools and machine components from carbon steels. Mild steel, because of its low carbon content, is not suitable for most heat treatment processes.

Hardening

Medium and high carbon steels can be hardened by heating the metal to red heat and then quenching it in cold water. Quenching (rapid cooling) gives added hardness but makes the metal rather brittle.

Tempering

Tempering removes the brittleness from hardened steel. The metal is reheated to a certain temperature and then quenched again. As the temperature increases, the colour changes from straw to red, brown, purple and blue. These colours are used to determine the quenching point.

A cold chisel is tempered by applying heat to the handle above the cutting edge, which is first polished with emery cloth. The chisel is then quenched when the cutting edge turns dark purple.

Annealing

Annealing softens metal to make it easier to machine. Steel is annealed by heating it to red heat and then allowing it to cool very slowly.

Case hardening

The heating and quenching process is used to give steel a hard wearing surface with a softer shock-resistant centre. Steels with a carbon content of 0.3 per cent or more can be case hardened by a factory process. Examples of case hardening include the teeth on gears and ball bearings.

A different process is necessary for case hardening mild steel with its low carbon content. It is heated to a bright red colour and then dipped in a carbon rich powder, which can be purchased from good tool shops. The metal is then reheated to bright red, so that the surface of the steel absorbs some of the carbon, and then quenched in clean cold water. The process can be repeated until the required degree of hardness is obtained. A file can be used to test the hardness of the surface.

Non-Ferrous Metals

There are many non-ferrous metals but few are used in the manufacture of horticultural machinery. Some non-ferrous alloys—a mixture of two or more non-ferrous metals—are also used to make horticultural equipment. The more important non-ferrous metals and alloys include the following:

Copper is a good conductor of heat and electricity. It can be welded, is easy to machine and resists corrosion. It is a soft metal but hardens with use or age. It can be annealed by heating to cherry red and then quenching in clean, cold water.

The more important uses of copper in horticulture include electrical components, water pipes and fittings.

Zinc is a corrosion resistant metal used to coat (galvanise) steel sheets, tanks and pipes.

Lead is used for making the plates in batteries for tractors and garden machinery. It is a very heavy metal with a bright appearance when new, but it soon becomes coated with lead oxide giving it a dull grey colour. This coating protects the lead from further corrosion. Solder is an alloy of lead and tin.

Aluminium is a very light metal of moderate strength used in a pure form for glasshouse construction, sheeting and irrigation equipment.

When alloyed with other metals including copper, zinc and silicon it can be used for castings. Examples include pistons, small engine cylinder blocks, irrigation equipment, etc.

Brass is an alloy of copper and zinc. The two main types are soft brass, which contains about 40 per cent zinc, and hard brass, which has more than 40 per cent zinc. Soft brass is easy to work and is used to make wire, strips and tubing. Hard brass is used for castings such as water supply fittings, sprayer parts, taps, bushes and some electrical components.

Metal Identification

Many non-ferrous metals are easily recognised by their colour and appearance, but the various types of iron and steel are more difficult to identify.

Iron and steel can be recognised by its hardness, the sparks produced when it is ground and whether it will bend. Hardness can be tested with a hacksaw or file.

Low carbon, or mild steel, is soft, easily bent and gives large quantities of white, spear shaped sparks with a few stars when held against a grinder.

Medium carbon steel is difficult to bend and will probably break. It is quite hard, and when ground produces star shaped sparks with a few spears, though fewer are produced than when grinding mild steel.

High carbon steel is very hard and tough and gives off a medium quantity of white star shaped sparks when ground.

Cast iron is easily recognised by its granular surface. It produces a small quantity of reddish sparks when held against a grinder.

Plastic

Plastic does not corrode, making it an ideal material for fertiliser spreaders and sprayers. It is also an electrical insulator. There are two main types of plastic:

Thermoplastic plastic will become pliable when subjected to gentle heat, but will distort if overheated. Polythene and PVC are two examples of thermoplastics. They are used for sheeting, water pipes and fittings, buckets, etc.

Thermosetting plastic is a hard material, which will not soften or melt if subjected to gentle heat, but will be destroyed if exposed to high temperatures. Bakelite is an example of this material used for electric switches, plug sockets, coil ignition distributor caps, etc.

Nuts and Bolts

Many types of nut and bolt can be found on horticultural equipment. *Hexagon headed bolts* need no description. *Carriage bolts* are used to fix timber to timber or steel. The square section under the head pulls into the wood to stop the bolt turning while the nut is tightened. *Plough bolts* are made in various patterns, and are used, for example, to fix a plough mouldboard onto the frog. The countersunk head with a flat top gives a smooth surface on the face of the mouldboard.

Machine screws may have round or countersunk heads. They are smaller in diameter than ordinary bolts and have thread right up to the head. *Set screws* are used in threaded holes to lock pulleys or gears on to their shafts. *Grub screws* serve a similar purpose but they are flush with the pulley hub when tightened.

Most nuts are hexagon shaped but some square nuts are used. *Wing nuts* can be tightened with the fingers and are useful where frequent adjustments are necessary. There are different types of *lock nuts.* Some of these have a soft metal or plastic insert built into the nut. The hole in the insert is slightly smaller than the bolt thread. As the nut is tightened, the bolt cuts a thread in the insert, and this is sufficient to lock the nut in position. Another type of lock nut is castle shaped, with a split pin which is passed through a hole in the bolt after it has been lined up with two slots in the castle nut. Where machine parts, such as a lever, must pivot on a bolt, two nuts, one tightened against the other, can be used.

The tendency for nuts to vibrate loose can also be overcome by putting a spring washer or a star washer under the nut. Sharp edges on the washer grip the nut to help keep it tight on the bolt. A tab washer (a flat washer with two or more locking tabs) may also be used. Some of the tabs are hammered

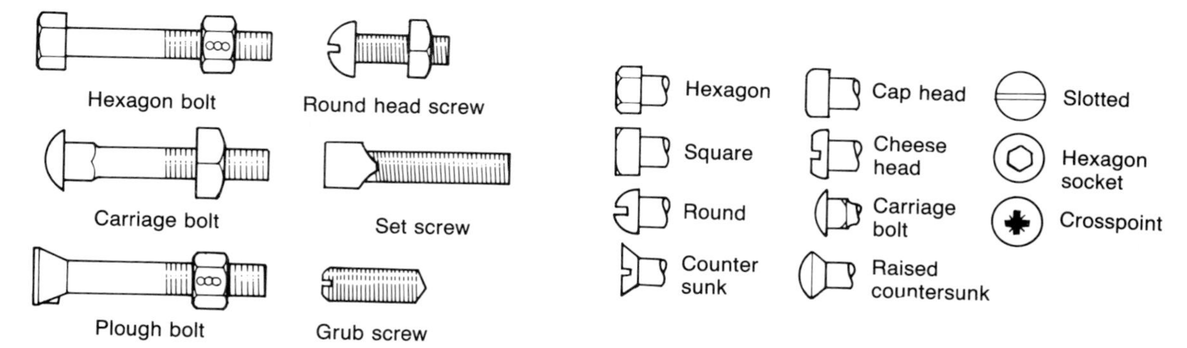

FIGURE 20.9 Types of nut and bolt.

upwards around one or more faces of the nut, while others are bent downwards over the edge of the component being held by the nut. Tab washers are often used to stop big end bearing nuts coming loose.

Workshop Safety

Unsafe hand tools, faulty electrical equipment, untidy benches and dirty floors can cause accidents in the workshop. A few simple precautions combined with general cleanliness will help prevent accidents.

The *workshop* should be warm and well lit with a concrete floor. Cold hands are easily bruised or cut when doing winter maintenance work.

Hand tools should be checked and any found to be unsafe should be repaired or discarded. Particularly check for hammers with loose heads, open ended spanners with stretched jaws and worn adjustable spanners.

Avoid wrapping your fingers around a spanner when undoing a nut or bolt. Apply pressure with the palm so that if the spanner slips, there is less risk of bruising the knuckles.

Power tools such as electric drills and grinders are a potential hazard if they are not well maintained. Do not use an electric power tool without a good earth or double insulation. This is most important when using electrical equipment in glasshouses or around the holding where there are cables trailing over damp ground or concrete. Check extension cables and scrap any which have damaged insulation. The use of a residual current operated circuit breaker is the only real safeguard when using mains electrical equipment in the workshop or around the holding. Better still, use cordless power tools.

Check plugs and cables frequently, report any damage, and do not use the equipment until it has been repaired. Use a fuse of the correct amp rating in fused plugs. Electricity cannot be seen but it can kill, so treat it with respect.

Jacks must be properly maintained. Do not work under a jacked up vehicle until it is supported with axle stands or strong wooden blocks. Never leave a jacked up vehicle without the extra support of an axle stand—someone may accidentally let it down.

Asbestos was widely used in the past for insulation purposes. This has now been proved a hazardous substance. Authorities advise that it should only be removed and taken away by contractors who are equipped for this work.

Some brake and clutch linings contain asbestos. Do not use compressed air to remove dust when servicing tractor brakes which have asbestos type linings, and do wear a dust mask.

A *first aid box* must be provided on all holdings in order to comply with Health and Safety Regulations. Information is available from the local health and safety inspector about the contents, which vary according to the number of people employed on the unit. It is also necessary for a member of staff to have some basic first aid training.

Personal Protection

Always wear goggles when using a grinder or portable electric tool which causes sparks. Goggles are also needed when using compressed air to blow away dust and dirt.

Ear defenders will safeguard your hearing. Above certain noise levels, the health and safety regulations require them to be worn.

Wear protective footwear. Shoes and trainers provide no protection against heavy objects dropping on your feet. Safety boots with steel toecaps should be the rule in the workshop. Boots with steel studs in the soles can also be a hazzard, especially on greasy concrete floors (which should, of course, never get into such a state)!

A boiler suit is the best thing to wear in a workshop or anywhere that maintenance work is done. Long flowing clothes and scarves should be avoided, especially when working with machinery.

Many people suffer from skin irritation after handling diesel fuel which can cause dermatitis, an unpleasant skin complaint. The use of a good quality barrier cream can help avoid this.

Horticultural spray chemicals present many hazards. Training, which places emphasis on personal protective clothing, is required for people who handle or apply spray chemicals.

Fire Prevention

All fire extinguishers should be readily accessible in an emergency. It is important to know which type of extinguisher can be used on a particular fire. A water based extinguisher must not be used on a fire caused by an electrical fault. When discovering such a fire, switch off the power supply, and call the fire brigade.

There are several types of fire extinguisher; make sure you can identify them and know when they are safe to use.

Dry powder extinguishers (coloured blue) and *carbon dioxide extinguishers* (coloured black) are suitable for use on live electrical fires and burning liquids such as oil, grease and paint. They can also be used to deal with small fires involving wood, paper and similar materials.

Foam extinguishers have yellow canisters. They must not be used to tackle live electrical fires because foam is an electrical conductor. Their main use is for fighting small fires involving wood, paper, etc.

Water extinguishers (coloured red) have a small cylinder of gas in the water filled canister, and when this gas is released it forces out a jet of water. *Soda acid extinguishers* are also water based. Both water and soda acid extinguishers are suitable for fires involving materials such as wood and paper but must not be used to tackle burning liquids or live electrical fires.

Other fire fighting aids include smother blankets, usually stored in a container or packet, and buckets of water or sand which should be kept at locations where there is a high fire risk.

Suggested Student Activities

1. Learn to identify the various types of spanner.
2. Practice using a hacksaw and a file. Use all the teeth on the file and on the saw blade and try to keep the work square.
3. Ask someone to show you how to sharpen a drill, and then practice this task until you get it right.
4. Get to know where the different types of fire extinguisher are located on your nursery or holding and make sure you know which types of fire they may be used against.
5. Attend a basic first aid course; the knowledge gained could be very useful in an emergency.

SAFETY CHECKS

Eyes and ears are at particular risk in the workshop. Wear ear defenders when using noisy power tools such as grinders. Always wear goggles when using a grinder, cutting bricks and concrete, cleaning equipment with compressed air or any other job where you are likely to harm your eyes.

Chapter 21

POWER FOR HORTICULTURE

There are numerous sources of power available in the horticultural industry. The internal combustion engine is the main source of power for grass care and land work. Electricity provides the power for motors, lighting, heating and cooling. Oil, solid fuels and gas are important providers of heat for glasshouses and other protected cropping areas.

INTERNAL COMBUSTION ENGINES

The power for tractors, grass cutting equipment, standby generators and various items of maintenance equipment like chain saws and concrete mixers is supplied by a petrol or diesel engine.

The use of internal combustion engines inside buildings presents problems because the poisonous exhaust fumes must be piped away. The air cleaner needs to be able to draw its air from outside if the engine is running in a dusty building. The installation must also comply with all regulations applying to stationary machinery.

Petrol and diesel engines are of equal importance in horticulture. Almost all pedestrian controlled machines, lawn tractors and domestic ride-on mowers have a petrol engine, most of them air cooled. The more powerful tractors and many ride-on mowers and compact tractors have diesel engines.

Power is defined as the rate of doing work, and the unit of engine power is usually stated in horsepower or kilowatts. Many machinery sales leaflets quote engine power in horsepower, while others use kilowatts. One horsepower is approximately equal to 750 watts or 0.75 kW. This means that an engine developing 10 hp can also be rated at 7.5 kW. To convert hp to kW, divide the horsepower by four and then multiply by three. To convert kW to hp, multiply the horsepower by 0.75.

One horsepower equals 33,000 ft lb per minute. In theory this means that a 1 hp engine can move a load of 330 lb through a distance of 100 feet in one minute.

One kilowatt equals 1,000 newton metres per second (nm/sec), and 10 newtons are approximately equal to a force of 1 kg (2.2 lb). Here again, in theory a 1 kW engine can move a load (force) of 1,000 newtons a distance of 1 metre in one second.

Tractor specifications, including those for compact models, may refer to three different horsepower outputs. They are:

Brake horsepower (BHP), also brake kW, is the power available at the engine flywheel.

PLATE 21.1 *Standby electric generator driven by a single cylinder four stroke engine. It has a maximum output of 2 kW.* (Sachs-Dolmar)

It is measured with an engine dynamometer on a test bed under controlled conditions and the results obtained are converted to brake horsepower.

Power take-off horsepower (PTO HP), also PTO kW, is less than the power developed at the engine flywheel because the gear train to the power shaft will absorb some power. It is measured by coupling a dynamometer to the power shaft and carrying out a similar set of calculations to those for BHP.

Drawbar horsepower (DBHP), also DBkW, is the power available at the drawbar to do work. It will be less than both pto and brake horsepower because of the power used to drive the transmission system and hydraulics and through rolling resistance and wheelslip. When pushing a loaded wheel barrow on soft ground, it is hard work because the wheel sinks into the soil. This is due to rolling resistance, and tractor front wheels have the same resistance to overcome when moving forwards. This is a lesser problem with four wheel drive because the front wheels help to pull the tractor along.

A typical two wheel drive tractor which develops 30.8 kW or 41.4 hp at the flywheel has 25.7 kW or 34.5 hp available at the power take-off. The power at the drawbar will be about 23.5 kW or 31.3 hp.

Some engine power outputs are given in ps or cv. These are metric horsepower ratings; 1 metric hp or 1 ps or 1 cv is equal to 1.014 hp.

SAE, DIN or BS

Tractor sales leaflets may state engine power as an SAE, DIN or BS rating. These abbreviations indicate the testing system used to calculate the power output at the engine flywheel.

DIN is widely used in Europe. DIN 70020 (Deutsche Industrie Normen) is a test standard which assumes that the engine is installed in a vehicle with an air cleaner, cooling fan, water pump and exhaust system fitted. These components use engine power, so the power at the flywheel will be a nett figure.

SAE is a North American rating. SAE (Society of Automotive Engineers) J270 121 is a test standard using a bare engine on a test bed with cooling, air cleaning, etc., absorbing none of the power produced at the flywheel. The SAE rating is a gross figure and will therefore be higher than the DIN rating for the same engine.

BS is the British Standard rating. BS AU 141a:1971 is basically intended for road vehicles but with less strict requirements for smoke emission when applied to tractors. This too is a gross figure and will be higher than the DIN figure for the same engine.

ELECTRICITY

Electricity is a major source of power for horticulture. It is used to provide heat and light and to operate electric motors for pumps, ventilation systems, etc.

Electricity is generated by a network of power stations which supply the main distribution lines (National Grid) at a pressure of 275,000 or 400,000 volts. This is reduced to 132,000 and then again to 33,000 volts by transformer sub-stations. A village may have a mains supply of 11,000 volts, which again is reduced to 415 volt three phase supply for business users, such as large nurseries, and to 240 volts for domestic use.

Mains electricity on 240 volt supply is transmitted by two cables. Some growers will have a three phase supply. This is a combination of three separate single phase supplies arranged to give a pressure of 415 volts. Three phase electricity is very expensive to install but has advantages where heavy electrical loads occur.

Alternating current (AC) from the mains cannot be stored in batteries, so power stations are geared to meet the fluctuating demands of consumers. Tractor engine alternators and dynamos supply direct current (DC), which can be stored in a battery until it is required by the starter motor or other electrical equipment.

Measurement of Electrical Power

The *watt* is the unit of electric power. There are 1,000 watts in 1 kilowatt.

The *volt* is the unit of measurement for the pressure of electricity in a circuit. Domestic supply from the mains is 240 volts; a tractor battery has a pressure of 12 volts.

The *amp* is the unit of measurement of current flow in a circuit. Circuits and electric appliances have fuses to protect them from damage through overloading. The size of the fuse (rated in amps) used in a circuit depends on the amount of current flowing in that circuit. This can be calculated by using the following formula:

$$\text{Amps} = \frac{\text{Watts}}{\text{Volts}}$$

For example, an electric heater has a power rating of 1,000 watts and is connected to a 240 volt supply. The current flow in amps is calculated in this way:

$$\text{Amps} = \frac{\text{Watts}}{\text{Volts}} = \frac{1{,}000}{240} = 4.16 \text{ amps}$$

When amps and volts are known, the formula can be turned round to find the maximum power output of the circuit in watts:

$$\text{Watts} = \text{Amps} \times \text{Volts}$$

For example, to find the maximum possible power output in a 240 volt circuit with a 5 amp fuse:

$$\text{Watts} = \text{Amps} \times \text{Volts} = 5 \times 240 = 1{,}200 \text{ watts.}$$

In practice, a 5 amp fuse will be required for a 1,000 watt appliance using a 240 volt supply.

Most portable electric equipment has a fused plug which should be made of hard rubber rather than the plastic material used for domestic plugs. It is very important to have the correct size fuse in the plug. The formula above can be used to calculate the current flow and to find the correct fuse rating provided that the voltage and wattage are known. This information will be found on a small data plate on the appliance. The following table gives suitable fuse ratings for some items of workshop and other equipment.

Fuse rating (Amps)	*Maximum loading* (Watts)	*Typical use*
2	500	Inspection lamp Soldering iron
5	1,200	Small bench grinder 1 kW heater
10	2,400	Heavy duty power tools Angle grinder Electric lawn mower
13	3,000	Small portable welder Water heater

Three phase power tools and equipment are wired direct to an isolator switch and are protected by a circuit breaker or a fuse.

Replacing a Fuse

When you need to fit a new fuse cartridge or rewire a fuse holder:

- Always disconnect the power supply first.
- Replace the fuse with a cartridge or wire of the same rating.
- Call in an electrician if the fuse blows again as soon as the power is turned on. Never try to solve the problem by using a heavier fuse.

Electrical Safety Devices

Many electrical installations have *Residual Current Devices (RCDs)* instead of rewirable fuses at the distribution board. When an overload or dangerous situation occurs, the RCD cuts off the supply in milliseconds to give instant protection to both user and appliance. When an RCD trips out, first disconnect the appliance and then reset the RDC. If it fails to reset or trips out again as soon as the appliance is reconnected, an electrician should be called.

RCDs should be checked at regular intervals; they have a test button for this purpose. When the button is pressed, the RCD should trip out immediately.

When using electrical appliances, such as a mower, grass trimmer, hedge cutter or workshop power tools, outdoors, it is very important to have the protection of an RCD. The safest system is to use a residual current circuit breaker (RCCB) unit. This will cut off the supply instantly if there is a fault in the equipment or if the cable is accidently cut by the mower or trimmer. It is good practice to check the RCCB every time outdoor electrical equipment is used.

Wiring a Plug

Always connect the three wires of an electric cable to the plug terminals in this way:

- *Green and yellow* to the terminal on the largest plug pin. This is the earth connection.
- *Brown* is connected to the live terminal, the one connected to the fuse holder in a fused plug.
- *Blue* is attached to the third, neutral terminal.

A few electric appliances may be found with the old colour coding. Green is earth, red is live and black is neutral. It is likely that cable with these colours will have deteriorated and should be replaced by a competent electrician.

To wire a three pin electric plug, follow this procedure:

- Strip about 50 mm of the outer covering from the cable. Do this carefully so that the insulation on the inner wires is not damaged.
- Secure the unstripped cable in the cable grip so that the stripped wires are just clear of the grip.
- Lay the three coloured wires over their terminals. Cut each wire about 10 mm beyond the terminal screw.
- Remove the cable from the grip and strip about 10 mm of insulation from each of the coloured wires. Where each wire is made up of several strands, twist them together to give the cable extra strength.
- Lay the stripped cable across the plug and secure the outer cable in the cable grip. Twist the wires clockwise around the terminal thread so that they will stay in position when tightening the terminal nuts. Some terminals have a hole for the wire with a screw which is tightened to hold it in position. The cable grip is designed to prevent the wires pulling from the terminals if the cable receives an accidental tug.
- Fit the correct fuse for the appliance and check for stray strands of wire out of place. Finally replace the plug cap.

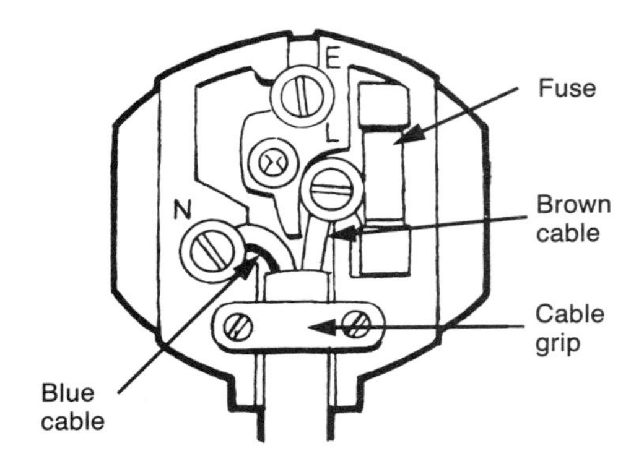

FIGURE 21.1 A three pin plug with the cap removed. The wiring indicates that it connected to a double insulated appliance because there is no earth wire.

Some electrical equipment has only two wires; one brown and the other blue. There is no green and yellow earth cable because the appliance is double insulated with a specially designed casing to protect the user if the appliance develops a fault. Such equipment has a double insulation mark consisting of a square with a smaller square inside it.

Electric Motors

Totally enclosed electric motors are normally used for equipment which works in dusty or damp conditions. Some electric motors are not totally enclosed, and they should be protected from excessive dust to ensure that the cooling vents do not become blocked. Attention to the following points will help extend the useful life of electric motors:

- Keep the motor dry. Store electric mowers, hedge trimmers etc., in a dry shed.
- Avoid overloading an electric motor, as this will cause it to overheat.

- Make sure the motor is securely mounted; vibration increases wear.
- Wipe any spilt oil from motors and all other electrical equipment. Oil will attack insulation and collect dust.
- Check that drive pulleys are correctly aligned. Misaligned belts and pulleys will result in excessive wear of belts and bearings.
- As a general rule, very little if any lubrication is required. Check the instruction book for lubrication requirements.

Buying Electricity

Consumers pay for electricity according to the number of units used. A unit of electricity is equivalent to one kilowatt of power used in one hour, so a 1,000 watt fire running for one hour will consume one unit. The cost of using electricity can be calculated in this way:

Total cost of electricity used = kilowatts × time (hours) × cost per unit

Example: To find the cost of running a 2 kW heater for 10 hours at 6.5 pence per unit:

Cost = Kilowatts used × time × cost per unit
= 2 kW × 10 hours × 6.5 p
= £1.30

The cost of running an electric motor is calculated by first converting horsepower to kilowatts. In theory, a one hp electric motor is equivalent to 750 watts, but in practice it should be assumed that a one hp motor will use one unit of electricity per hour. This allows for extra power demand when the motor is started and for losses in efficiency while it is running.

The mains electricity supply passes through a meter which measures the power consumed in units. Domestic properties pay for their electricity according to the number of units used plus a fixed quarterly charge. The unit cost is cheaper during off-peak periods. Many growers prefer to pay for electricity on a block tariff system. This means that the first block, or number of units used during each quarter, is charged at a very high rate, after which the price per unit becomes very much less. This is a particular advantage to the grower who uses a great deal of electricity.

The Safe Use of Electricity

You cannot hear, see or smell electricity, but it can kill very quickly. The following points should be remembered when using or working near electric cables and equipment:

- Always use the correct size fuse for the appliance.
- Keep cables, extension leads and plugs in good condition. Avoid laying cables across floors and other surfaces where they might be damaged. Scrap cut or damaged cables, as they can prove particularly dangerous.
- Protect light bulbs from dirt and moisture. Use waterproof light fittings in glasshouses and similar moisture-laden areas.
- It is much safer to use low voltage or cordless equipment where possible.
- Always use an RCD protected supply when operating outdoor equipment such as mowers and hedge trimmers.
- Take great care when working near or under overhead cables with metal ladders, irrigation pipes and similar long metal items. Make sure there is plenty of clearance before taking high loads or machinery beneath overhead wires. Do not operate a tipping trailer or allow others to tip loads from a vehicle underneath overhead wires. Should a trailer or lorry body ever tangle with overhead cables, make sure the driver jumps clear of the cab. By climbing down in the usual way, the driver will earth out the vehicle and may well be electrocuted.
- Do not attempt to make temporary repairs to electrical equipment or circuits; call in a qualified electrician.

Electric Shock

If you find a person who has received an electric shock, the first action is to turn off the power supply. Where this is not possible, try to free the person from the electrical contact by using his clothing to drag him clear. As an alternative put a dry cloth or other insulating material over the person's skin before pulling him clear. Never touch the victim's skin until the power has been turned off or until the casualty has been dragged clear of the live cable or appliance, as you are likely to receive an electric shock if you do so. When the danger has gone, carry out first aid procedures, including artificial respiration if breathing has stopped. (If you do not know how to do this, you should attend a first aid course, as it could save a life.)

Oil and Gas

Oil is an alternative to electricity for heating. Many growers use electricity for soil warming cables and air heaters, but others prefer oil or gas. Oil presents few problems provided that the burners are serviced regularly and storage regulations are observed. One major disadvantage is its fluctuating price level. Fuel may thicken in very cold weather unless winter grade fuel is used.

Many growers use Liquid Petroleum Gas (LPG), usually propane, as a source of power for heating and for internal combustion engines on forklift trucks.

Direct fired mobile heaters which burn propane are used for heating and for frost protection in glasshouses, tunnels and other buildings. Permanent installations use gas fired, warm air heaters or radiant tube heaters. Propane heating offers the added advantage of producing carbon dioxide as a by-product, which promotes rapid plant growth.

The handling and transport of produce is an important activity in horticulture, and many holdings use propane powered forklift trucks which can be used inside and outside buildings. Their engines are relatively quiet and produce a minimum of exhaust fumes.

Propane is delivered and stored under pressure in outside tanks which are normally the property of the gas supplier. The fuel is delivered at regular intervals by a road tanker. The gas can be piped from the storage tank to permanent heaters, or it can be supplied in smaller containers for use by forklift trucks, portable heaters, etc.

Care must be taken to avoid leakages of gas. If a leak is suspected, the supply must be turned off at once, and all flames and naked lights must be extinguished immediately.

Transmission of Power

The power developed by electric motors and internal combustion engines can be transmitted by means of belts, chains, gears and shafts.

Belt Drives

Vee-belts are widely used to transmit drive from electric motors. They are also used on a wide range of mowers and other horticultural machinery. Power is transmitted by the sides of a vee-belt which grip against the sides of the pulley. For this reason, the pulley and belt must match so that the belt cannot run in the bottom of the pulley. Single vee-belts are made in various widths and a wide range of lengths. Pulleys are made to suit each belt width.

Where it is necessary to transmit a lot of power, vee-belts are sometimes used on multiple pulleys with two or three belts running side by side. When they are worn, all belts must be replaced as a set, and if one of the belts breaks, it is still necessary to replace the full set because those which remain intact will be stretched.

Vee-belts are tensioned either with an idler or a jockey pulley, or by moving the driving or driven pulley in relation to the other, according to the design of the machine. The drive between two pulleys can be reversed

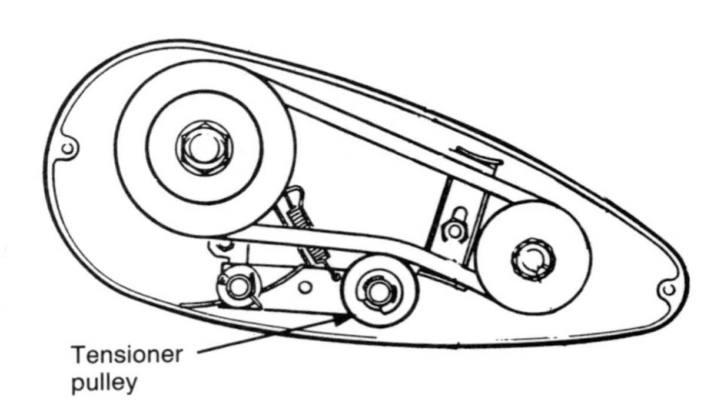

FIGURE 21.2 Vee-belt drive with a spring loaded tensioner or jockey pulley.

by crossing the belt in the shape of a figure of eight.

In many applications the fact that a vee-belt will slip if overloaded is an added advantage. An example of this is the vee-belt drive on a rotary mower.

Flat belts are suitable for transmitting power over a greater distance than is possible with vee-belts. Flat belt pulleys often have a slight ridge or crown in the centre of the pulley face to help keep the belt in correct alignment. Drive can be reversed by crossing the belt.

Toothed belts run on ribbed pulleys and provide a non-slip drive. They are becoming more popular than chain drives as less maintenance is required.

Chain Drives

Chain drives are used where belt slip is unacceptable; this is important where the driven parts are timed with each other. One example is an engine timing chain which is arranged to open valves and provide a spark at the right moment in the four stroke cycle. A slip clutch is used on some machines with a chain drive to protect it from damage through overloading.

Roller chains are available in both high and low speed versions. High speed roller chains have short links made to close limits, and are used as engine timing chains. Low speed roller chains have much longer links and are only suitable for machinery where relatively low power is transmitted at equally low speed. The toothed wheels which carry roller chains are called sprockets.

Roller chains require frequent lubrication. Both chain and sprockets wear at a similar rate. A worn sprocket has hooked teeth and worn roller chains will have a lot of movement between the links.

Roller chains are normally tensioned with an idler or jockey sprocket in a similar way to a vee-belt. When the chain becomes stretched, it may be necessary to remove a link if all available adjustment with the idler sprocket has been used.

Unlike belts, roller chains cannot be crossed to reverse the direction of drive.

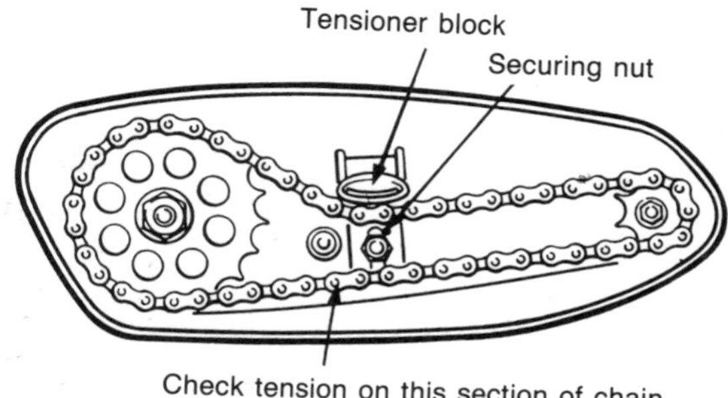

FIGURE 21.3 Roller chain drive with a chain tensioner block.

Shaft Drives

Shafts provide a simple method of power transmission over a distance. When the two ends of the shaft are not contained within a rigid frame, allowance for misalignment must be made with some form of flexible coupling.

The most common method is the use of universal joints. A tractor power take-off shaft has two of these, making it possible for the implement input shaft to be moved through quite a wide angle in relation to the tractor power shaft.

Gear Drives

Gears are used to achieve a positive drive between two adjacent shafts. The speed of the two shafts will be directly related to the number of teeth on the pair of gears. Shaft speed can be increased, decreased or remain unchanged depending on the arrangement of the gears.

Spur gears are used to transfer power from one shaft to another parallel to it. They may have either straight teeth or helical (angled) teeth (see Chapter 5). A gear turning in a clockwise direction on one shaft will drive a gear mated with it on an adjacent shaft in an anti-clockwise direction.

Bevel gears are used to transmit drive through an angle from one shaft to another. The

PLATE 21.2 *Cylinder mower chain drives. One chain is tensioned with an idler sprocket, the other (bottom) has an adjustable tensioner block. The clutch is attached to the large sprocket.*
(Ransomes, Sims and Jefferies)

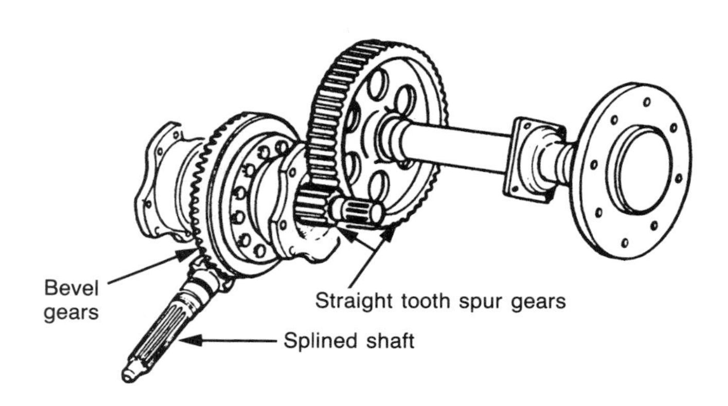

FIGURE 21.4 Gear transmission system. *(Kubota)*

crown wheel and pinion in the tractor transmission system is an example of a pair of bevel gears transmitting drive through a right angle. Not all bevel gears, which may have straight or helical teeth, are designed to run at a right angle.

Gear and Pulley Speeds

The speed of a driven gear can be calculated if the speed of the driving gear and the number of teeth on both gears is known. In the same way, the speed of the driving gear can be calculated to give a required driven gear speed. This is the formula to use:

$$\frac{\text{Teeth on gear A}}{\text{Teeth on gear B}} = \frac{\text{rpm of gear B}}{\text{rpm of gear A}}$$

The same formula can be used to calculate belt pulley speeds by using pulley diameter instead of number of teeth on the gears.

Example: To find the speed of a 12 teeth gear (B) driven by an 18 teeth gear (A) which turns at 60 rpm:

$$\frac{\text{Teeth on A}}{\text{Teeth on B}} = \frac{\text{rpm B}}{\text{rpm A}}$$

$$\frac{\text{18 teeth}}{\text{12 teeth}} = \frac{\text{rpm B}}{\text{60 rpm}}$$

$$\frac{18 \times 60}{12} = \text{rpm B} = \text{90 rpm.}$$

Example: To find the speed of a 400 mm diameter pulley driven by a 600 mm pulley running at 1,000 rpm:

$$\frac{\text{Diameter A}}{\text{Diameter B}} = \frac{\text{rpm B}}{\text{rpm A}}$$

$$\frac{\text{600 mm}}{\text{400 mm}} = \frac{\text{rpm B}}{\text{1,000 rpm}}$$

$$\frac{600 \times 1{,}000}{\text{400 rpm}} = \text{rpm B} = \text{1,500 rpm.}$$

Safety Devices

Chain drives and vee-belt drives on some machines are protected by safety overload clutches which will slip if overloaded. Timed chain drives do not have overload clutches; instead some have a shear bolt holding the sprocket onto its hub which will break if severe overloading occurs. When the bolt is replaced, timing is restored.

One type of slip clutch has two serrated metal faces held together by spring pressure. When overloading occurs the driving pulley continues to turn but the shaft does not. The movement of the serrated faces against each other makes a clattering sound.

Another type of clutch has friction lined plates held between smooth metal faces by spring pressure. The metal plates are connected to the driving shaft and the friction plates are attached to the output shaft. When overloading occurs, the metal plates on the driving shaft continue turning but the friction plates on the output shaft remain stationary.

There is no noise from the clutch when it slips, but continuous slipping will overheat the clutch and may burn out the friction discs.

Some power take-off driven machines have an over-run clutch built into the power shaft. This allows the implement to slow down at its own speed after the power shaft is disengaged. An over-run clutch works on a similar principle to the freewheel on a bicycle.

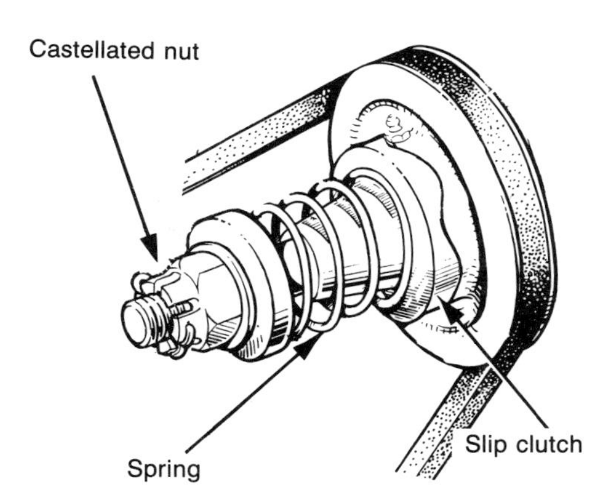

FIGURE 21.5 A vee-belt drive protected from damage through overload by a serrated face slip clutch.

Suggested Student Activities

1. Study sales leaflets to find the differences between BHP and PTO HP for different horticultural tractors.
2. Look for the different types of belt and chain drive. Find out what type of safety overload device is used on different horticultural machines with chain or belt drives.
3. Practice wiring a three pin plug.
4. Find out where the fuse board and the residual circuit breakers are located around the holding.
5. Attend a basic first aid course—the knowledge gained could be very useful in an emergency.

Safety Checks

Get to know the location and purpose of the different types of fire extinguisher at your place of work.

Use a dry powder extinguisher when fighting a fire caused by a live electrical fault. Never use a water based extinguisher because water is a conductor of electricity. When possible, turn off the power at the mains before using any type of extinguisher.

METRIC CONVERSION TABLES

British to Metric		Metric to British	
Length			
1 inch (in)	= 2.54 cm or 25.4 mm	1 millimetre (mm)	= 0.0394 in
		1 centimetre (cm)	= 0.394 in
1 foot (ft)	= 0.30 m	1 metre (m)	= 1.09 yd
1 yard (yd)	= 0.91 m	1 kilometre (km)	= 0.621 miles
1 mile	= 1.61 km		
Conversion Factors			
inches to cm	× 2.54	centimetres to in	× 0.394
or mm	× 25.4	millimetres to in	× 0.0394
feet to m	× 0.305	metres to ft	× 3.29
yards to m	× 0.914	metres to yd	× 1.09
miles to km	× 1.61	kilometres to miles	× 0.621
Area			
1 sq inch (in^2)	= 6.45 cm^2	1 sq centimetre (cm^2)	= 0.16 in^2
1 sq foot (ft^2)	= 0.093 m^2	1 sq metre (m^2)	= 1.20 yd^2
1 sq yard (yd^2)	= 0.836 m^2	1 sq metre (m^2)	= 10.8 ft^2
1 acre (ac)	= 4047 m^2 or 0.405 ha	1 hectare (ha)	= 2.47 ac
Conversion factors			
sq feet to m^2	× 0.093	sq metres to ft^2	× 10.8
sq yards to m^2	× 0.836	sq metres to yd^2	× 1.20
acres to ha	× 0.405	hectares to ac	× 2.47
Volume (liquid)			
1 fluid ounce (1 fl oz) (0.05 pint)	= 28.4 ml	100 millilitres (ml or cc)	= 0.176 pints
1 pint	= 0.568 litres	1 litre	= 1.76 pints
1 gallon (gal)	= 4.55 litres	1 kilolitre (1000 litres)	= 220 gal
Conversion factors			
Pints to litres	× 0.568	litres to pints	× 1.76
gallons to litres	× 4.55	litres to gallons	× 0.220

Weight

1 ounce (oz)	= 28.3 g	1 gram (g)	= 0.053 oz
1 pound (lb)	= 454 g	100 grams	= 3.53 oz
	or 0.454 kg	1 kilogram (kg)	= 2.20 lb
1 hundredweight (cwt)	= 50.8 kg	1 tonne (t)	= 2204 lb
1 ton	= 1016 kg		or 0.984 ton
	or 1.016 t		

Conversion factors

ounces to g	× 28.3	grams to oz	× 0.0353
pounds to g	× 454	grams to lb	× 0.00220
pounds to kg	× 0.454	kilograms to lb	× 2.20
hundredweights to kg	× 50.8	kilograms to cwt	× 0.020
hundredweights to t	× 0.0508	tonnes to tons	× 0.984
tons to kg	× 1016		
tons to tonnes	× 1.016		

INDEX